Lecture Notes in Computer Science 8487

Commenced Publication in 1973
Founding and Former Series Editors:
Gerhard Goos, Juris Hartmanis, and Jan van Leeuwen

T0223381

Song Guo   Jaime Lloret
Pietro Manzoni   Stefan Ruehrup (Eds.)

Ad-hoc, Mobile, and Wireless Networks

13th International Conference, ADHOC-NOW 2014
Benidorm, Spain, June 22-27, 2014
Proceedings

 Springer

Volume Editors

Song Guo
The University of Aizu
School of Computer Science and Engineering
Fukushima, Japan
E-mail: sguo@u-aizu.ac.jp

Jaime Lloret
Universitat Politècnica de València
Integrated Management Coastal Research Institute (IGIC)
Valencia, Spain
E-mail: jlloret@dcom.upv.es

Pietro Manzoni
Universitat Politècnica de València
Department of Computer Engineering (DISCA)
Valencia, Spain
E-mail: pmanzoni@disca.upv.es

Stefan Ruehrup
FTW - Telecommunications Research Center Vienna
Vienna, Austria
E-mail: ruehrup@ftw.at

ISSN 0302-9743   e-ISSN 1611-3349
ISBN 978-3-319-07424-5   e-ISBN 978-3-319-07425-2
DOI 10.1007/978-3-319-07425-2
Springer Cham Heidelberg New York Dordrecht London

Library of Congress Control Number: 2014939293

LNCS Sublibrary: SL 5 – Computer Communication Networks
and Telecommunications

Typesetting: Camera-ready by author, data conversion by Scientific Publishing Services, Chennai, India

Printed on acid-free paper

Springer is part of Springer Science+Business Media (www.springer.com)

Preface

The International Conference on Ad-Hoc Networks and Wireless (ADHOC-NOW) is one of the most well-known venues dedicated to research in wireless networks and mobile computing. Since its creation and first edition in Toronto, Canada, in 2002, the conference celebrated 12 other editions in 6 different countries. Its 13th edition in 2014 was held in Benidorm, Spain, during 22 to 27 June.

The 13th ADHOC-NOW attracted 78 submissions. A total of 33 papers were accepted for presentation after rigorous reviews by Program Committee members, external reviewers, and discussions among the program chairs. Each paper received at least three reviews; the average number of reviews per paper was around 4. The accepted papers covered various aspects of mobile and ad hoc networks, from the physical layer and medium access to the application layer, as well as security aspects, and localization.

ADHOC-NOW does not restrict its scope to either experimental or purely theoretical research, but tries to provide an overall view on mobile and ad hoc networking from different angles. This goal was reflected in the 2014 program, which contained a variety of interesting topics. Moreover, the 13th ADHOC-NOW was accompanied by a workshop program covering selected topics related to ad hoc networks, which led to a lively exchange of ideas and fruitful discussions.

Many people were involved in the creation of these proceedings. First of all, the review process would not have been possible without the efforts of the Program Committee members and the external reviewers, who provided their reports under tight time constraints. We also thank Springer's team for their great support during the review and proceedings preparation phases. Last, but not least, our special thanks goes to the Organization Committee for preparing and organizing the event and putting together an excellent program.

June 2014

Song Guo
Jaime Lloret
Pietro Manzoni
Stefan Ruehrup

Organization

General Chairs

Jaime Lloret   Universitat Politècnica de València, Spain
Ivan Stojmenović   University of Ottawa, Canada

Program Chairs

Song Guo   University of Aizu, Japan
Pietro Manzoni   Universitat Politècnica de València, Spain

Submission Chair

Miguel Garcia   Universitat Politècnica de València, Spain

Proceedings Chair

Stefan Ruehrup   FTW – Telecommunications Research Center Vienna, Austria

Publicity Chairs

Paul Yongli   Deakin University, Australia
Gongjun Yan   Indiana University, USA
Sandra Sendra   Universitat Politècnica de València, Spain

Web Chair

Milos Stojmenović   Singidunum University, Serbia

Technical Program Committee

Flavio Assis   UFBA – Federal University of Bahia, Brazil
Michel Barbeau   Carleton University, Canada
Jose M. Barcelo-Ordinas   UPC, Spain
Zinaida Benenson   FAU, Germany
Matthias R. Brust   Louisiana Tech University, USA

Volker Turau   Hamburg University of Technology, Germany
Vasos Vassiliou   University of Cyprus, Cyprus
Cheng Wang   Tongji University, China
Konrad Wrona   SAP, France
Yulei Wu   Chinese Academy of Sciences, China
Weigang Wu   Sun Yat-sen University, China
Qin Xin   University of the Faroe Islands, Faroe Islands
Stella Kafetzoglou   NTUA, Greece

External Reviewers

Daniel Bimschas
Nicolas Bonichon
Walter Bronzi
Timm Buhaus
João Caldeira
German Castignani
Thierry Derrmann
Dejan Drajic
Sebastian Ebers
Giuseppe Fedele
Markus Forster
Stefano Galzarano
Nenad Gligoric
Antonio Guerrieri
Christiana Ioannou
Stevan Jokic
Aggelos Kapoukakis
Marek Klonowski

Milan Lukic
Nicola Marchetti
Florian Meier
Julian Ohrt
Pasquale Pace
Tomasz Radzik
Vladan Rankov
Xiaojiang Ren
Peter Rothenpieler
Charalambos Sergiou
Marc Stelzner
Francisco Vazquez-Gallego
Lin Wang
Wenzheng Xu
Siqian Yang
Zhang Yi
Zinon Zinonos

Table of Contents

MAC and Physical Layer

Mobile Ad Hoc, Sensor and Robot Networks

Routing II

Localization and Security

Vehicular Ad-Hoc Networks

Combined Mobile Ad-Hoc and Delay/Disruption-Tolerant Routing

Christian Raffelsberger and Hermann Hellwagner

Institute of Information Technology, Alpen-Adria Universität Klagenfurt,
Klagenfurt, Austria
{craffels,hellwagn}@itec.aau.at

Abstract. The main assumption of many routing protocols for wireless mobile ad-hoc networks (MANETs) is that end-to-end paths exist in the network. In practice, situations exist where networks get partitioned and traditional ad-hoc routing fails to interconnect different partitions. Delay/disruption-tolerant networking (DTN) has been designed to cope with such partitioned networks. However, DTN routing algorithms mainly address sparse networks and hence often use packet replication which may overload the network. This work presents a routing approach that combines MANET and DTN routing to provide efficient routing in diverse networks. In particular, it uses DTN mechanisms such as packet buffering and opportunistic forwarding on top of traditional ad-hoc end-to-end routing. The combined routing approach can be used in well-connected networks as well as in intermittently connected networks that are prone to disruptions. Evaluation results show that our combined approach can compete with existing MANET and DTN routing approaches across networks with diverse connectivity characteristics.

Keywords: mobile ad-hoc networks, disruption-tolerant networks, routing, simulation.

1   Introduction

The majority of state-of-the-art routing protocols [1] for wireless mobile ad-hoc networks (MANETs) assume the existence of an end-to-end path between source and destination pairs. These protocols fail to deliver packets if such an end-to-end-path does not exist. However, in real application scenarios, ad-hoc networks may not be fully connected since disruptions cause the network to get partitioned. In practice, many ad-hoc networks will provide well-connected regions but still suffer from partitioning which prevents end-to-end communication between a subset of the nodes. A reason for such disruptions are link failures caused by obstacles or the mobility of nodes. Diverse connectivity characteristics impose challenges on the communication network, especially on the routing protocol. Hence, there is a need for hybrid routing protocols that exploit multi-hop paths to efficiently route packets in well-connected parts of the networks and permit inter-partition communication by storing packets that cannot be

S. Guo et al. (Eds.): ADHOC-NOW 2014, LNCS 8487, pp. 1–14, 2014.

routed instantly. One example for networks that are prone to partitioning are hastily formed ad-hoc networks for emergency response operations. These networks may be diverse in terms of connectivity and networking equipment. The connectivity may range from well-connected networks, where nearly all nodes are interconnected, to sparse networks, where most nodes are disconnected. In between these two extremes, the network may also be intermittently connected and provide several 'islands of connectivity'. For instance in disaster response scenarios, which are a promising application domain for mobile ad hoc networks, members from the same search and rescue team may be interconnected as they tend to work near each other. However, there may be no end-to-end paths available between different teams or the incident command center and teams that are spread on the disaster site. MANET protocols that try to find end-to-end paths will not work satisfactorily in such emergency response networks that provide diverse connectivity characteristics [6].

Routing algorithms for Delay- or Disruption-Tolerant Networking (DTN) [8] do not assume the existence of end-to-end paths but allow nodes to store messages until they can be forwarded to another node ,or delivered to the final destination. This mechanism is called store-and-forward or store-carry-forward routing and increases robustness in the presence of network disruptions. However, many DTN routing algorithms use packet replication to improve delivery probability and delivery delay. Whereas this mechanism is beneficial in sparse networks that provide few contact opportunities, it may dramatically decrease performance in dense networks, since it introduces high overheads in terms of transmission bandwidth and storage.

The contributions of this paper are as follows. We introduce a combined MANET/DTN routing approach called CoMANDR that extends end-to-end MANET routing with mechanisms from DTN routing. CoMANDR is designed to cover a broad range of connectivity characteristics, from intermittently connected to well-connected networks. The combined routing approach makes no assumption about the existence of end-to-end paths. It can deliver packets instantly if end-to-end paths exist or select custodian nodes opportunistically to bridge islands of connectivity. We evaluate our approach in several scenarios and compare it with other state-of-the-art routing approaches from the MANET and the DTN domain.

The remainder of the paper is structured as follows. Section 2 introduces the routing protocols that are used in the evaluation. Section 3 describes the design of CoMANDR. Section 4 presents the evaluation setup including a scenario description and the used metrics. The simulation results are discussed in Section 5. Finally, Section 6 concludes the paper and discusses possible future work.

2   Related Work

This section briefly describes the protocols that are used in the evaluation. PROPHET [5] is a flooding-based DTN routing protocol that uses the so called delivery predictability metric to decide to which nodes a message should be forwarded. The delivery predictability is a measure of how likely it is that a node can deliver a message to its destination. It is based on the assumption that nodes that have met frequently in the past, are also likely to meet again in the future. Whenever two nodes meet, they exchange and update their delivery predictability values and exchange all messages for which the other node has a higher delivery probability. For this evaluation, CoMANDR uses PROPHET's delivery predictability metric in its utility calculation function (see Section 3.2 for details). CoMANDR is dependent on a MANET routing protocol that finds end-to-end paths in the network. We did not choose a specific MANET routing protocol for the evaluation. Instead, we use a generic link state protocol, referred to as MANET, that finds the shortest end-to-end paths in the network and is capable of routing packets in connected parts of the network. The MANET routing protocol implementation supports to limit the maximum length of end-to-end paths that are reported. By limiting the path length it is possible to simulate imperfections of MANET routing protocols in real networks. Without this limitation, the packet delivery ratio of MANET represents the upper bound for all protocols that rely on end-to-end paths. The same MANET routing protocol is used in CoMANDR to build the routing tables and route packets if a path is available. Some recent approaches that combine MANET and DTN routing have added packet buffering to a MANET routing protocol [2][7]. These approaches buffer packets instead of dropping them if no end-to-end path is available. We added packet buffering to the optimal MANET routing protocol to simulate this kind of approach. The resulting protocol is called MANET store-and-forward (MANET-SaF) and is one example for hybrid MANET/DTN routing. Additionally, the evaluation includes the Epidemic routing protocol [9]. Epidemic routing floods the whole network. In particular, whenever two nodes meet, Epidemic routing exchanges all messages that the other node has not already buffered. If transmission bandwidth and buffers are unlimited, Epidemic would utilize all available routes and optimize delivery delay and packet delivery ratio. Hence, Epidemic sets the upper bound for the performance of any routing algorithm. However, Epidemic's high resource usage negatively affects its performance in resource-constraint environments.

3   Combined MANET/DTN Routing

Combined MANET/DTN Routing (CoMANDR) works like a traditional routing protocol for MANETs when end-to-end paths are available. It uses the routing table that is calculated by the MANET protocol to route packets that can be reached instantly over a multi-hop end-to-end path. Thus, CoMANDR will exactly work like the underlying MANET routing protocol if the network is fully connected. To cope with disruptions, CoMANDR utilizes two mechanisms from delay/disruption-tolerant networking: packet buffering and utility-based forwarding. If the routing table contains no valid entry for a packet's destination, CoMANDR buffers the packet instead of discarding it. The rationale behind this behavior is that a buffered packet may be sent later when a route becomes available (i.e., sender and receiver are in the same connected component). There may be situations where an end-to-end path between sender and receiver will never be available. To handle such situations, CoMANDR may also forward packets to nodes that are assumed to be closer to the destination. The decision to which node a buffered packet should be forwarded is based on a utility function. One interesting aspect is that the utility function can re-use information that is collected by the MANET routing protocol (e.g., information from the routing table or link-state announcements). However, it is also possible to collect additional information to calculate utility values for other nodes in the network. The utility values are used to determine an alternative path if no end-to-end path has been found. Nodes with higher utility values are more likely to deliver packets to the destination. In general, CoMANDR first tries to send a packet via available MANET routes. If this is not possible, the packet is sent to the neighbor with the highest utility value for that packet. While this procedure is repeated, the packet is sent hop-by-hop towards the destination. The following pseudo code describes the basic algorithm to combine MANET and DTN routing:

> **procedure** ROUTEPACKET(p)
> $nextHop = $ **queryRoutingTable**(p)
>   **if** $nextHop \neq NULL$ **then**
>   **sendTo**$(nextHop, p)$
>   return
>   **end if**
> $nextHop = $ **getMaxUtilityNode**(p)
>   **if** **getUtiliy**$(this, p) < $ **getUtility**$(nextHop, p)$ **then**
>   **sendTo**$(nextHop, p)$
>   **else**
>   **bufferPacket**(p)
>   **end if**
> **end procedure**

3.1   Packet Buffering

In order to provide store-carry-forward routing, packets need to be buffered when no end-to-end path is available. Additionally, it is checked if a routing table entry is valid. The packet delivery ratio can be increased if the validity of routes is checked and packets are buffered in case of stale routes [7]. An entry is invalid if its next hop entry is currently not available (i.e., there is no wireless link to the one hop neighbor that is advertised by the entry). Such stale route entries may be an effect of link outages caused by the mobility of nodes or by obstacles and MANET routing protocols need some time to detect and handle such events. To identify if a next hop is available, MANET routing control traffic can be monitored [7]. Additionally, information from other layers may be used. For instance, information about the status of links that are provided by the underlying link layer protocol.

Apart from deciding when to buffer a packet, it is also important to decide when a buffered packet can be sent. In case of temporary link outages, packets may be sent as soon as the link is available again or a proactive MANET routing protocol has provided an alternative path that includes a valid next hop. However, there may be cases where no end-to-end path will be found since the destination of a buffered packet is in another partition. In these cases it is beneficial to forward the packet to a node that is more likely to deliver the message. This mechanism is called utility-based forwarding and is described in the next section. It is important to note that the evaluated version of CoMANDR only forwards a single copy of every packet since every node deletes a packet that it has forwarded to another node. This saves transmission bandwidth and storage but may negatively affect routing performance in sparse networks.

3.2   Utility-Based Forwarding

The utility of a node describes the node's fitness to deliver a packet towards its destination. In general, a node will hand over a packet to another node if the other node has a higher utility value. The utility may be dependent or independent of the destination [8]. A destination-independent utility function is based on characteristics of the potential custodian node, such as its resources or mobility. On the other hand, destination-dependent utility functions are based on characteristics concerning the destination, such as how often a node has met the destination or if a node and the destination belong to the same social group.

The combined use of a utility table and a MANET routing table allows nodes to route packets in both connected and disrupted networks. The MANET routing table represents some sort of spatial information (i.e., which nodes are currently in the vicinity of a node). Combining routing table information with utility functions that contain historic data (e.g., information about previous states of the routing table), effectively calculates spatio-temporal clusters of nodes. This information allows a node to determine to which other node a packet should be sent, when there is currently no end-to-end path to the destination available.

The performance of a utility function is influenced by the characteristics of the scenario. Hence, it is important to choose a utility function or a combination of functions that fits the specifics of the intended application scenario. We have chosen a utility function that uses routing table entries to calculate meeting probabilities based on the well-known PROPHET routing algorithm [5] for DTNs. Although we performed some experiments with different parameters for the utility calculation, to empirically determine parameters that suit the scenarios, it is important to note that the purpose of this work is not to find an optimal utility function. Instead, this work intends to show that a combination of MANET routing and DTN routing (i.e., applying mechanisms such as packet buffering and utility-based forwarding on top of a MANET routing protocol) is beneficial in some scenarios.

To limit the calculation efforts for the utility function and the amount of data to be exchanged, every node should limit the number of nodes it keeps in its utility table. One possibility is that every node only keeps the n highest entries in its utility table. Another possibility is to remove nodes if their utility value drops under a certain minimum threshold. The second option was used for the evaluation (i.e., nodes are removed from a utility table if their utility value drops below 0.2). If nodes are not in the utility table of another node, they will not be used as custodian nodes. This prevents nodes from forwarding packets to custodians that only offer a low chance to deliver the packet to its destination. Otherwise, a lot of transmissions would be performed without significantly increasing the delivery probability.

CoMANDR uses a modified version of the PROPHET meeting probability calculation function to calculate the utility of a node. In contrast to the PROPHET protocol, that only considers when two nodes directly meet (i.e., there is a direct link between the nodes), CoMANDR also considers multi-hop information from the routing table. When a node i has a routing table entry for another node j (with a distance less than infinite), CoMANDR considers node i and j to be in contact. This allows nodes to exploit multi-hop paths to determine contacts with other nodes.

As the MANET routing protocol regularly updates the routing table entries, the meeting probabilities and thus the utility values for other nodes need to be updated as well. To be precise, every node i manages one utility value for every node j that it knows. The set of known nodes includes all destinations for which a routing entry exists or has existed previously (i.e., disconnected nodes that are still kept in the cluster). So if a route to node j is known, node i will update the utility value for node j (denoted as U_{ij}) using PRoPHET's probability update function:

$$U_{ij} = U_{ij} + (1 - U_{ij}) * \alpha \tag{1}$$

On the other hand, for a node k that is not in the routing table but has a utility value, the utility value U_{ik} is reduced:

$$U_{ik} = U_{ik} * \gamma \tag{2}$$

The parameter α determines how fast the utility converges to 1 if there is a contact between two nodes, whereas γ determines how fast it converges to 0 if there is no contact. Both parameters need to be in the range between 0 and 1.

Every node needs to regularly broadcast all calculated utilities (the utility table) to its direct neighbors. When a node receives the utility table of another node, it can use this information to update its own utility table. If a node i is in contact with node j that has a utility value for node k, node i can transitively update its utility value for node k. For instance using PROPHET's transitive update function:

$$U_{ik} = max(U_{ik\ old}, U_{ij} * U_{jk} * \beta) \tag{3}$$

β is used to control the impact of transitivity. It is worth noting that the transitive update function is general and not tied to the use of PROPHET's meeting probability.

4   Evaluation Setup and Scenarios

The overall goal of the evaluation is to show that CoMANDR performs well in a broad range of connectivity settings. The Opportunistic Network Environment

(ONE) simulator [4] is used to evaluate all protocols. The ONE is mainly intended to evaluate DTN routing protocols. It focuses on the network layer and does not implement physical characteristics of the transmission. Although this imposes a lack of realism, we believe that it is still possible to make a fair comparison between MANET, DTN and our hybrid MANET/DTN approach.

We needed to implement multi-hop MANET routing within the ONE. In particular, we implemented a link state protocol that uses Dijkstra's shortest path algorithm to calculate the shortest paths in the network. This link state MANET routing protocol is also used by CoMANDR to route packets in the connected parts of the network. Hop count is used as route metric, as the ONE does not provide any information about the quality of links. All nodes have the same view on the network. Thus, the implemented MANET routing protocol is optimal. In reality, routing protocols have to cope with imperfect information about the network (e.g., information about links is missing or wrong). Hence, routing protocols have problems to find end-to-end paths in mobile scenarios. In particular, paths that comprise many hops may not be found. Additionally, the throughput of end-to-end paths drastically decreases with the hop count [3]. Thus, we restrict the maximum length of paths that are reported by the MANET routing algorithm to simulate these problems. If not denoted otherwise, the routing table only includes routes with a maximum end-to-end path length of five hops for all experiments. This restricts the maximum path length that MANET can exploit to five. All other protocols may still exploit longer paths as they do not only use end-to-end paths but also store-and-forward routing to deliver packets.

4.1 Scenarios

All protocols are evaluated in several scenarios that offer different connectivity characteristics. In a first set of experiments we varied the transmission range and simulation area size to get a diverse set of networking scenarios, ranging from well-connected to sparse networks. We calculated the connectivity degree for all scenarios (see 4.2) and selected three scenarios offering different levels of connectivity. In particular, we selected three scenarios that use the same transmission range of 100 m but have a different simulation area size. The selected scenarios include a well-connected scenario, a sparse scenario and an intermittently connected scenario that lies between the other two.

The mobility model that is used in all scenarios is the random walk model as implemented in the ONE. In particular, a node selects its next destination by randomly selecting a direction, speed and distance, after waiting for a random pause time. Since the maximum distance between two consecutive waypoints is limited, nodes moving according to this model tend to stay close to each other for a longer time, compared to the random waypoint model. It is important to note that random mobility rather puts PROPHET and CoMANDR at a disadvantage because both protocols assume that the future encounter of nodes is predictable. However, we argue that the low movement speed of nodes (i.e., the max speed is 2 m/s) and the fact that consecutive waypoints are close to each other, mitigate

Table 1. Simulation parameters

Mobility model	
No. nodes	100
Model	Random Walk
Movement speed	1 to 2 m/s
Pause time	0 to 60 s
Distance (min,max)	0 to 50 m
Wireless settings	
Transmission range	100 m
Transmission rate	4 Mbps
Traffic model	
Packet creation interval	500 to 2500 s
Packet creation rate	1 msg every 30 s (per node)
Packet size	100 kB
Packet buffer size	700 MB (per node)
Parameter for PROPHET routing/CoMANDR	
$P_{init}(=\alpha)$	0.9
β	0.7
γ	0.995

the effects of random mobility to some extent. For instance, two nodes that have met recently are also more likely to meet each other again than two nodes that are far away from each other. Moreover, it has been shown that PROPHET is still able to perform reasonably well in scenarios with random mobility [5].

The total simulation time is 4500 s and all experiments are repeated 23 times using different seeds for the mobility model and the traffic generator. All scenarios include 100 nodes. Traffic is generated by creating a new packet with random source and destination every 0.3 s. Hence, a node generates a new packet every 30 s on average. No traffic is generated after 2500 s to allow the routing protocols to deliver buffered packets before the simulation ends. All packets have an infinite time to live. Important simulation parameters are listed in Table 1.

4.2   Metrics

The first metric that is used to evaluate the routing approaches is the *packet delivery ratio* (PDR). It shows the ratio between successfully received packets at the destination and the number of created packets. The *hop count* shows how many nodes a packet has passed from source to destination. The *transmission cost* metric denotes the ratio between transmitted packets and successfully received packets. For single-copy schemes such as MANET routing and CoMANDR, the transmission cost is proportional to the average hop count of all successfully received messages. For the other schemes, the transmission cost is mainly influenced by the number of message replicas. The *latency* represents the time that is needed to transfer a packet from the source to the destination. Latency includes the buffering time and the transmission time for all nodes along the path.

A metric that is often used for evaluating mobile ad-hoc routing protocols is the routing control overhead (i.e., the traffic overhead for finding end-to-end routes). However, different MANET routing protocols greatly vary in the amount of control overhead they introduce [10]. As this study only includes a generic MANET protocol, it is not feasible to directly measure control overhead. As CoMANDR and MANET-SaF are extensions of the generic MANET protocol, the overhead for these three protocols is comparable. It is also fair to assume that the control overhead of the underlying MANET routing protocol is significantly lower than the data overhead introduced by the multi-copy schemes that are evaluated in this paper. Hence, we argue that not taking control overhead into account should not hinder a fair comparison of the evaluated protocols.

Three additional metrics are used to characterize the network connectivity of the simulation scenarios. The *connectivity degree* CD is the probability that two randomly selected nodes are in the same connected component at a given point in time (i.e., an end-to-end path between the two nodes exists). A connectivity degree of 1 denotes a fully connected network, whereas 0 denotes a network were all nodes are isolated. The connectivity degree at a given point in time t is calculated as follows:

$$CD_t = \sum_{P_i \in P_t} \frac{|P_i|}{|N|} * \frac{|P_i| - 1}{|N| - 1}, \tag{4}$$

where N denotes the set of all nodes in the network and P_t denotes the set of partitions that comprise the network at a given time t. $|P_i|$ denotes the number of nodes in one particular partition and $|N|$ the total number of nodes in the network. As the connectivity degree changes over time, the average connectivity degree for the duration of the simulation has to be calculated as follows:

$$CD = \frac{1}{T} * \sum_{t}^{T} CD_t, \tag{5}$$

where T denotes the number of samples that have been taken and CD_t the connectivity degree for one sample. For the scenarios in this paper, CD denotes the mean value of 4500 samples (i.e., one sample per second). Another metric that describes the connectivity of a network is the *largest connected component* (LCC). The LCC denotes the number of nodes that are located in the largest partition. The third metric used to characterize the scenario in terms of connectivity is the *number of partitions* with at least two nodes. Hence, the number of partitions does not include isolated nodes. Table 2 lists the connectivity characteristics for the evaluation scenarios.

5   Results

This section includes the evaluation results for CoMANDR, Epidemic, PRO-PHET, MANET and MANET-SaF. Unless otherwise stated, figures show mean values of all simulation runs and error-bars denote the 95% confidence interval.

Table 2. Scenario characteristics in terms of network connectivity

Size of area (in m x m)	Avg. connectivity-ning degree CD	Largest connected component (avg)	Avg. no of partitions
700x700	0.882	92.886	1.915
800x800	0.634	74.95	3.853
1000x1000	0.157	30.276	17.982

The packet delivery ratio for all evaluated protocols in the three scenarios is shown in Figure 1a. Traditional end-to-end MANET routing is clearly outperformed by the other protocols and achieves the lowest PDR in all scenarios. Epidemic routing can deliver most packets in all scenarios. This is due to the fact that the link bandwidth is very high and nodes can store all packets in their buffers, which is the ideal case for Epidemic. No packets are dropped because of full buffers which maximizes Epidemic's performance. PROPHET can achieve a similar PDR in well-connected and intermittently connected scenarios. The performance results of CoMANDR and MANET-SaF are comparable in the well-connected scenario. The reason for this is that source and destination are very likely to be in the same connected component at some point in time and the packets can be delivered via an end-to-end path. Hence, MANET-SaF works similarly to CoMANDR in this scenario and both protocols achieve nearly the same PDR. However, CoMANDR outperforms MANET-SaF in the other two scenarios. In the sparse scenario, CoMANDR could deliver nearly 50% more packets than MANET-SaF. This performance gain is achieved by the utility-based forwarding scheme of CoMANDR that forwards packets towards the destination. Thus, CoMANDR can deliver packets to destinations that are never in the same connected component as the source, which improves its performance compared to MANET-SaF.

The protocols are diverse in terms of transmission cost as shown in Figure 1b. Due to its aggressive replication scheme, Epidemic nearly performs 100 packet transmissions to deliver one packet. Although PROPHET can reduce this number by not forwarding packets to neighbors that have a lower delivery predictability, it still replicates packets extensively. MANET produces the lowest transmission cost as it only delivers packets via the shortest available end-to-end path. As the path has to be available instantly, it drops packets if it fails to find such an end-to-end path. MANET-SaF has a higher transmission cost than MANET as buffering packets instead of dropping them allows it to deliver more packets, especially via longer paths. CoMANDR has a higher transmission cost if the connectivity is low. However, compared to the multi-copy schemes Epidemic and PROPHET, its transmission cost is still very low. Thus, CoMANDR offers the best trade-off between packet delivery ratio and transmission cost among all protocols. We believe that this is a very important feature of CoMANDR as resources are often scarce in mobile networks. Reducing the number of transmissions and hence reducing the wireless channel utilization and battery consumption, while still providing a good packet delivery ratio, is an important issue in many scenarios.

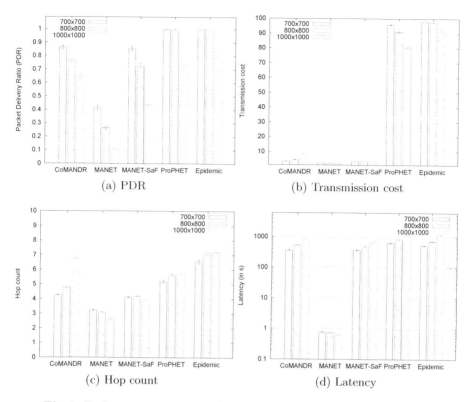

Fig. 1. Performance comparison for scenarios with different connectivity

The hop count is shown in Figure 1c. In general it can be said that, for the same scenario, the hop count is correlated with the packet delivery ratio. In particular, the protocols that achieve a higher packet delivery ratio achieve this mainly by utilizing longer paths which increases the average hop count. Since MANET only delivers packets via end-to-end paths, its hop count is limited by the fact that end-to-end paths do not comprise many hops, especially in the sparse scenario. Additionally, as long end-to-end paths have been removed from the routing table to simulate imperfections of MANET protocols in real networks, the maximum hop count is limited. We also performed some experiments with a higher hop limitation for end-to-end paths. With higher hop limitations, MANET also utilizes longer paths and the average hop count is higher. Due to space constraints, we cannot present detailed results about hop count for these experiments. As mentioned before, MANET-SaF can deliver more packets via longer paths as it stores packets if no end-to-end path is available, or the end-to-end path breaks while the packet is on its way to the destination. Similarly, the multi-copy schemes Epidemic and PROPHET have a higher hop count as they are able to deliver more packets via long paths. The hop count of CoMANDR is similar to the one of MANET-SaF for the well-connected and intermittently connected scenarios. In the sparse scenario, CoMANDR's utility-based forwarding technique finds more paths but also needs more hops. However, as only one message copy is passed in the network, this does not cause a high transmission cost.

Latency is shown in Figure 1d. Since MANET only uses instantly available end-to-end paths, it has the lowest latency. However, at the cost of a low PDR. The other protocols have a significantly higher latency due to packet buffering. Similar to the hop count, the latency is correlated with the PDR.

We also performed experiments with a varying hop count limit for the end-to-end paths. As mentioned before, MANET routing protocols often fail to find multi-hop paths including many hops, especially in mobile scenarios. For the previous experiments, we limited the maximum path length to five which is a rather conservative estimation and limits the performance of MANET and protocols depending on it (i.e., CoMANDR and MANET-SaF that use MANET to route packets in the connected parts of the network). Figure 2 shows how the PDR is affected by the length limitation of end-to-end paths. An interesting finding is that the store-and-forward mechanism of MANET-SaF and CoMANDR is a good means to increase the PDR, when the MANET protocol does not find longer multi-hop paths. For instance, in the intermittently connected scenario (see Fig. 2b), CoMANDR with a relatively strict maximum end-to-end path length of four achieves a higher PDR than MANET with practically no restriction (i.e., hop limit 20). Even in the well-connected scenario, idealistic MANET routing has a lower PDR than CoMANDR and MANET-SaF for hop limits greater than five (see Fig. 2a). This is an indication that CoMANDR may also perform better than traditional MANET protocols in well-connected but quickly changing networks, where traditional MANET protocols fail to find end-to-end paths because of the mobility of nodes.

We also performed experiments to assess the performance of CoMANDR using different values for α, β and γ. Due to space constraints we cannot present the results in detail. However, results show that the aging factor γ has a higher impact on routing performance than α and β. Especially in scenarios with low connectivity, γ should be set to a high value as this increases the PDR, without increasing the transmission cost significantly. The values listed in Table 1 offered the best performance over all scenarios.

In the given scenarios, the packet delivery ratio of CoMANDR is always better than or equal to the delivery ratio of MANET and MANET-SaF routing. This shows that the mechanisms applied by CoMANDR on top of MANET routing, namely packet buffering and utility-based forwarding, are beneficial. In contrast to MANET routing, CoMANDR achieves packet delivery ratios that are comparable to state-of-the-art DTN routing algorithms in the intermittently and low connected scenarios. It is worth noting that sufficiently large buffers were provided in all scenarios. This is very beneficial for Epidemic and PROPHET since the packet delivery ratio is not negatively affected by packet drops caused by full buffers. On the other hand, CoMANDR is much more efficient. Thus, the

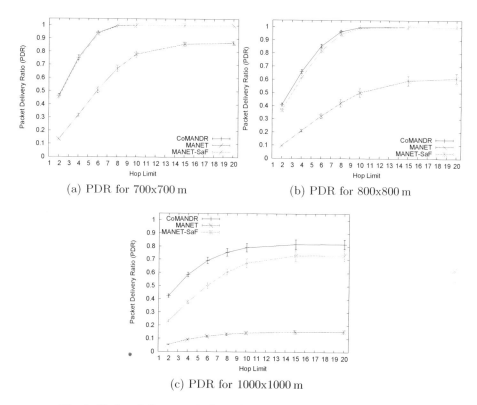

(a) PDR for 700x700 m

(b) PDR for 800x800 m

(c) PDR for 1000x1000 m

Fig. 2. Packet delivery ratio for different end-to-end path hop limits

performance of CoMANDR will obviously be less affected by limited resources. This shows that CoMANDR is well suited for a broad range of networks.

6   Conclusion

CoMANDR combines MANET and DTN routing in order to ensure good performance across a broad range of networks. In well-connected networks, it works similar to traditional MANET routing. Additionally, it uses mechanisms to store and opportunistically forward packets to custodian nodes if no end-to-end path exists. Evaluation results show that our approach can compete with or outperform other state-of-the-art routing protocols both from the MANET and DTN domain. One important advantage of CoMANDR is that it offers a good trade-off between packet delivery ratio and transmission cost. As the intended application scenarios of CoMANDR include networks consisting of resource-constrained mobile devices, using resources efficiently is a very important feature of the protocol.

For this evaluation CoMANDR was implemented as single-copy scheme. However, it would be interesting to assess its performance if packet replication were used. This should improve CoMANDR's performance in very sparse networks. However, as the intended application domain of CoMANDR are diverse networks, it is important to design a replication scheme that does not introduce too high overheads concerning storage and bandwidth which would decrease performance in well-connected parts of the network and waste possibly scarce resources such as transmission bandwidth, battery or storage. Other topics for future work are to look into additional utility functions and evaluate CoMANDR in realistic scenarios such as emergency response operations, that are an interesting application domain for this kind of routing approach.

Acknowledgments. The research leading to these results has received funding from the European Union 7th Framework Programme (FP7/2007-2013) under grant agreement no 261817, the BRIDGE project, and was partly performed in the Lakeside Labs research cluster at Alpen-Adria-Universität Klagenfurt.

References

1. Boukerche, A., Turgut, B., Aydin, N., Ahmad, M., Bölöni, L., Turgut, D.: Routing protocols in ad hoc networks: A survey. Computer Networks 55(13), 3032–3080 (2011)
2. Delosieres, L., Nadjm-Tehrani, S.: Batman store-and-forward: the best of the two worlds. In: Proc. Int. Conf. Pervasive Computing and Communications Workshops (PerCom Workshops 2012), pp. 721–727. IEEE (2012)
3. Johnson, D., Hancke, G.: Comparison of two routing metrics in OLSR on a grid based mesh network. Ad Hoc Networks 7(2), 374–387 (2009)
4. Keränen, A., Ott, J., Kärkkäinen, T.: The ONE simulator for DTN protocol evaluation. In: Proc. 2nd Int. Conf. Simulation Tools and Techniques (SIMUTools 2009), pp. 55:1–55:10. ICST (2009)
5. Lindgren, A., Doria, A., Schelén, O.: Probabilistic routing in intermittently connected networks. SIGMOBILE Mob. Comput. Commun. Rev. 7, 19–20 (2003)
6. Raffelsberger, C., Hellwagner, H.: Evaluation of MANET routing protocols in a realistic emergency response scenario. In: Proc. Int. Workshop on Intelligent Solutions in Embedded Systems (WISES 2012), pp. 88–92. IEEE (2012)
7. Raffelsberger, C., Hellwagner, H.: A hybrid MANET-DTN routing scheme for emergency response scenarios. In: Proc. Int. Conf. Pervasive Computing and Communications Workshops (PerCom Workshops 2013), pp. 505–510. IEEE (2013)
8. Spyropoulos, T., Rais, R.N., Turletti, T., Obraczka, K., Vasilakos, A.: Routing for disruption tolerant networks: Taxonomy and design. Wireless Networks 16(8), 2349–2370 (2010)
9. Vahdat, A., Becker, D.: Epidemic routing for partially-connected ad hoc networks. Tech. Rep. CS-2000-06, Duke University (July 2000)
10. Viennot, L., Jacquet, P., Clausen, T.H.: Analyzing control traffic overhead versus mobility and data traffic activity in mobile ad-hoc network protocols. Wireless Networks 10(4), 447–455 (2004)

A Multipath Extension for the Heterogeneous Technology Routing Protocol

Josias Lima Jr.[1], Thiago Rodrigues[1], Rodrigo Melo[1], Gregório Correia[1],
Djamel H. Sadok[1], Judith Kelner[1], and Eduardo Feitosa[2]

[1] Federal University of Pernambuco (UFPE), Recife, Brazil
{josias,trodrigues,rodrigodma,gregorio,jamel,jk}@gprt.ufpe.br
[2] Federal University of Manaus (UFAM), Manaus, Brazil
efeitosa@icomp.ufal.com.br

Abstract. In recent years we have witnessed the emergence of new access techniques that use both wireless technologies and self-organizing features. Their combination eliminates the need for using pre-defined wired structures and prior configurations. In this paper, we propose an extension by enabling multipath routing over our Heterogeneous Technologies Routing (HTR) Framework. HTR Multipath routing offers several benefits such as load balancing, fault tolerance, routing loop prevention, energy-conservation, low end-to-end delay, congestion avoidance, among others. This work performs a comparative analysis of the proposed HTR extension, with the baseline HTR, and the widely-used Optimized Link State Routing (OLSR) protocol. The evaluation is validated through the simulation of heterogeneous technologies such as WiMAX, 3GPP LTE and Wi-Fi. Results show that our proposal effectively improves the data delivery ratio and reduces the end-to-end delay without major impact on network energy consumption.

Keywords: Wireless Mobile Communication, Mobile Ad hoc Networks (MANET), Heterogeneous technologies, multipath routing, WiMAX, Wi-Fi, LTE, simulation.

1 Introduction

The popularity of mobile communication is constantly increasing. Tasks once handled by wired communication can now be performed by mobile devices equipped with several network interfaces, which may be of both wireless and cellular access technologies. With such diversity, the fundamental goal [1] is to render the existence of heterogeneous networks transparent. Furthermore, efficiently selecting the most appropriate technology to use is crucial for obtaining the levels of performance required by future networks.

Recently, we designed a new routing framework to interconnect devices in a heterogeneous ad hoc network environment [2]. It creates an enclosed heterogeneous mobile ad hoc network (MANET) [3], or a multi-hop ad hoc wireless network where nodes can move arbitrarily leading to rapid and unpredictable infrastructure changes.

S. Guo et al. (Eds.): ADHOC-NOW 2014, LNCS 8487, pp. 15–28, 2014.
© Springer International Publishing Switzerland 2014

Such MANETs can be set up quickly in diverse environments and can be composed of different communication technologies such as Bluetooth, Wi-Fi, 6LowPAN, Zigbee and Ethernet.

HTR (Heterogeneous Technologies Routing) also provides self-organizing support to bootstrap its nodes, through the self-configuration of network interfaces requiring minimum human interaction. For energy awareness, HTR employs the HTRScore special metric to help the path computation process and interface selection. This is essential to mitigate excessive energy consumption since such networks are generally composed of nodes with constrained capabilities (e.g. energy level, processing capacity and so forth). Our work extends the HTR framework baseline, as it is typically susceptible to routing loops, with a new multipath routing scheme preventing routing loops and enhancing load balancing as well as energy-conservation. The presence of these loops was associated to the HTRScore computation, which, in some cases, was producing a divergent view of the network because each node computes and propagates its own perception about the network energy consumption to reach others.

Contributions of this paper are as follows: First, an integrated solution for heterogeneous ad hoc communication scenarios, which is typically lacking from industry and academic research, is given. As a second contribution, network performance is effectively increased under the multipath extension. Finally, to the best of our knowledge, although many methodologies to compare ad hoc network protocols have been published, none of them provide a comparison methodology for use with heterogeneous technologies. Thus, our evaluation methodology sets a direction for the analysis of heterogeneous wireless MANETs and their protocols.

The paper is organized as follows: Section 2 introduces related works. Section 3 describes the proposed framework. Section 4 presents the scenario, simulation environment and the methodology used. The metrics and results taken from our experiments are presented in Section 5. Finally, Section 6 provides concluding remarks and directions for future work.

2   Related Works

Several ad hoc routing protocols have been developed and can be classified according to different criteria. In [4,5], a state-of-the-art review and a set of classification criteria for typical representatives of mobile ad hoc routing protocols are presented. However, none of them consider heterogeneous environments.

2.1   Heterogeneous Ad Hoc Networks

Recently, a number of approaches dealing with interoperability among heterogeneous ad hoc networks have emerged. While some have focused on the interoperation among multiple wireless domains, adopting high level architectures, and having merely sketched the required components (e.g. the translation of different naming spaces) [6-8], others have addressed the heterogeneous routing below the IP layer (underlay level) [9-11].

Ana4 [9] defines a generic layer 2.5 as part of an ad hoc architecture, which relies on the concept of a virtual ad hoc interface. This interface is a logical entity, which abstracts a set of network devices into a single and addressable network component. By designing an ad hoc proposal at layer 2.5, it becomes possible to provide end-to-end communication, regardless of the number of network interfaces in each node. In spite of the similarity with the proposed HTR framework, Ana4 does not offer routing based on context information (e.g. residual energy), nor does it support MANET auto-organization. [10] uses an approach based on MPLS (Multiprotocol Label Switching) [12] to permit forwarding of packets over various heterogeneous links; however, it lacks support for the handling of logical sub-networks and a self-configuration mechanism.

The 3D-Routing protocol [11] makes it possible to compose a fully connected heterogeneous MANET using a 2.5 layer approach. An interesting 3D-Routing concept is the use of roles. The idea is to allow nodes to have one or more roles associated with them. Despite its advantages, 3D-Routing introduces considerable overhead, increasing network bandwidth usage. The reasons for this are two-fold: the lack of a mechanism to control packet flooding, as well as the use of a complex representation scheme for node information and policies. Moreover, it does not describe the mechanism used to compute its routing table.

2.2 Multipath Routing Protocol for Ad Hoc Networks

Several multipath routing protocols were proposed for ad hoc networks [13]. Most of them are based on existing single-path ad hoc routing protocols. In [14], a multipath extension to the well-known AODV [15] protocol is introduced. In [16], a new QoS-aware multipath source routing protocol called MP-DSR, and based on DSR [17] is designed and it focuses on a new QoS metric to provide increased stability and reliability of routes. [18] introduces a multipath Dijkstra-based algorithm to obtain multiple routes and it uses a proactive routing protocol based on OLSR [19]. [20] recommends a scheme that uses two disjoint routes for each session using an on-demand approach.

Notwithstanding the important contributions of these existing solutions, there is still a need for an efficient multipath scheme to connect heterogeneous devices seamlessly.

3 HTR Overview

HTR offers a proactive cross layer routing protocol. Being proactive, HTR ensures that a path is computed as early as possible. The control messages of HTR are sent utilizing MAC layer datagrams. HTR was developed for mobile ad hoc network; it abstracts multi and heterogeneous interfaces to construct a self-organized heterogeneous communicating ad hoc network. An HTR node may have many interfaces, with similar or different technologies such as Wi-Fi, Bluetooth, WiMAX and LTE, but must have only one IP address. To guarantee address uniqueness in an evolving environment, HTR adopts the Network Address Allocation Method proposed in [21].

HTR is based on the OLSR protocol and includes HELLO messages and Topology Control (TC) messages, as well as MPR for reduction of traffic flooding. However, HTR uses additional metrics such as link quality information and node device capabilities to choose MPR nodes. In addition to the features previously mentioned, HTR uses control messages to piggyback service propagation and policies that are based on human roles.

The majority of routing approaches use only the hop count during routing path computation [4]. In contrast, HTR uses a cost metric called HTRScore that is defined considering factors such as the awareness of link conditions and power efficiency in order to perform path computation. The HTRScore formula can be seen in (1).

$$HTRScore\ (i,j) = \frac{e_{i,j}^{\alpha}}{(1 - \rho_{i,j})^{\beta}} \cdot \frac{E_i^{\gamma}}{R_i^{\theta}} \tag{1}$$

Where i is the source node; j is the destination neighbor; $e_{i,j}$ is the transmission energy required for node i to transmit an information unit to its neighbor j; $\rho_{i,j}$ is the probability to lose a packet sent from i to j; R_i is the residual energy of node i; and E_i is the initial battery energy of node i.

The symbols α, β, γ, and θ represent nonnegative weighting factors for each described parameter. Note that if all weights are equal to zero, then the lowest-cost path is the shortest path, and if only γ and θ are equals to zero, then the lowest cost path is the one that will require the least energy consumption, considering retransmission or not, regarding the value of β. If γ is equal to θ then normalized residual energy is used, while if only θ is equal to zero then the absolute residual energy is used. In case all three parameters α, γ and θ are equal to zero, then only the paths with best link stability are emphasized. The stability of a link is given by its probability to persist for a certain time span.

Two modules compose the HTR framework [2]. The former is the bootstrap module wherein the startup and configuration of a node (i.e. assignment of IP address and link layer adaptive configuration) is provided. The latter designates how the routing module takes over path computation.

3.1   Routing Module

The HTR Routing Module adopts the Dijkstra Algorithm to perform path computation, however it utilizes the HTRScore metric to weigh the edges of the network graph and then based on this, calculates the best (i.e. with the smallest cost) routing path for each available destination. In the initial approach, only one route was constructed for each probable destination. Similarly to [18,22], multipath routing is set to increase the network performance by decreasing the congestion, end-to-end delay, and packet loss. Added benefits include the mitigation of routing loops due to the multipath nature of our MANETs.

3.2   Multipath

Unlike the single path strategy, with a multipath approach different paths are computed between source and destination. Multipath routing could offer several benefits [23] such as load balancing, fault tolerance, routing loop prevention, higher aggregate bandwidth, energy-conservation, lower end-to-end delay, security, bottlenecks and congestion avoidance [13,24].

During the HTR evaluation, we noticed the occurrence of routing loops, which were significantly decreasing the performance of the protocol. The presence of these loops was associated to the HTRScore computation, which, in some cases, was producing divergent views of the network. Considering that backup routes could be used to prevent or effectively decrease the occurrence of routing loops, a new process for path computing was developed. This new process uses a multipath routing algorithm based on the algorithm introduced in [18], the MultipathDijkstra, to mitigate the routing loops. Our modified version of the MultipathDijkstra algorithm is shown as follows:

```
MultiPathDijkstra (s, d, G, N)
    HTRScore₁ ← HTRScore
    G₁ ← G
    for i ← 1 to N do
        SourceTreeᵢ ← Dijkstra (Gᵢ, s)
        Pᵢ ← GetPath(SourceTreeᵢ, d)
        if e or Reverse(e) is in Pᵢ then
            HTRScoreᵢ₊₁← f_p (HTRScore ᵢ (e))
        else if Head(e) is in Pᵢ  then
            HTRScoreᵢ₊₁ (e) ← f_e (HTRScoreᵢ(e))
        else
            HTRScoreᵢ₊₁ (e) ← HTRScoreᵢ(e)
        end if
    end for
```

The general principle of this algorithm occurs at step i which looks for the shortest path Pi to the destination d. Then the edges of Pi or pointing to Pi have their cost increased in order to prevent the next step from taking this same path again. The fp function is used to increase the cost of arcs belonging to the path Pi. This encourages future paths to use different arcs but not different vertices. fe is used to increase the cost of the arcs that would lead to the vertices of the previous path, Pi. The fp and fe functions are used to get link-disjoint paths or node-disjoint routes as necessary. The Dijkstra (Gi, s) is the standard Dijkstra's algorithm which provides the source tree of shortest paths from vertex s in graph G; GetPath(ST,d) is the function that extracts the shortest path to d from the source tree ST; the function $Reverse(e)$ gives the opposite edge of e; $Head(e)$ provides the vertex edge e points.

To avoid the creation of similar paths, the fp and fe functions consider vertices and edges used in previous paths. In contrast to the approach presented in [18], which updates the scores by multiplying them by a constant, our proposal multiplies them (the HTRScores) by the number of times each vertex/edge was used.

To prevent routing loops, the approach adds a header to the packet whereupon every node of the path records its identification, creating a list of nodes that has already processed a packet. When received, the node inspects the header of the packet and searches the routing table for the next node to which it is possible to forward the packet, considering the backup paths defined, but excluding the nodes already on the list. If there is no alternative path available, the packet is discarded to avoid unnecessary use of resources. The header is removed when the packet is received by the destination node. The multipath routing algorithm computes a default value of ten distinct paths to any given destination, this value may not be the optimal one; however, it drastically decreases the amount of routing loops to a minimum margin. The analysis of the multipath HTR routing table computation process is presented in Section 5.

4   Methodology and Simulation Model

This section presents the adopted methodology. Firstly, the scenario is described in detail as well as its topology and behavior. Next, the configuration and execution of the simulation are shown. The metrics collected in this scenario and their results are analyzed and discussed in the third subsection.

Fig. 1 shows the chosen topology scenario. It is composed of heterogeneous technologies and the main idea is to send traffic from nodes in the extremities regions (A). Nodes in (A) and (B) have only Wi-Fi interfaces, nodes in the (C) area are bridges responsible for changing Wi-Fi to WiMAX or LTE and vice-versa. Finally, those nodes communicate with the other line of bridges through the tower (D). Therefore, the routes made between nodes in (A) have to pass through, at least, one node from (B), one node from (C), a tower (D), a node from (C), another node from (B) to, finally, arrive in destination node in (A). Thereby, it is possible to have several routes connecting extremity nodes.

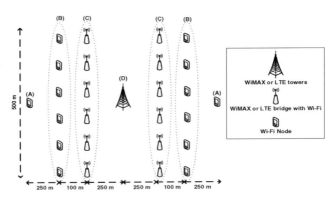

Fig. 1. Simulated scenario with heterogeneous technologies

To evaluate this scenario we used the ns-3 (Network Simulator 3), and the configuration shown in Table 1. To ensure the heterogeneity of the network, the setup included three distinct technologies (Wi-Fi, WiMAX, and LTE) where two of them are used at each time, i.e., each protocol is executed in the scenario twice, one using Wi-Fi and WiMAX and a second time using Wi-Fi and LTE.

Table 1. Configuration parameters of simulation

Parameters	Values
Simulator	ns-3
Routing Protocol	OLSR / HTR / Multipath HTR
Simulation Area	1200m x 500m
Simulation Time	1010s
Applications	CBR
Application Packet Size	512 bytes
Transmission Interval	0.1s
CBR start-end	10 – 1010s
Transport Protocol	UDP
Network Protocol	IPv4
IP Fragmentation Unit	2048
Data Rate	300 / 600 / 1000 Kbps

Table 2 details the values of the configuration parameters for each technology used in the simulation, whereas Table 3 shows the configuration for each propagation loss model in use.

Table 2. Technologies configuration

Parameters	Technologies		
	Wi-Fi	WiMAX	LTE
Physical layer model	PHY 802.11g	PHY 802.16	PHY 3GPP LTE
Wireless channel frequency	2.4 GHz	5 GHz	1929-1980 Mhz (Uplink) / 2110-2170 Mhz (Downlink)
Propagation loss model	Two Ray Ground	COST-Hata	Friis
Transmission power	35-39 dBm	30 dBm	10 dBm (UE) / 30 dBm (eNb)
Modulation	OFDM 64 QAM	OFDM 16 QAM	OFDM QPSK / 16 QAM / 64 QAM
Coding rate	3/4	1/2	1/2; 2/3; 3/4

The initial energy level of each node was established in order to guarantee that at the end of the simulation all nodes would maintain at least 25% of resilient energy. This definition allows for better control of the HTRScore variation and prevents the deactivation of nodes. Each node uses a power source based on the Panasonic CGR18650DA [25] Li-Ion battery and for each technology in use, a corresponding model for its energy consumption was defined based on real radio chips [26-28]. Table 4 shows the radio chip consumption model and its parameters for each technology, as it draws power from the battery source.

Table 3. Propagation loss model configuration

Propagation Model	Default Parameters	Values
COST-Hata	Center Frequency	2.3 GHz
	Base Station	50m
	Mobile Station Antenna Height	3m
	Minimum Distance	0.5m
Friis	Wave Length	58.25mm
	System Loss	1
	Minimum Distance	0.5m
Two Ray Ground	Wave Length	58.25mm
	System Loss	1
	Minimum Distance	0.5m
	Height above Z	1m

Table 4. Radio chip Energy Models

Technology	Radio Chip	Parameters	Values
Wi-Fi	Texas Instru-ments CC2420	Supply voltage	2.5 V
		TX current	17.4 mA
		RX current	19.7 mA
		IDLE, CCA BUSY, SWITCHING current	0.426 mA
WiMAX	Atmel AT86RF535B	Supply voltage	3.3 V
		TX current	315 mA
		RX current	270 mA
		IDLE current	2.5 mA
		SCANNING current	135 mA
LTE	Infineon BGA777L7	Supply voltage	2.8 V
		TX current	4.2 mA
		RX current	4.2 mA
		IDLE current	0.53 mA

5   Results and Discussion

5.1   Statistic Methods

The statistical methodology applied in this work is: (1) for each metric, an initial sample with size 50 was collected (the size 50 was chosen based on the central limit theorem), (2) the samples were tested for normality using the Shapiro-Wilk's W statistic test [30]; (3) 95% is the confidence interval considered.

5.2   Throughput

This metric represents the average data delivery bitrate at a given destination and takes into account only the received data packets, disregarding the control messages. Figures 2 and 3 illustrate the comparison between the Multipath HTR (i.e. with the proposed multipath extension), HTR (i.e. the original single-path protocol), and OLSR regarding throughput. Fig. 2 shows the results for LTE and Wi-Fi nodes and Fig. 3 for the combined use of WiMAX and Wi-Fi. One notes that our proposal achieves very similar throughput results to OLSR and HTR at a 300 Kbps traffic rate. Nonetheless, in contrast, our extension shows higher delivery rates at 600 Kbps, an increase of more than half (approximately 57%) and a twofold increase (225%) for LTE and Wi-Fi when comparing with OLSR and HTR, respectively. With the same bitrate, it shows an increase of 11% and 14% for WiMAX and Wi-Fi when compared with OLSR and HTR. At 1 Mbps, the gain reaches around 120% and 476% for LTE and Wi-Fi respectively and only 2% and 13.6% for WiMAX and Wi-Fi respectively.

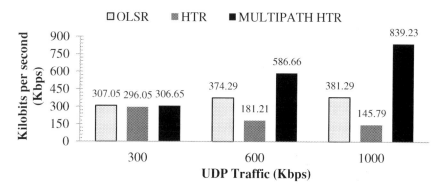

Fig. 2. Throughput of LTE and Wi-Fi devices

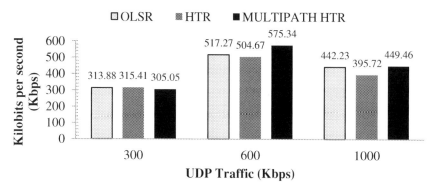

Fig. 3. Throughput of WiMAX and Wi-Fi devices

5.3   Expended Energy (EE)

Fig. 4 shows the expended energy results for WiMAX and Wi-Fi and Fig. 5 shows results for the LTE and Wi-Fi scenario. This metric indicates the energy consumption of the network and is obtained by summing the initial energy level of all network nodes and then subtracting this value by the sum of the resilient energy. In both cases, our proposal shows minor impact on network energy consumption, it increased nearly 4.4% for the LTE and Wi-Fi scenario and 8.3% for the WiMAX and Wi-Fi scenario when compared to OLSR. In the case of our multipath extension, the overhead is due to the additional header information included for routing loop control. Furthermore, our routing strategy achieves a higher throughput, which also understandably impacts energy consumption.

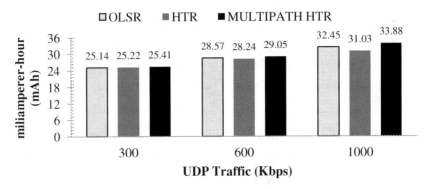

Fig. 4. Expended Energy for the scenario using LTE and Wi-Fi devices

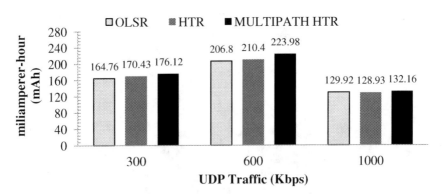

Fig. 5. Expended Energy for the scenario using WiMAX and Wi-Fi devices

5.4   Packet Loss Ratio (PLR)

Figures 6 and 7 show the results of PLR. In Fig. 6 (PLR with LTE and Wi-Fi), our proposal shows a lower packet loss ratio, in which OLSR had 3, 10 and 3.7 times more packet loses than our proposal at 300, 600 and 1000 Kbps rate, respectively. Furthermore, the HTR had a higher PLR when comparing with its correspondent multipath version: about 7, 18 and 5 times higher, respectively, since routing loops occurred.

In Fig. 7 (PLR with WiMAX and Wi-Fi), our proposal had the same PLR when compared with OLSR and 4 times less PLR than the single-path HTR at 300 Kbps rate, and had nearly 4 times less PLR than the OLSR and the single-path HTR at 600 Kbps, however at 1000 Kbps had almost the same PLR as the OSLR and single-path HTR.

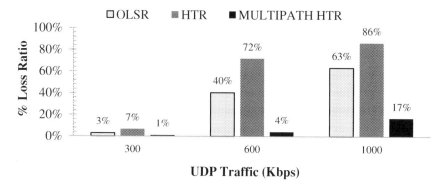

Fig. 6. PLR for the scenario using LTE and Wi-Fi devices

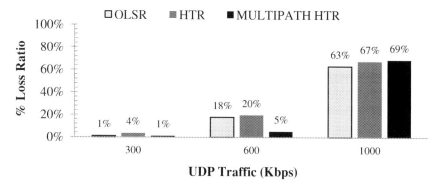

Fig. 7. PLR for the scenario using WiMAX and Wi-Fi devices

5.5   End-to-End Average Delay

Figures 8 and 9 show the results of end-to-end average delay. In Fig. 8 (end-to-end delay with LTE and Wi-Fi technologies), our proposal shows a lower delay, a decrease of 44.6%, 51% and 35.4% when compared with OLSR and 47.5%, 51% and 30.2% when compared with the HTR at 300, 600 and 1000 Kbps, respectively.

In Fig. 9 (end-to-end delay with WiMAX and Wi-Fi technologies), the end-to-end average delay was about the same at 300 Kbps, but at 600 and 1000 Kbps our approach shows a lower end-to-end, a decrease of 96% and 77.2% respectively when compared with OLSR and single-path HTR.

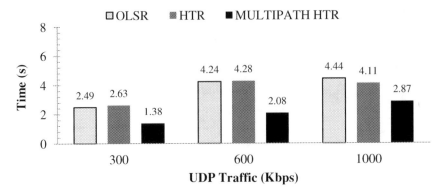

Fig. 8. End-to-end average delay for the scenario using LTE and Wi-Fi devices

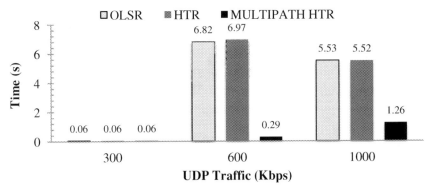

Fig. 9. End-to-end average delay for the scenario using WiMAX and Wi-Fi devices

5.6   Lessons Learned

Our results show that in terms of delivery data ratio, end-to-end average delay and energy expended, our proposal improved the network performance with minimal impact on energy consumption. Moreover, we surpass the data delivery and reduce significantly the end-to-end delay when comparing against the single-path HTR and the well-known OLSR.

Additionally, we observe that the application of different heterogeneous technologies results in different network efficiencies. Moreover, our results indicate that the use of LTE rather than WiMAX leads to better network improvement since first achieved, with minor end-to-end delay and packet losses, a better throughput and improved energy consumption. Furthermore, we note the decrease of throughput and increase of packet loss ratio when using WiMAX technology at 1 Mbps. This behavior can be related to the maximum date rate expected for the technology when using the selected modulation scheme, coding rate and propagation loss model. In [29], the author provides a prediction (2.55 Mbps for uplink using our parameters) of the WiMAX MAC layer throughput when varying the modulation scheme, packet length, coding rate and number of users and concludes that the throughput decreases for smaller packet sizes and for a larger number of users. Thus, for this reason we conclude that the WiMAX base station was overloaded at 1 Mbps traffic rate.

6   Concluding Remarks and Future Works

In this work, we addressed the problem of routing loops on HTR, a protocol for connecting heterogeneous devices based on OLSR, by proposing a multipath extension that offers several benefits such as load balancing, routing loop prevention, energy-conservation, low end-to-end delay, congestion avoidance, among others.

Additionally, we performed a comparative analysis of our proposal with the single-path HTR and the widely used OLSR protocol. The evaluation was validated through simulation (using ns-3) and makes use of heterogeneous technologies such as WiMAX, LTE and Wi-Fi. Our results show that the proposed solution provides a more responsible protocol that effectively improves network performance by increasing data delivery and reducing the end-to-end delay without a major impact on network energy consumption.

For future work we consider implementing an analytic model to compute the optimal quantity of distinct paths necessary for the multipath scheme. We would also be interested in including QoS requirements, improving routing context-awareness based on human roles or other context information and, finally, adding security features.

References

1. Li, J., Luo, S., Das, S., McAuley, T., Patel, M., Staikos, A., Gerla, M.: An Integrated Framework for Seamless Soft Handoff in Ad Hoc Networks. In: MILCOM 2006, pp. 1–7. IEEE (2006)
2. Souto, E., Aschoff, R., Lima Junior, J., Melo, R., Sadok, D., Kelner, J.: HTR: A framework for interconnecting wireless heterogeneous devices. In: 2012 IEEE Consumer Communications and Networking Conference (CCNC), pp. 645–649 (2012)
3. Calafate, C.M.T., Garcia, R.G., Manzoni, P.: Optimizing the implementation of a MANET routing protocol in a heterogeneous environment. In: Proceedings of the Eighth IEEE Symposium on Computers and Communications, ISCC 2003, pp. 217–222. IEEE Comput. Soc. (2003)
4. Liu, C., Kaiser, J.: A Survey of Mobile Ad Hoc network Routing Protocols (2003)
5. Abolhasan, M., Wysocki, T., Dutkiewicz, E.: A review of routing protocols for mobile ad hoc networks. Ad Hoc Networks 2, 1–22 (2004)
6. Crowcroft, J.: Mobile Ad-hoc Intern-domain Networking. In: Annual Conference of ITA (2007)
7. Chau, C., Crowcroft, J., Lee, K., Wong, S.H.Y.: IDRM: Inter-Domain Routing Protocol for Mobile Ad Hoc Networks. Computer (2008)
8. Schmid, S., Eggert, L., Brunner, M., Quittek, J.: Towards autonomous network domains. In: Proc. IEEE 24th Annu. Jt. Conf. IEEE Comput. Commun. Soc., pp. 2847–2852 (2005)
9. Boulicault, N., Chelius, G., Fleury, E.: Ana4: a 2.5 Framework for Deploying Real Multi-hop Ad hoc and Mesh Networks. Science (80-X), 1–24 (2006)

10. Untz, V., Heusse, M., Rousseau, F., Duda, A.: Lilith: an interconnection architecture based on label switching for spontaneous edge networks. In: The First Annual International Conference on Mobile and Ubiquitous Systems: Networking and Services, MOBIQUITOUS 2004, pp. 146–151. IEEE (2004)
11. Jacinto, B., Vilaça, L., Kelner, J.: 3D routing: a protocol for emergency scenarios. In: Proc. 6th Int. Wirel. Commun. Mob. Comput. Conf., pp. 519–523 (2010)
12. Rosen, E., Viswanathan, A.: RFC 3031 - Multiprotocol Label Switching Architecture (2001), http://tools.ietf.org/html/rfc3031
13. Tarique, M., Tepe, K.E., Adibi, S., Erfani, S.: Survey of multipath routing protocols for mobile ad hoc networks. J. Netw. Comput. Appl. 32, 1125–1143 (2009)
14. Marina, M.K., Das, S.R.: On-demand multipath distance vector routing in ad hoc networks. In: Proceedings Ninth International Conference on Network Protocols, ICNP 2001, pp. 14–23. IEEE Comput. Soc. (2001)
15. Perkins, C., Royer, E., Das, S.: RFC 3561 - Ad hoc On-Demand Distance Vector (AODV) Routing (2003), http://tools.ietf.org/html/rfc3561
16. Leung, R., Poon, E., Chan, A.-L.C.: MP-DSR: a QoS-aware multi-path dynamic source routing protocol for wireless ad-hoc networks. In: Proceedings LCN 2001 26th Annual IEEE Conference on Local Computer Networks, pp. 132–141. IEEE Comput. Soc. (2001)
17. Maltz, D., Hu, Y., Johnson, D.: RFC 4728 - The Dynamic Source Routing Protocol (DSR) for Mobile Ad Hoc Networks for IPv4 (2007), http://tools.ietf.org/html/rfc4728
18. Yi, J., Adnane, A., David, S., Parrein, B.: Multipath optimized link state routing for mobile ad hoc networks. Ad Hoc Networks 9, 28–47 (2011)
19. Clausen, T., Jacquet, P.: RFC 3626 - Optimized Link State Routing Protocol, OLSR (2003), http://www.ietf.org/rfc/rfc3626.txt
20. Lee, S.-J., Gerla, M.: Split multipath routing with maximally disjoint paths in ad hoc networks. In: ICC 2001, IEEE International Conference on Communications, Conference Record (Cat. No.01CH37240), pp. 3201–3205. IEEE (2001)
21. Aschoff, R., Souto, E., Kelner, J., Sadok, D.: (WO2011009177) Network Address Allocation Method (2011)
22. Devi, R., Rao, S.: QoS Enhanced Hybrid Multipath Routing Protocol for Mobile Adhoc Networks. Int. J. Distrib. Parallel Syst. 3, 89–105 (2012)
23. Yi, J., Cizeron, E., Hamma, S., Parrein, B.: NET 08-2 - Simulation and Performance Analysis of MP-OLSR for Mobile Ad Hoc Networks. In: 2008 IEEE Wirel. Commun. Netw. Conf., pp. 2235–2240 (2008)
24. Mueller, S., Tsang, R.P., Ghosal, D.: Multipath routing in mobile ad hoc networks: Issues and challenges. In: Calzarossa, M.C., Gelenbe, E. (eds.) MASCOTS 2003. LNCS, vol. 2965, pp. 209–234. Springer, Heidelberg (2004)
25. Panasonic: Panasonic CGR18650DA (Lithium Ion Batteries: Individual Data Sheet) (2007)
26. Texas Instruments: CC2420 2.4 GHz IEEE 802.15. 4/ZigBee-ready RF Transceiver (2004)
27. Atmel: Atmel AT86RF535B (3.5 Ghz WiMAX Transceiver) (2007)
28. Infineon: Infineon BGA777L7 Data Sheet (2009)
29. Pareit, D., Petrov, V., Lannoo, B., Tanghe, E., Joseph, W., Moerman, I., Demeester, P., Martens, L.: A Throughput Analysis at the MAC Layer of Mobile WiMAX. In: 2010 IEEE Wireless Communication and Networking Conference, pp. 1–6. IEEE (2010)
30. Shapiro, S.S., Wilk, M.B.: An Analysis of Variance Test for Normality (Complete Samples). Biometrika 52, 591–611 (1965)

Anticipation of ETX Metric to Manage Mobility in Ad Hoc Wireless Networks

Sabrine Naimi[1], Anthony Busson[2], Véronique Vèque[1],
Larbi Ben Hadj Slama[3], and Ridha Bouallegue[3]

[1] Laboratory of Signals and Systems Université Paris Sud, Supélec, CNRS
[2] University Claude Bernard Lyon 1, LIP (ENS Lyon, INRIA, CNRS, UCBL)
[3] Innov'COM Laboratory Higher School of Communication, Tunisia

Abstract. When a node is moving in a wireless network, the routing metrics associated to its wireless links may reflect link quality degradations and help the routing process to adapt its routes. Unfortunately, an important delay between the metric estimation and its inclusion in the routing process makes this approach inefficient. In this paper, we introduce an algorithm that predicts metric values a few seconds in advance, in order to compensate the delay involved by the link quality measurement and their dissemination by the routing protocol. We consider classical metrics, in particular ETX (Expected Transmission Count) and ETT (Expected Transmission Time), but we combine their computations to our prediction algorithm. Extensive simulations show the route enhancement as the Packet Delivery Ratio (PDR) is close to 1 in presence of mobility.

Keywords: ETX, Metric, Ad Hoc, Wireless Network, Mobility, Anticipation.

1   Introduction

Recent developments in wireless communication technologies and the emergence of mobile units (laptop, smart-phone, etc.) have made possible access to the network anywhere at any time. As the node may be mobile while staying connected, mobility management has to be efficiently performed to ensure a good quality of experience to the users. In cellular networks or IEEE 802.11 infrastructure mode networks, mobile nodes are associated with the best base station or access point according to signal strength. Prediction algorithms [1], based on the quality of the wireless links (signal strength, SINR, etc.), may be used to anticipate the association to the next infrastructure point. When the node is moving, these complex handover mechanisms [2] allow the node to associate transparently from an infrastructure point (Base Station or Access Point) to another. If the mobile node is no more in the same IP sub-network, Layer 3 mechanisms such as Mobile IP [3] are provided to operate IP address changes.

These handover mechanisms and the IP mobility management do not hold in mobile ad hoc networks. Indeed, instead of managing a single link, each node

S. Guo et al. (Eds.): ADHOC-NOW 2014, LNCS 8487, pp. 29–42, 2014.
© Springer International Publishing Switzerland 2014

senses the wireless links with all nodes within its radio range. Moreover, there is no IP address change to manage. Nodes mobility is then managed by the routing process rather than layer 2 handovers and Mobile IP.

An efficient mobility management can then be performed in two ways: an efficient link sensing algorithm may prevent link failure and invalidates links before they are actually unusable [4], or metrics associated to each link may reflect link degradations due to nodes mobility. An increase of these metrics will involve a path change from the routing process that will switch from degrading links to more stable links. Unfortunately, as link sensing is performed at the network layer, there is always a bias between the real link states and the estimated ones. This bias is two-fold: a time shift of the measured quality exists due to frequency of the probes used to manage the link, and a measurement error is caused by the lack of information from lower layers. Metrics obtained from link states information suffer from these two problems. Moreover, metrics dissemination by the routing protocol can cause an appreciable delay between the time at which the link states are estimated and their inclusion in the route calculations. All these uncertainties lead to high packet loss rate in presence of mobility.

In this paper, we introduce a new method to anticipate the values of routing metrics. Our main idea consists in using a prediction algorithm applied to received signal strengths to estimate future values of the metrics in order to compensate the delay involved by the link quality measurements and their dissemination by the routing protocol. This technique is used in infrastructure based networks, but to our knowledge, has not been proposed to anticipate link metrics in ad hoc networks. Our contributions deal with a new metric calculation method to manage the mobility problem. We consider classical metrics, in particular ETX [5] (Expected Transmission Count) and ETT [6] (Expected Transmission Time), but we combine their computation to our prediction algorithm. We focus on mobile ad hoc networks but it does not prevent its use in other contexts. Simulation results of our approach show that when our algorithm is well configured the Packet Delivery Ratio (PDR) is close to 100% in presence of high mobility.

The paper is organized as follows. In Section 2, related works on prediction algorithm in infrastructure networks and the main routing metrics used in ad hoc networks are presented. Section 3 highlights the problem statement. Section 4 introduces our prediction method to anticipate the future routing metrics values. In Section 5, we detail the method used to predict the signal strength and its performance. Simulation scenarios and results are described in Section 6. We conclude in Section 7.

2   Related Works

Prediction algorithm. Mobility prediction approaches can detect the next base station for resource reservation prior to the actual handover. This approach has attracted several research interests [1], [7], [8], [9]. The handoff integration in Layer 2 and Layer 3 has been proposed in [10]. The authors describe a predictive handoff approach for Mobile IP to support fast or smooth handoff. They use a real-time mobility estimation to predict a future handoff. In [11], the Received Signal Strength Indicator (RSSI) and the direction of mobile node are also used as input parameters to a fuzzy inference system to predict the handover decision. The paper presents Mobility and Signal Strength-Aware Hand-off Decision (MSSHD) approaches to predict the handover decision. A new machine learning based prediction system is considered in [12]. The authors use information available in cellular networks as Channel State Information (CSI) and handover history, and by embedding Support Vector Machines (SVMs) into an efficient pre-processing structure to resolve prediction. All these mechanisms have been designed for cellular networks, or infrastructure based networks, where a mobile only needs to select the best link (associated to the best infrastructure point). These techniques are thus not suitable in ad hoc networks where all links may be potentially used.

Also, many approaches using mobility prediction algorithm have been proposed in the context of ad hoc networks. A mobility prediction based on neural network and a recording the geographical location of nodes is introduced in [13]. In [14], an adaptive mobility prediction method is proposed. It predicts the future distance of 2 neighboring nodes using automaton. A mobility prediction approach based on a hello protocol is developed in [15]. Each node and its neighborhood predict their own position through an autoregression-based mobility model. All these techniques have been designed for ad hoc networks with the aim to predict future nodes location or the distance between them. But generally, nodes do not know their positions as they are not equipped with geolocation devices, and mobility is often unpredictable (due to directions change for instance). Moreover, the variations of the link qualities are not only linked to the distance between nodes, but also to the environment where the node evolves (city, building, etc.) and complex radio phenomena (fading, shadowing, etc.). Therefore, we think a more appropriate link quality prediction method should be based on signal strength rather than geographical locations.

Metrics. Routing protocols use metrics to determine the shortest-path from a source to a destination. A metric is a numerical value associated to each link. When additive metrics are considered, the shortest-path is the one for which the sum of the metrics of its links is minimal.

Metrics are generally not specifically defined to manage mobility but there are a few exceptions. Some metrics try to reflect the mobility of the nodes, for instance the average number of link breaks [16], or the Link Duration [17] (LD) that corresponds to the time where two nodes are within the transmission range of each other. These metrics aim at capturing the nodes mobility, and favor stable routes. But they are strongly dependent on the mobility patterns, useless when all nodes are mobile, and may be counterproductive when all nodes are fixed. Other metrics, not particularly related to mobility could nevertheless be efficient to manage mobility. These metrics are supposed to reflect link quality and increase when a node recedes from a neighbor. The routing protocol may find better routes and switch from receding to more stable links. Therefore, these metrics have two important benefits: to manage mobility efficiently and to favor efficient link according to a performance criteria like bandwidth, loss rates, or delay. A lot of metrics have been proposed in literature [18]. In this paper, we focus on the most popular metrics, ETX [5] and ETT [6], which have been shown among the most efficient [18]. Moreover, we present the LD (Link Duration) metric in detail as it will be used in our simulations.

In order to take into account the link quality, more complex metrics have been defined as ETX [5]. Calculation of ETX uses the forward delivery ratio (**df**) and the reverse delivery ratio (**dr**) of the link. **df** is the probability that a data packet reached the destination successfully. **dr** is the probability that a data packet is received successfully. ETX is formally defined as follows:

$$ETX = \frac{1}{df * dr} \tag{1}$$

ETX evaluates the mean number of transmissions and retransmissions needed to send a data packet. Therefore, ETX promotes links with low loss rates. But ETX does not distinguish links with different capacities. To solve this problem, the ETT (Expected Transmission Time) metric [6] was proposed. The ETT calculation is based on ETX but takes into account both data packets length L and transmission rates B:

$$ETT = ETX * \frac{L}{B} \tag{2}$$

ETT can be seen as the time expected to transmit a packet of length L. It multiplies the number of attempts (the ETX metric) to send the packet by the time to transmit the packet on the link ($\frac{L}{B}$).

The Link Duration metric [19] is calculated by measuring the life duration of a link between two nodes. In case of mobility, the authors of [20] have shown that LD can be used as a management tool for mobility. For a given source-destination pair, the path with the maximum path duration is selected as the best route. The path duration is the minimum LD of all the links forming the path. The idea is to consider that routes with long path duration tend to be more stable. In our simulations, we shall show that it is very dependent on mobility models and wrong in practice.

3   Problem Statement

As we claim, ETX and ETT metrics could be efficient to manage mobility. Indeed, when a node is moving away from a neighbor the quality of the link degrades and the Frame Error Rate increases, as well as ETX and ETT metrics. This link metric becoming high, the routing protocol should find paths with lower metrics (if available) and packet loss should be avoided. But a certain number of phenomena prevent the realization of this ideal scenario.

In Fig. 1, we plotted the Frame Error Rate function for two nodes depending on the Signal on Noise Ratio (SNR). In this simulation, SNR is strictly decreasing

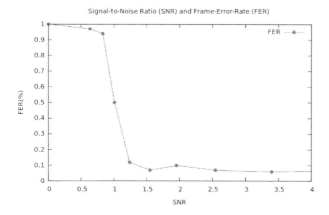

Fig. 1. FER for 2 nodes versus SNR

with the distance according to a path-loss function. It has been obtained with the Network Simulator NS-3 [21]. We observe that FER behaves as a step function. It switches suddenly from 100% to $5 - 10\%$ when the SNR is approximately equal to 1.11 and stays constant otherwise. It means that if a link is receding due to mobility, FER will be close to $5 - 10\%$ until a certain distance where it will increase very sharply. This phenomenon is not particular to our model but it is the FER behavior that we observe in general. ETX or ETT will reflect the degradation only when the FER would have reached a high FER. This lossy link will be still used for a while leading to packet losses.

Moreover, calculation of these metrics is generally processed at the network layer and is based on the reception or non-reception of some specific control packets (Hello packets for instance). It introduces a bias in the df and dr estimations used in ETX or ETT with regard to the instantaneous Frame Error Rate. Also, routing protocols introduce a delay between the ETX/ETT calculations and the time at which they are taken into account in the route computations. These delays are due to the metrics dissemination and the delay to process information and compute routes. All these phenomena add a major delay between the FER degradation and its consideration in the path taken by the data packets. ETX and ETT are then quite inefficient to manage mobility.

The use of information from the physical and MAC layers can help us to estimate instantly and accurately FER, link quality and to prevent link degradation. The main idea of our algorithm is to anticipate the values of link metrics a few seconds in advance. This anticipation allows us to compensate the different delays induced by the FER estimation and the routing protocol. Our mechanism allows these metrics to be also suitable in the context of mobility. Ideally, when our algorithm is perfectly configured, it allows the routing protocol to consider link states in real time and to avoid losses.

4   Metric Anticipation Algorithm

4.1   Relationship between FER and Signal Strength

Anticipation is made through a prediction mechanism based on the history of link quality measurements. ETX and ETT metrics are mainly based on the Frame Error Rate of the wireless link. Therefore, it is this quantity that we try to anticipate. But, we do not use previous FER measures since it behaves as a step function. The received signal strength is strictly decreasing/increasing as the nodes are moving and is directly related to FER. It is thus easier to forecast a trend for signal strengths than for FER.

We collect and store the signal strength of the packets exchanged between neighbors for each link. This quantity is generally available through the Received Signal Strength Indicator (RSSI) with 802.11 and 3G/LTE cards. These measures are collected only for control packets as the routing daemon may not be informed of the reception of data packets. We assume that each node has a mapping table to match signal strength to FER. This table can be obtained by the network card manufacturer or inferred through a set of measures performed before the ad hoc network deployment. A function, denoted $FER(.)$, is then used to map a signal strength to a FER. We assume also that FER predictions are exchanged in the Hello packets of the routing protocol. It will allow a node to predict FER for the two directions of a link. The choice of this parameter will be confirmed as shown in our experimentation in Section 5.2.

4.2   Metric Anticipation

Our proposal consists in using the classical ETX or ETT metrics when the link quality is good, and in increasing artificially and gradually metric values when links start to deteriorate.

The algorithm is presented in Table 1. It considers the ETX metric, but we could consider ETT as well. The new metric is denoted ETX_ANT. The algorithm is applied at each control packet reception. At each reception, the receiver measures the $LINK\_QUALITY$, either RSSI, SINR (Signal on Interference plus Noise Ratio), or any quantity reflecting the link quality.

Table 1. ETX_ANT ALGORITHM

```
IF ( LINK_QUALITY > TH_Q )
        ETX_ANT=COMPUTE_ETX()
ELSE
        SIGNAL_PRED=SIGNAL_PREDICTION()
        FER_PRED=FER(SIGNAL_PRED)
        ETX_ANT= 1/(FER_PRED * d_f)
ENDIF
```

The *LINK_QUALITY* variable is compared to a predefined threshold denoted TH_Q. We anticipate the value of ETX, only if the quality is lower than this threshold. First, the signal strength is predicted with the *SIGNAL_PREDICTION(.)* function using a past measurement history of signal strength values. One possible method to anticipate signal strength is described in the next section. Then, we map this future signal strength to an expected FER with the *FER(.)* function. It allows the node to estimate the d_r parameter of the ETX metric (see Equation (1)). d_f is computed on the other end of the link (predicted or not depending on the link quality), and obtained through routing control packets. The *ETX_ANT* metric is then computed from these anticipated values.

5   Signal Strength Prediction Algorithm

Several methods have been proposed to predict signal strength [22]. Most of these methods assume a particular radio environment, through given path-loss or fading. We considered the linear regression method proposed and evaluated in [23] as it has a low complexity and does not rely on a particular radio context.

5.1   Linear Regression Method

With a linear regression method, an history of signal measurements, denoted $S = \{s_1, s_2, .., s_n\}$, is stored in a table of the last n measurements updated at each new reception. The different times at which they are collected are stored in the vector $T = \{t_1, t_2, ..., t_n\}$. A linear regression model with the form $S = a + bT$ is used to fit the data. The future signal strength is then estimated as:

$$\hat{s}_{n+p} = a + bp \tag{3}$$

where p is the step of the prediction. a and b are given by:

$$b = \frac{\sum_i s_i t_i - \bar{S}.\bar{T}}{\sum_i t^2 - n\bar{T}^2} \tag{4}$$

$$a = \bar{S} - b\bar{T} \tag{5}$$

$\bar{S}$ (respectively $\bar{T}$) is the mean of vector S (respectively T).

The authors of [23] used a dynamic window (dynamic size of the vectors S and T) to adapt to abrupt changes. During sudden changes, "old" data would not reflect the current trend. The reduction of the window size is thus necessary to discard measurements that are no more pertinent. The size of the history window is initially set to a default value, and changes according to errors observed in the prediction process. If the prediction error is greater than a threshold, the size of the window will be immediately reduced. However, if the error does not exceed the threshold, the size of the window is increased until it reaches its maximum size.

5.2   Experimentations

We evaluated the prediction algorithm described in the previous Section through
a set of experimentations. We configured two laptops in ad hoc mode under Linux
(Ubuntu 12.04 LTS). They were equipped with two extern USB wireless cards.
We performed this evaluation in the outdoor garden of our laboratory. The two
laptops were in line of sight of each other. First, both computers moved in opposite directions at a pedestrian constant speed, then stay more or less at the same
distance before approaching a few seconds and moving away for a second time.
It aims at evaluating the prediction algorithm when there are clear movements
in the signal strength (nodes approaching or moving away) and when the signal
oscillates (the distance between the two PCs does not vary much). In Fig. 2,
we plotted the measured signal and the estimated signal strengths (2 seconds
in advance). It shows that this mechanism successes to predict signal strengths
when nodes are approaching and moving away. As expected, when the signal
strength oscillates, the prediction is less accurate. Several similar experiments
have been performed and led to the same behavior.

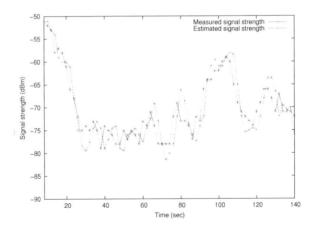

Fig. 2. Comparison of the measured signal strength and the prediction (made 2 seconds
in advance)

6   Simulations

Our next evaluation uses simulation in order to consider more complex environment and multihop routes.

6.1   Parameter Settings

Before performing the evaluation of our proposal, we need to tune the parameters
$TIME$ and TH_Q. We denote $TIME$ (expressed in seconds) the variable that

corresponds to the prediction time (the number of seconds we anticipate the signal strength/metric). *TIME* is related to the variable p in Equation 3. It must be set according to the time needed to disseminate a new metric values in the network and the time to process it (compute routes, etc.). Therefore, it strongly depends on the used routing protocol. In our simulations, we set *TIME* equal to 2 seconds.

The other parameter, TH_Q, must be set in such a way that the prediction begins at least *TIME* seconds before a link breakage. Therefore, a fine tuning of this parameter assumes the knowledge of the radio environment and the maximum relative speed of the mobile nodes. In our scenarios, this information is known. In case of a real deployment, they may be previously inferred through experimentations.

Under the assumption that the physical reception threshold is reached at least *TIME* seconds after the *LINK_QUALITY* threshold, we get:

$$TH_{Rec} \geq PL\left(PL^{-1}(TH_Q) + S_{rel} \times TIME\right) \tag{6}$$

Where $PL(distance)$ is the path-loss function that gives the expected receiving power with regard to the distance, TH_{Rec} is the physical threshold for the frame reception, and S_{rel} the relative speed between nodes. We get:

$$TH_Q \geq PL\left(PL^{-1}(TH_{Rec}) - S_{rel} \times TIME\right) \tag{7}$$

6.2   Simulation Scenarios

We performed simulations using NS-3. We used the Optimized Link State Routing Protocol (OLSR) [24] because it is a proactive link-state protocol for which it was easy to take into account different routing metrics. Link states information (links and metrics) is disseminated periodically through Topology Control messages (TC). We change the TC message format to include metrics values. Also, we changed the algorithm used to update the routing table. We implemented the Dijkstra algorithm instead of the default algorithm of OLSR, optimized for hop count metric. The rest of the OLSR routing protocol has not been modified. The new metric is disseminated only every *TC_INTERVAL* seconds, when the node sends its new TC.

We compare our approach (ETX_ANT and ETT_ANT) with the classical metrics ETX, ETT and LD. LD metric is defined by the difference between the last time that the link was symmetric as managed by OLSR and the time at which the TC is sent. The fixed parameters values used for the simulations are given in table 2.

We consider three scenarios: a chain of nodes, a grid network and a completely mobile ad hoc network. Details of these scenarios are described below. In all scenarios, we use the Log Distance Path Loss Model [25] as the radio propagation model. Therefore, we combine this function with a random variable S modeling the channel randomness. S follows a log-normal distribution with zero mean and

Table 2. Simulation Parameters

Parameters	Numerical Values
Simulation time	62 [sec]
Window size (Chain scenario)	$1200 \times 200\ [m^2]$
Window size (Mesh and ad hoc scenario)	$300 \times 300\ [m^2]$
Number of nodes (Chain scenario)	12
Number of nodes (Grid scenario)	15
Number of nodes (Ad hoc scenario)	25
Wireless technology	IEEE 802.11a
Traffic type	Constant Bit Rate (CBR)
Packet size	1024 bytes
TH_Q	2.23 mWatt
HELLO_INTERVAL (OLSR)	0.25 [sec]
TC_INTERVAL (OLSR)	2 [sec]

different standard deviations σ (σ^2 ranges from 0 to 25). The received signal strength is then $PL(d) + S$ (expressed in dBm).

6.3   Simulation Results

In all curves, each point is the average of 20 simulations and is shown with a confidence interval of 95%. It may appear that the confidence interval lengths are smaller than the symbols used in the plots, making them barely visible.

☐ Chain Scenario

The first topology considered is a chain of 11 nodes which are separated from a distance of 100 meters A further mobile node moves at constant speed along the chain of fixed nodes. This topology is a trivial case that allowed us to test the efficiency of our algorithm, and where results are easy to interpret. The source is the mobile node. It transmits data packets at regular interval (CBR traffic). The destination is the first node in the chain. We make varying the speed of the mobile from $10km/h$ to $70km/h$. The source moves and the application starts after a few seconds to let the time to OLSR to fill the routing tables. The simulation stops when the mobile node arrives at the end of the chain.

The *Packet Delivery Ratio* (PDR) is the ratio between the number of packets received by the destination over the number of packets sent by the source. Fig. 3(a) shows the PDR when using different metrics (Hop Count, LD, ETX, ETT, ETX_ANT and ETT_ANT) and varying the speed of the mobile node, and without fading ($\sigma^2 = 0$). The PDR with ETX_ANT is 100% for all speeds. It empirically proves that nodes mobility is perfectly managed for this scenario. For higher speed, performance of ETT_ANT decreases, but are greater than 98.5% and outperforms the classical metrics. The lower performance of ETT_ANT is due to the Wi-Fi rate (parameter

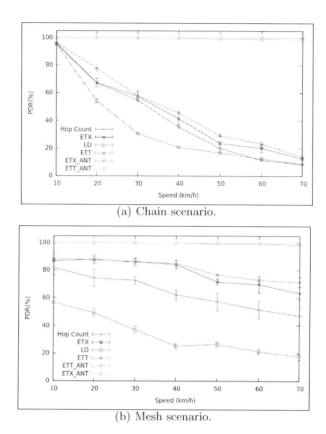

(a) Chain scenario.

(b) Mesh scenario.

Fig. 3. Packet Delivery Ratio in chain and mesh scenarios

B in Equation (2)) that is not predicted in our algorithm, and may lead to inaccurate predictions. LD selects paths with the longest duration which explains why it is the worst metric. Indeed, rather than privileging new links, LD metric chooses the oldest.

Mesh Network Scenario

In this scenario, we consider a mesh network modeled through a grid topology of 9 fixed nodes. 6 nodes are mobile following the Random Waypoint Model (RWP) [26] with constant speed. We plotted the PDR for the different metrics in Fig. 3(b) when varying the speed. We do not consider fading ($\sigma^2 = 0$). Results show that ETX_ANT and ETT_ANT are 100% for all speeds, proving that our mechanism manages perfectly the mobility in this scenario.

Ad hoc Network Scenario

In the third scenario, we considered an ad hoc network where all nodes are mobile and are randomly distributed in an area with size $300 \times 300\ m^2$. Mobility of nodes follows the RWP model.

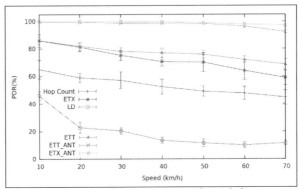

(a) Mobile ad hoc scenario without fading.

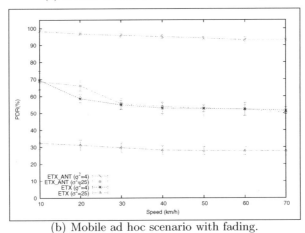

(b) Mobile ad hoc scenario with fading.

Fig. 4. Packet Delivery Ratio in ad hoc scenario with and without fading

In Fig. 4(a), we can observe that anticipated ETX metric is still very close to 100%. The PDR with ETX_ANT does not drop below 97% and ETT_ANT does not exceed 92%. LD metric presents the worst performances: the PDR drops to 11.5%.

To show the impact of fading, we plotted in Fig. 4(b) the metrics ETX and ETX_ANT for different fading value (we vary the standard deviation σ^2). We observe that performances degrade when the variance increases whatever the metric. With small fading ($\sigma^2 = 4$), PDR with ETX_ANT remains greater than 95% showing that the mobility is well managed. For $\sigma^2 = 25$, the PDR is at most 70%. Approximately 25% of the packet losses are due to the fading itself and not to mobility (it is the percentage that we observed when nodes were not mobile). This corresponds to a worst case scenario where fading is very important, and where by definition such randomness is unpredictable. But, PDR for the ETX_ANT metric stays more or less constant with the mobile speeds showing that even with high fading it still predicts properly the metrics. By lack of space, these results have not been shown for the previous scenarios, but they present exactly the same behavior.

7   Conclusion

In this paper, we have presented an algorithm to anticipate the value of routing metrics based on signal strength prediction. This technique is based on a cross-layer approach. This anticipation allows the routing protocol to compensate the delay due to the computation and dissemination of the metrics. We considered ETX [5] (Expected Transmission Count) and ETT [6] (Expected Transmission Time), but we combine their computation to our prediction algorithm. Through a large set of simulations and scenarios, we have shown that our method improves significantly the PDR. When the network is a Mesh network, where the connectivity is ensured, a proper design of our algorithm leads to a PDR of 100%. For other scenarios, packet losses are provoked by network disconnections and fading, but even in these cases our algorithm outperforms the classical metrics.

This work may be extended in different ways. Particular wireless managers have to be considered in order to combine the FER prediction algorithm and the computation of the future link rate. Also, we plan to deploy our solutions on a testbed to evaluate its efficiency in a real environment.

References

1. Menezes, S.L.: Optimization of Handovers in Present and Future Mobile Communication Networks. Doctor of philosophy in computer science, University of Texas at Dallas (2010)
2. Shayea, I., Ismail, M., Nordin, R.: Advanced handover techniques in LTE- Advanced system. In: International Conference on Computer and Communication Engineering (ICCCE), pp. 74–79 (2012)
3. Perkins, C., Johnson, D., Arkko, J.: Mobility Support in IPv6. Internet Engineering Task Force (IETF), RFC 6275 (Proposed Standard) (2011)
4. Ali, H.M., Naimi, A.M., Busson, A., Vèque, V.: An efficient link management algorithm for high mobility mesh networks. In: 5th ACM International Workshop on Mobility Management and Wireless Access (MobiWac), pp. 42–49 (2007)
5. De Couto, D.S.J., Aguayo, D., Bicket, J., Morris, R.: A high-throughput path metric for multi-hop wireless routing. In: 9th Annual International Conference on Mobile Computing and Networking (MobiCom), New York, USA, pp. 134–146 (2003)
6. Esposito, P.M., et al.: Implementing the Expected Transmission Time Metric for OLSR Wireless Mesh Networks. In: 1st IFIP Wireless Days. IEEE (2008)
7. Francois, J.M.: Performing and Making Use of Mobility Prediction. Doctor thesis, University of Lyte (2007)
8. Yavas, G., Katsaros, D., Ulusoy, Ö., Manolopoulos, Y.: A Data Mining Approach for Location Prediction in Mobile Environments. Data Knowl. Eng. 54, 121–146 (2005)

9. Bergh, A.E., Ventura, N.: Prediction Assisted Fast Handovers for Mobile IPv6. In: IEEE MILCOM, Washington DC (2006)
10. Zhang, D.-W., Yao, Y.: A Predictive Handoff Approach for Mobile IP. In: International Conference on Wireless and Mobile Communications (ICWMC), p. 78 (2006)
11. Sadiq, A.S., Abu Bakar, K., Ghafoor, K.Z., Gonzalez, A.J.: Mobility and Signal Strength- Aware Handover Decision in Mobile IPv6 based Wireless LAN. In: International MultiConference of Engineers and Computer Scientists, Hong Kong (2011)
12. Chen, X., Mériaux, F., Valentin, S.: Predicting a User's Next Cell With Supervised Learning Based on Channel States. CoRR (2013)
13. Kaaniche, H., Kamoun, F.: Mobility Prediction in Wireless Ad Hoc Networks using Neural Networks. CoRR (2010)
14. Mousavi, S.M., Rabiee, H.R., Moshref, M., Dabirmoghaddam, A.: Model Based Adaptive Mobility Prediction in Mobile Ad-Hoc Networks. In: International Conference on Wireless Communications, Networking and Mobile Computing (WiCom), pp. 1713–1716 (2007)
15. Li, X., Mitton, N., Simplot-Ryl, D.: Mobility Prediction Based Neighborhood Discovery in Mobile Ad Hoc Networks. In: Domingo-Pascual, J., Manzoni, P., Palazzo, S., Pont, A., Scoglio, C. (eds.) NETWORKING 2011, Part I. LNCS, vol. 6640, pp. 241–253. Springer, Heidelberg (2011)
16. Qin, L., Kunz, T.: Mobility Metrics to Enable Adaptive Routing in MANET. In: IEEE International Conference on Wireless and Mobile Computing, Networking and Communications (WiMob), pp. 1–8 (2006)
17. Sadagopan, N., Bai, F., Krishnamachari, B., Helmy, A.: PATHS: analysis of PATH duration statistics and their impact on reactive MANET routing protocols. In: 4th ACM International Symposium on Mobile Ad Hoc Networking and Computing (MobiHoc), Maryland, USA, pp. 245–256 (2003)
18. Campista, M.E.M., et al.: Routing Metrics and Protocols for Wireless Mesh Networks. IEEE Network 22, 6–12 (2008)
19. Gerharz, M., de Waal, C., Frank, M., Martini, P.: Link Stability in Mobile Wireless Ad Hoc Networks. In: 27th Annual IEEE Conference on Local Computer Networks (LCN), Washington, USA, p. 0030– (2002)
20. Bai, F., Sadagopan, N., Helmy, A.: IMPORTANT: a framework to systematically analyze the Impact of Mobility on Performance of Routing Protocols for Adhoc Networks. In: Twenty-Second Annual Joint Conference of the IEEE Computer and Communications (INFOCOM), pp. 825–835. IEEE Societies (2003)
21. Network Simulator 3, http://www.nsnam.org/
22. Duel-Hallen, A., Hu, S., Hallen, H.: Long Range Prediction of Fading Signals: Enabling Adaptive Transmission for Mobile Radio Channels. IEEE Signal Processing Magazine 17, 62–75 (2000)
23. Long, X., Sikdar, B.: A Real-Time Algorithm for Long Range Signal Strength Prediction in Wireless Networks. In: Wireless Communications and Networking Conference (WCNC), pp. 1120–1125. IEEE (2008)
24. Clausen, T., Jacquet, P.: Optimized Link State Routing Protocol (OLSR). RFC 3626 (Experimental) (2003)
25. Erceg, V., et al.: An Empirically Based Path Loss Model for Wireless Channels in Suburban Environments. IEEE Journal on Selected Areas in Communications 17 (1999)
26. Hyytiä, E., Virtamo, J.: Random waypoint mobility model in cellular networks. Wireless Networks 13, 177–188 (2007)

O-SPIN: An Opportunistic Data Dissemination Protocol for Folk-Enabled Information System in Least Developed Countries*

Riccardo Petrolo[1], Thierry Delot[1,2], Nathalie Mitton[1],
Antonella Molinaro[3], and Claudia Campolo[3]

[1] Inria, France
[2] LAMIH, University of Valenciennes, France
[3] DIIES, Universitá Mediterranea di Reggio Calabria, Italy

Abstract. Without universal access to the Internet, Least Developed Countries are left by the wayside of the digital revolution. Research is underway to overstep the barrier to the development of information technology services in these areas. In this context, the *Folk-IS* (Folk-enabled Information System) is a new *fully decentralized* and *participatory* approach, in which, each individual can transparently perform data management and networking tasks through highly secure, portable, and low-cost storage and computing personal devices, as physically moving, so that global services can finally be delivered by crowd. In this paper, we propose *Opportunistic SPIN* (O-SPIN), an information dissemination protocol that augments the well-known data-centric energy-aware SPIN (*Sensor Protocols for Information via Negotiation*) protocol to enable networking facilities for Folk-nodes, by exploiting opportunistic contacts among users. Performance of the proposed solution has been evaluated through simulations carried out in the OMNeT++ framework under different settings. Achieved results demonstrate its effectiveness and efficiency in the information dissemination process.

1 Introduction

Today, citizens in developed countries receive payslips, banking statements, medical records, and other personal data through the Internet. The advantages are numerous, to name a few, e-services are faster, more convenient, and can be used to save time, strengthen the markets, improve the quality of life, etc. Unfortunately, this is not yet possible in *Least Developed Countries (LDCs)*, where the capillary diffusion of *Information and Communication Technology (ICT)* is still hindered by several social, technical, and economical barriers.

According to Non-Governmental Organizations (NGOs), four main requirements must be met to build a practical technical solution in LDCs: privacy protection, immediate personal benefit, self-sufficiency, and very low deployment

---

* This work has been partially supported by the PalmaRES Project, funded by MIUR - Cooperlink initiative and by CPER NPdC/FEDER CIA, PREDNET LIRIMA and the FP7 VITAL.

S. Guo et al. (Eds.): ADHOC-NOW 2014, LNCS 8487, pp. 43–57, 2014.
© Springer International Publishing Switzerland 2014

cost. To comprehensively address the mentioned issues, a paradigm called *Folk-enabled Information System (Folk-IS)* has been preliminarily introduced in [1]. Folk-IS promotes the idea of an *infrastructure-less* and *participatory platform*, where each individual implements a small subset of the complete information system, thanks to emerging highly secure, portable, and low-cost storage as well as computing personal devices, called *Smart Tokens*. Within Folk-IS, people transparently and opportunistically performs data management and networking tasks as they physically move, thus enabling *human networking*. Human networks are a special kind of opportunistic networks, in which people exchange information when they meet. Opportunistic networking is, indeed, claimed to be the only affordable way to help bridging the digital divide by providing intermittent Internet connectivity to rural and developing areas [2,3]. However, the design of simple and efficient opportunistic routing and data dissemination strategies is, generally, a complicated task due to the lack of knowledge about the topological evolution of the network. Routes are built dynamically while messages are *en route* between the sender and the destination(s), and any node can opportunistically be used as a next-hop, provided it is likely to bring the message closer to the final destination.

In order to contribute to this issue, in this paper, an opportunistic data dissemination protocol is proposed to enable the Folk-IS paradigm. The proposed *Opportunistic SPIN (O-SPIN)* takes advantages of both *data-centric* and *opportunistic* routing philosophies. O-SPIN borrows the main features of SPIN (Sensor Protocols for Information via Negotiation), a family of data-centric routing protocols used to efficiently disseminate information in Wireless Sensor Networks (WSNs) [4,5]. The rationale behind our design choice is as follows:
- Being deployed for WSNs, SPIN is designed to keep the energy consumption as low as possible. This is an attractive feature for Folk-IS nodes equipped with resource-constrained battery-powered Smart Tokens.
- Targeting the delivery of sensor data, SPIN is optimized for the single exchange of small-sized packets. This is the typical data exchange mainly targeted in Folk-IS scenarios, like public utility data (e.g., drought warnings, humanitarian aids advertisements), e-administration data (e.g., recording of births and deaths).

To fit the requirements of Folk-IS scenarios, experiencing intermittent connectivity due to a variety of reasons, O-SPIN is enhanced to exploit the opportunistic contacts among personal devices. To serve this purpose, *periodical advertisement* that supports the dissemination process and *neighbourhood information exchange* are additionally foreseen.

The remainder of the paper is organized as follows. Section 2 presents the Folk-IS paradigm and the Smart Token, the requirements of the data dissemination protocol and use cases. Section 3 describes the proposed O-SPIN protocol. Performance comparison between O-SPIN and a variant of SPIN are reported in Section 4. Section 5 concludes the paper and provides hints for future work.

2   Preliminaries: The Folk-IS Paradigm and Architecture

Alongside good governance, technology is considered among the greatest enablers for improving quality of life. However, the majority of its benefits have been concentrated in industrialized nations and are thus limited to a world's population fraction. E-services could be considered superfluous luxury for LDCs, where population makes less than \$2,000 per year [6]. On the contrary, several reports [7–9] make evident that ICT are called to play a catalytic role in these countries. ICT can help to achieve universal primary education, promote gender equality and empower woman, reduce child mortality, combat diseases or ensure environmental sustainability. However, making e-services practical and afford-able is still challenging because of lack of infrastructure, high initial investment, difficulty to maintain the system operational, reluctance to use the system due to security concern, etc.

The Folk-IS paradigm aims to address four main requirements [1]:

• *Ethic & security:* the lack of a strict legal framework regulating people's privacy leads to recognize an LDC as an especially hostile context. Privacy abuses, data corruption and denial of service attacks, often driven by financial or political in-terests or ethnic disputes, are usually not deterred by sufficiently coercive laws. Thus, privacy protection and security are considered by NGOs as two strong prerequisites before deploying any solution in the field;
• *User's benefit:* providing a global interest for the community is not a sufficient incentive. The solution must provide a direct benefit to each user to be effec-tively used in the field;
• *Self-sufficiency:* relying on the expected improvement of the technical infras-tructure or on upcoming governmental programs or laws implies a major risk of failure since not perennial;
• *Sustainability:* the deployment cost of the solution must be very low (a few dollars per user) and proportional to the life's cost of those areas, without huge initial investment. The maintenance cost must also be minimal and should ide-ally generate a source of revenues for new local jobs linked to the solution.

Folk-IS builds upon the emergence of highly secure, portable and low-cost storage and computing devices called *Smart Tokens* (Figure 1(a)). These devices combine the tamper-resistance of a secure micro-controller with high storage capacity external NAND flash chips and short-range communication capabilities. The Smart Token considered in this paper is very similar to a smart token product, provided by Gemalto, used by Inria in a field experiment [10]. This kind of smart token is available for a few dollars, in a SIM card form factor (plugged in a USB key casing), and simply needs to be extended with a fingerprint reader, in a similar way to traditional secure USB keys. In our context, the Smart Token is more complex, since it embeds a battery (and the required means to charge it) and powers a wireless communication element (Bluetooth, led-light communications, etc.). It is also equipped with I/O resources to allow basic users interacts autonomously.

Folk-IS is characterized by the following features:

• *Infrastructure-less environment:* nodes cooperatively establish the network in-dependently of any fixed base station infrastructure.

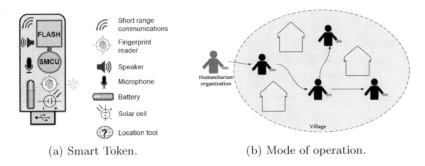

(a) Smart Token.   (b) Mode of operation.

Fig. 1. Folk-IS paradigm

- *Distributed communication:* each individual implements a small subset of the complete information system; at the same time, a node can act as a user of the system and also as a human network node. There is no central coordinator for the communications among humans that are typically ad-hoc.
- *Mobility:* nodes move inside the area more or less randomly so, the topology is highly dynamic. Moreover, Smart Tokens carried by pedestrians are not equipped with GPS-like receivers due to cost/energy requirements. Hence, human communications should take advantage from recurrent mobility patterns (e.g., a bus) to improve message delivery performance.
- *Resource-constrained devices:* Smart Tokens are devices with several limitations: *(i) Low-cost:* the deployment cost must be very low (a few dollars each) in order to be suitable for those countries; *(ii) Low-energy:* the devices must be able to power the wireless communication component and the other functionalities while consuming less battery as possible; *(iii) Hardware constraints:* the devices have small capacity, like a tiny RAM (about 64 kB), a small micro-controller and a small internal stable storage. Thus, it is fundamental to well manage the information stored in the device, removing the old one in order to free up space.

Many applications can benefit of the Folk-IS paradigm, provided that they are compliant with high-latency asynchronous data exchanges. Let us imagine a humanitarian organization in a LDC that wants to inform the population about the date for medical checks and vaccinations. As shown in Fig. 1(b), it transmits this information through the ad-hoc *Folk-enabled Network (Folk-Net)*. Every Folk-node associated to an individual acts as a human network node and forwards the message to its neighbours.

2.1   Data Dissemination in Folk-Net

To match the peculiarities of Folk-IS scenarios, routing should be:

- *data-centric*: all the interest is in the data, not in the location/identity of the node. This is because it is not possible to exactly track position of individuals since Smart Tokens are not equipped with a GPS.

• *energy-efficient*: personal devices are battery-powered and might be not rechargeable often because of the lack of charging points in LDCs.

• *opportunistic*: Communications can follow a carry-and-forward approach. Given the lack of infrastructure and the highly dynamic nature of Folk-Net, people will transparently perform networking tasks as they physically move, by exploiting the short-lived contacts with encountered users.

Several protocols in the literature may match the aforementioned requirements. Various data dissemination schemes have been proposed for delay-tolerant and opportunistic networks and energy saving protocols for resource-constrained WSNs [11], [12]. Among the two most representative energy-efficient data-centric approaches, i.e., Directed Diffusion [13] and SPIN [4], the latter, proposed by Heinzelman et al. in [4] and [5], is a family of adaptive protocols that disseminate the sensed information to every interested node in the network by using meta-data advertisement and negotiation to favour aggregation, redundancy elimination and energy saving [11]. The conceived data dissemination protocol, O-SPIN, borrows its main tenets from the SPIN protocol proposed for WSNs. By doing so, it provides data-centric and energy-efficient communications, and it is augmented to leverage opportunistic contacts among users to disseminate data.

3   The Proposed *Opportunistic SPIN*

We now describe the designed O-SPIN protocol, by emphasizing aspects in common and differences with the SPIN-2 protocol in [4] (hereafter shortened as SPIN).

3.1   O-SPIN in a Nutshell

Opportunistic data delivery in O-SPIN occurs in the following three steps: *Data Advertisement*, *Data Request*, and *Data Delivery*, as shown in Figure 2(a). The organization in three steps is shared with traditional SPIN, but the message format and the behavior details of each phase are different. Each O-SPIN node keeps the following data structures: the *Neighbours Table*, storing information about encountered neighbours, and the *Data Table*, storing retrieved data and related meta-data that describes the data content.

Data Advertisement. In traditional SPIN, a sensor node advertises new data only when they are sensed by the node itself or received by neighbouring nodes. So the broadcasting of advertisement (ADV) messages is *asynchronous*. The first main difference between SPIN and O-SPIN consists in the introduction of *periodical advertisements*. Since human networking context is mobile and sparse, when a node transmits the ADV, it may not find any neighbour. The introduction of periodical advertisements (every *ADV Message Interval*) provides a more efficient information dissemination, as demonstrated in Section 4, on the other hand, it implies higher energy consumption.

Data Request. When a node receives an ADV Message, it checks whether it already holds the corresponding Data; otherwise, it sends a Request (REQ)

Message to ask for it. REQ is sent if the following conditions hold: the destination of the packet is the node itself, *OR* the destination of the packet is the entire community, *OR* somehow the node knows how to reach the destination, *AND* the residual energy in the node allows it to complete the operation.

Data Delivery. When the advertising node receives a REQ Message, it sends back the DATA Message. Data can then be stored by the receiving node and further forwarded hop-by-hop to the destination.

(a) Basic phase.   (b) Ehnanced phase.

Fig. 2. O-SPIN basic (a) and enhanced (b) phases

3.2   Neighbour Discovery

Contrarily to a static typical WSN scenario where neighbourhood information is easy to get, in the scenario under analysis, the neighbourhood of each Folk-Node can change more or less rapidly. Especially if the network is sparse, the risk is that source nodes and potential forwarders (or destination) hardly meet or even if they meet they do not realize it (e.g., because they miss the ADV message). O-SPIN makes nodes broadcast HELLO Messages every *Hello Message Interval* regardless of the availability of Data to help neighbour discovery and enjoy opportunistic contacts among users. When a node receives a HELLO Message, it updates its Neighbours Table. If a node has new data to transmit, it piggybacks it in its HELLO Message and decreases the *Hello Message Interval* to the *ADV Message Interval* . These new HELLO messages also contain the ADV for the Data (Fig. 2(b)). If a node does not receive a REQ for the Data advertised in the HELLO Message, for a time equal to four times the *ADV Message Interval*, it decreases the *Hello Message Interval* to the *ADV Message Interval*. All in all, O-SPIN changes the Advertisement step of traditional SPIN by *(i)* making the advertisement periodic, *(ii)* adding new messages (HELLO), and *(iii)* updating the Neighbours Table at every interaction between nodes.

3.3   Energy Conservation

O-SPIN implements a simple energy-conservation heuristic similar to SPIN-2 [5]. When nodes are energy plentiful, they work normally by following the three-way handshake protocol (ADV-REQ-DATA), as described in the previous sections. Otherwise, if a node observes that its energy reaches a low threshold, it reacts by reducing its participation in the protocol. Specifically, a node will only partici-pate in a protocol stage if it can complete all the successive stages without going below the low-energy threshold. This conservative approach implies that, if a node receives some new data, it only initiates the ADV phase if it has enough energy to participate in the full three-stage protocol with all its neighbours. Similarly, if a node receives an ADV, it does not send out a REQ unless it has enough energy to both send a REQ and receive Data. This approach does not prevent a node from spending energy in receiving ADV or REQ messages even when it is below its low-energy threshold. It does, however, prevent the node from ever handling a DATA message when it runs out of energy.

3.4   O-SPIN Packets and Data Structures

O-SPIN uses 4 types of messages (Fig. 3): HELLO, ADV, REQ, and DATA.

HELLO Packet. A HELLO packet (Fig. 3a) contains the user identifier (USER ID), unique in the entire network, and the home town (HOME TOWN). When a node u receives a HELLO Message from v, it stores information about v in its Neighbours Table and, eventually, updates the number of times (NUMBER OF MEETINGS) and the last time (LAST MEETING) v has been met, which allow removal of expired entries and additional criteria for the next-hop selection. More details are given in Section 3.5.

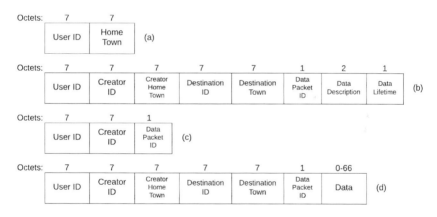

Fig. 3. O-SPIN messages: HELLO (a), HELLO+ADV (b), REQ (c), DATA (d)

HELLO+ADV Packet. When a node u gets new data from a node v, it adds the Data in the HELLO message issuing a HELLO+ADV message (Fig. 3b), which actually is an ADV sent periodically. The fields of this HELLO+ADV packet are:

- USER ID, the identifier of u;
- CREATOR ID, the identifier of v;
- CREATOR HOME TOWN, the town of residence of v;
- DESTINATION ID, the identifier of the ultimate destination node; it can be a broadcast address if the message is to be spread among all the population;
- DESTINATION TOWN, the destination town of the DATA packet, if specified;

• DATA PACKET ID, the identifier of the Data Packet (unique);
• DATA DESCRIPTION, is an optional meta-data description of the data content;
• DATA LIFETIME, it represents the lifetime of a data packets, once obsolete, these packets should not be considered any more.

REQ Packet. When a node receives a HELLO+ADV message, it first updates its Neighbours Table and then checks whether it already has the Data, otherwise, it sends a REQ message (Figure 3c) to specify the Data it is interested into.

DATA Packet and Data Table. When a node u receives a REQ message, first it updates its Neighbours Table, and then sends the DATA message (Fig. 3d). Once the DATA message is received, u stores the following information in its Data Table: DATA PACKET ID; CREATOR ID; CREATOR HOME TOWN; DESTINATION ID (representing the identifier of the destination user); DESTINATION HOME TOWN; DATA (representing the information to be disseminated); RECEIVED TIMES (date of Data reception). If the destination of the received message is u itself, it consumes the Data; otherwise, it starts to send HELLO+ADV Messages to advertise the new Data.

3.5   Next-Hop Selection Criteria

Some fields of the O-SPIN packets like DESTINATION ID, DESTINATION TOWN, HOME TOWN, NUMBER OF MEETINGS and LAST MEETING are fundamental in the routing strategies. Some possible usage examples are identified in this section; the different routing strategies depend on the application type. For example, public utility applications require data to be broadcast to the largest population in the shortest time (e.g., vaccination advertisements); e-administration applications usually target a specific village (e.g., birth recording); and so on. When a node receives a HELLO+ADV message, it may send a REQ message (and hence candidate itself as a possible forwarder) if one of these conditions is met, for example: the destination of the packet is its direct neighbour; the destination of the packet is its most-frequently met neighbour; the destination of the packet is a neighbour that has been very recently met; the destination town of the packet is its residence town; the destination town of the packet is the residence town of one of its neighbours.

In the simulation study presented in the next section we have implemented just few of the possible routing policies by focusing on the broadcasting of a message that is useful for the entire community.

4   Performance Evaluation

4.1   Simulation Settings

To evaluate the performance of the proposed O-SPIN protocol we used OMNeT++[1] and the simulation framework Castalia[2]. We consider a village as an

---

[1] http://www.omnetpp.org
[2] http://castalia.research.nicta.com.au/index.php/en/

area of $500 \times 500 \ m^2$ in which N pedestrian nodes, equipped with a *Smart To-ken*, move according to the Random Waypoint Model, with an average speed of 1 m/s. Nodes are generated inside the village by using a uniform distribution. Simulations have been conducted under two different population densities: *Low density*, with $N = 50$ nodes (200 persons per km^2); *High density*, with $N = 100$ nodes (400 persons per km^2).

For each simulated scenario, we varied the *ADV Message Interval* from 5 to 40 s; *Hello Message Interval* from 10 to 80 s. The initial energy budget of a Smart Token is set to 18720 Joules, which is the typical energy of two AA batteries. Only the radio component consumes energy for transmission, reception and carrier sensing. The Smart Tokens's *RF Transmission Power* is a varying parameter in our simulations; it is set to 0 dBm and -10 dBm, corresponding to a power consumption of 57.42 mW and 36.3 mW, respectively. Table 1a reports the power transmission matrix with information on the power spent by each device in each status. In this paper the Duty Cycle, the fraction of time that the node listens to the channel and does not sleep, is set to 0.005, so nodes sleep for 1990 ms. The nodes are not aligned in their sleeping schedules, so a train of beacons is sent before each data transmission to wake up potential receivers. Table 1b summarizes further Smart Token parameters, based on the Texas Instruments CC2400 datasheet.

Table 1. Simulator parameters

#	Rx	Tx	Sleep
Rx	-	62	62
Tx	62	-	62
Sleep	1,4	1,4	-

(a) Power transition matrix (mW).

Parameter	Value
Data Rate	250 kbps
Modulation	PSK
Bits per symbol	4
Bandwidth	20 MHz
Noise Bandwidth	194 MHz
Noise Floor	-100 dBm
Sensitivity	-95 dBm

(b) TI CC2400 datasheet.

In the scenario under study, a node in a village creates a data packet of public utility; the goal is to spread the information in the entire village and inform the highest number of persons before the Smart Tokens run out of energy. The following metrics have been computed to compare the performance of the two protocols, in terms of effectiveness and efficiency of the data dissemination process:

• *Percentage of informed nodes (PIN):* representing the number of reached nodes, computed like: $PIN = \left(\frac{Number of Nodes That Received Data}{N} \right) * 100$.

• *Energy Consumption (ECons):* representing the average energy consumption per node, computed as $ECons = \left(\frac{\sum_{i=0}^{N}(EnergyConsumed_i)}{N} \right)$.

Simulation results are reported with the 99% confidence intervals.

Table 2. Main features of benchmarked protocols

Protocol	Hello Exchange	ADV Exchange	Energy conservation heuristic
SPIN-Enhanced	N/A	periodical	supported
O-SPIN	periodical	periodical with adaptive repetition interval	supported

4.2 Simulation Results

A first analysis showed us that the original version of SPIN, traditionally designed for a static WSN scenario, performs very poorly, i.e., after 1 hour of simulation, only two nodes received the Data packet, regardless of the transmission power (0 or -10 dBm). Therefore, for the sake of equity and fairness, we have introduced some simple improvements to SPIN that we call *SPIN-Enhanced* with our O-SPIN. In *SPIN-Enhanced* the advertisement process is regularly repeated at a fixed interval once new Data is created or received, and the same energy conservation heuristic as SPIN is implemented. For the sake of clarity, the main features of the benchmarked solution are summarized in Table 2.

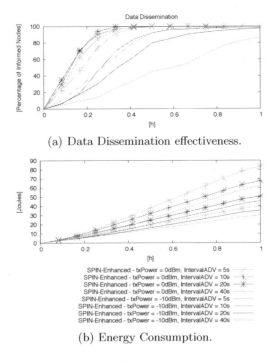

(a) Data Dissemination effectiveness.

(b) Energy Consumption.

Fig. 4. SPIN-Enhanced performance (low density)

SPIN-Enhanced (low density population). As shown in Figure 4(a), SPIN-Enhanced with Transmission Power of 0 dBm presents better performance than the case with Transmission Power of −10 dBm. In the former case, all nodes inside the area receive the packet after about 30 minutes; this is because the transmission range of the Smart Token is larger. Moreover, it is possible to observe that the performance improves when the ADV Message Interval is shorter, except for the case with ADV Message Interval equal to 5 s. Intuitively, when the interval is shorter the probability to exploit opportunistic contacts increases; however, if this interval is too short the number of transmitted ADV messages creates congestion on the channel. Finally, the effect of ADV Message Interval on data dissemination is more visible when the Transmission Power is −10 dBm.

Figure 4(b) shows the relative Energy Consumption when changing Transmission Power and ADV Message Interval. When the Transmission Power is 0 dBm the energy consumption is higher. Also, when ADVs are sent more frequently the energy consumption increases. Using SPIN-Enhanced we need to find a good trade-off between the parameters. So we use the ADV Message Interval equals to 10 s to achieve good performance when the Transmission Power is 0 and when it is −10 dBm.

O-SPIN (low density population). Figure 5(a) shows the trend of Data Dissemination effectiveness when using our proposal. It is possible to observe also in this case that when the Transmission Power is set to 0 dBm the performance is better.

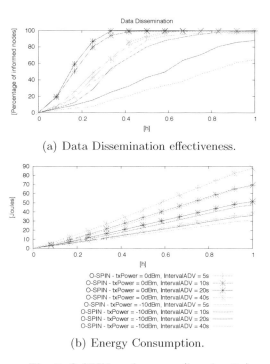

(a) Data Dissemination effectiveness.

(b) Energy Consumption.

Fig. 5. O-SPIN performance (low density)

Also in this case, the performance decreases when the ADV Message Interval is equal to 5s. Figure 5(b) shows the Energy Consumption; again sending the ADV Messages with higher frequency corresponds to higher battery consumption.

SPIN-Enhanced vs. O-SPIN (low density population). We compare the performance of O-SPIN and SPIN-Enhanced under the simulation assumptions considered above, but excluding the case with ADV Message Interval of 5 s because it is a low-performing case for both protocols. Both protocols are able to inform all the network nodes in one hour (Figure 6(a)). In most cases, O-SPIN is slightly slower to diffuse the information, but we can observe that when the ADV Message Interval is set to 10 s, both protocols achieve similar performance. Figure 6(b) compares the Energy Consumption between the two routing protocols. O-SPIN spends less energy than SPIN-Enhanced. Figure 7 shows the network lifetime using Transmission Power to 0 dBm and the Interval ADV to 10 s. The assumption is that a single node creates a Data Packet each hour. The curves show that Smart Tokens running O-SPIN have 1 day more of autonomy compared with devices which run SPIN-Enhanced, which is a significant improvement since devices might be not recharged often.

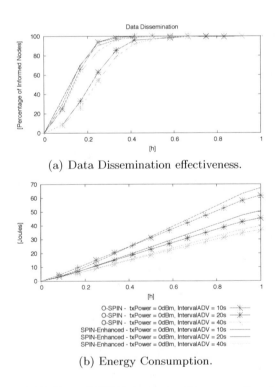

(a) Data Dissemination effectiveness.

(b) Energy Consumption.

Fig. 6. O-SPIN vs. SPIN-Enhanced performance (low density)

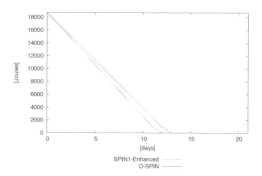

Fig. 7. O-SPIN vs. SPIN-Enhanced Network lifetime

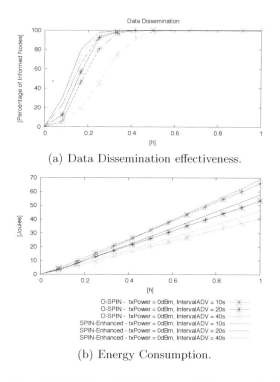

(a) Data Dissemination effectiveness.

(b) Energy Consumption.

Fig. 8. O-SPIN vs. SPIN-Enhanced (high density)

SPIN-Enhanced vs. O-SPIN (high density population). In this section, we compare the performance of SPIN-Enhanced and O-SPIN protocols in the high density scenario. Figure 8(a) shows that in 30 min all nodes receive the Data packet. O-SPIN performance increases when the ADV Interval is smaller. Figure 8(b) shows the Energy Consumption; O-SPIN saves more energy than SPIN-Enhanced.

5   Conclusion

In this paper a new routing technique to enable human networking in LDC was presented. Opportunistic contacts between people equipped with a personal device, empowered with communication and storing capabilities, allow information dissemination (e.g., data of public utility, like the presence of drinkable water, healthcare service, humanitarian aids, etc.) in an efficient and effective way. An important feature for these newly developed personal devices is the battery consumption. Indeed, it is fundamental to design communication and data exchange protocols that preserve the devices' battery. At the same time, it is also required that these devices are able to disseminate information to as many people as possible. We proposed a new routing protocol that enhances SPIN by exploiting the opportunistic contacts among personal devices.

We evaluated performance in different density scenarios, under different settings for transmission power and repetition intervals of messages. The results show that the proposed O-SPIN outperforms the straightforwardly enhanced version of SPIN, in terms of effectiveness and efficiency of the dissemination process with beneficial effects on the network lifetime. Transmission Power is the most important parameter that influences the performance metrics; indeed if it is high then both data dissemination and battery consumption are faster. Another important parameter is the ADV Message Interval, which indicates the periodicity of the ADV Messages. By increasing this interval, both data dissemination and battery consumption are slower.

In the future, we plan to study more sophisticated criteria for the next hop selection, and to introduce the adaptive setting of the ADV Message Interval based on traffic load, scenario characteristics, and node capability. We expect that the proposed Folk-IS vision may help to bridge the digital divide while pointing out research directions and solutions for efficient and effective information dissemination in hostile environments.

References

1. Anciaux, N., Bouganim, L., Delot, T., Ilarri, S., Kloul, L., Mitton, N., Pucheral, P.: Folk-IS: Opportunistic Data Services in Least Developed Countries. In: VLDB 2014 - 40th International Conference on Very Large Data Bases, pp. 2014–2040 (September 2014)
2. Pentland, A., Fletcher, R., Hasson, A.: DakNet: rethinking connectivity in developing nations. Computer 37(1), 78–83 (2004)
3. Doria, A., Uden, M., Pandey, D.P.: Providing connectivity to the Saami nomadic community. In: Development by Design Conference (2002)
4. Heinzelman, W.R., Kulik, J., Balakrishnan, H.: Adaptive protocols for information dissemination in wireless sensor networks. In: Proceedings of the 5th Annual ACM/IEEE International Conference on Mobile Computing and Networking, MobiCom 1999, pp. 174–185. ACM Press, New York (1999)
5. Kulik, J., Heinzelman, W., Balakrishnan, H.: Negotiation-based protocols for disseminating information in wireless sensor networks. Wireless Networks 8(2/3), 169–185 (2002)

6. Brewer, E., Demmer, M., Du, B., Ho, M., Kam, M., Nedevschi, S., Pal, J., Patra, R., Surana, S., Fall, K.: The case for technology in developing regions. Computer 38(6), 25–38 (2005)
7. Coceres, R., Belding, E., Parikh, T., Subramanian, L.: Information and Communication Technologies for Development (Guest editors' introduction). IEEE Pervasive Computing 11(3), 12–14 (2012)
8. ITU: The Role of ICT in Advancing Growth in Least Developed Countries - Trends, Challenges and Opportunities
9. Rossi, G., Murugesan, S., Godbole, N.: IT in Emerging Markets. IT Professional 14(4), 2–3 (2012)
10. Allard, T., Anciaux, N., Bouganim, L., Pucheral, P., Thion, R.: Pervasive and Smart Technologies for Healthcare. IGI Global (March 2010)
11. Watteyne, T., Molinaro, A., Richichi, M.G., Dohler, M.: From MANET To IETF ROLL Standardization: A Paradigm Shift in WSN Routing Protocols. IEEE Communications Surveys & Tutorials 13(4), 688–707 (2011)
12. Akkaya, K., Younis, M.: A survey on routing protocols for wireless sensor networks. Ad hoc Networks 3(3), 325–349 (2005)
13. Intanagonwiwat, C., Govindan, R., Estrin, D.: Directed diffusion: a scalable and robust communication paradigm for sensor networks. In: Proceedings of the 6th Annual International Conference on Mobile Computing and Networking, pp. 56–67. ACM (2000)

Probing Message Based Local Optimization of Rotational Sweep Paths

Florentin Neumann, Christian Botterbusch, and Hannes Frey

Institute for Computer Science, University of Koblenz-Landau, Germany
{fneumann,cbotterbusch,frey}@uni-koblenz.de

Abstract. In localized geographic routing, the beaconless recovery problem with guaranteed delivery refers to a nodes' task of selecting the next hop on a local minimum recovery path without use of beaconing. The state-of-the-art solution, the Rotational Sweep algorithm (RS), solves this problem using only three messages per hop. However, immediate data forwarding, as part of the next-hop selection process, causes problems concerning routing efficiency. Within transmission range of a single forwarding node the RS recovery path may consist of multiple nodes each of which transmits the data. In this paper, we suggest and evaluate a simple, yet promising extension of RS to avoid this. Instead of immediate data forwarding, a lightweight probing packet is used to explore the RS recovery path and to determine a locally optimal hop for data transmission (w.r.t. some user-defined optimization function). Our simulations suggest that independent of the node density, our extension can lead to significant reductions of data packet transmissions, and reduced energy consumption, compared to RS, which helps, e.g., to save valuable energy resources of nodes.

Keywords: Reactive greedy recovery, beaconless geographic routing, localized algorithms, energy efficiency, wireless sensor networks.

1 Introduction

Wireless sensor networks consist of small, usually battery powered, devices which are equipped with low range radios for wireless communication. One typical application is environmental monitoring which may involve routing of large-sized data packets via multiple hops to far away data sinks. While energy resources are limited, one possibility to increase the network's overall lifetime—a goal which is strongly desirable—is to decrease energy consumption, e.g., by optimizing energy efficiency of data routing. This study focuses on improving existing approaches for localized multi-hop georouting towards energy efficiency and more generally, towards local optimization of routing decisions with respect to some user defined optimization criteria.

The problem of *localized geographic routing with guaranteed delivery* may be considered as solved under the following assumptions [1]: Network nodes know

S. Guo et al. (Eds.): ADHOC-NOW 2014, LNCS 8487, pp. 58–71, 2014.

their geographic positions in the plane, are provided with a priori knowledge on destinations' positions, and have uniform wireless transmission ranges, i.e., the underlying network graph obeys the unit disk graph model, where any two nodes share a bidirectional edge if their Euclidean distance is at most some fixed constant R. We adopt this model throughout this work.

However, most of the geographic routing algorithms base their routing decisions on the nodes' complete neighborhood information (adjacencies in the network graph), and therefore require the use of *beaconing*.

Beaconless geographic routing algorithms avoid the wasteful beaconing process by using contention mechanisms: The forwarder (a node that wants to route a message) notifies its neighbors by sending an Request-to-Send (RTS). These start a delay-based competition and the first node to answer with a Clear-to-Send (CTS) is selected by the forwarder and is being sent the data packet, which disables other nodes' delay timers and prevents sending of further CTS messages.

Using this scheme, (distance based) *Greedy routing* [2], the task of forwarding the data to the neighbor minimizing the distance towards the destination, as well as *recovery from local minimum situations* can be solved using only constant number of message transmissions per hop (see [3,4] and references given therein). A node is said to be in a local minimum situation, when no neighbor is closer to the destination than the node itself and Greedy routing thus fails. For guaranteed delivery, recovery from a local minimum requires to forward the message to a node which is closer to the destination than the local minimum node. In case beaconing is not allowed, this problem is very challenging and referred to as the *beaconless recovery problem*.

The currently best solution to the beaconless recovery problem is the *Rotational Sweep algorithm* (RS) by Rührup and Stojmenović [4] (see Sect. 2 for details). In combination with Greedy routing their algorithm guarantees delivery while using only three messages per routing step, including the data packet. It has also been shown that the recovery paths produced by this algorithm are potentially very short regarding Euclidean path length [5]. Yet, a major drawback of this algorithm is that within transmission range of a single forwarding node the recovery path may consist of arbitrarily many hops [5] (see Fig. 1b for an example). That is, considering hop-count as a criterion for energy efficiency, RS recovery paths can be fairly inefficient because a data packet is repeated several times by nodes within transmission range of a single forwarder.

In this work, we introduce and evaluate a generic extension of the RS algorithm, called *Rotational Sweep Probing* (RSP), which facilitates better or even locally optimal recovery path routing decisions, with respect to some user defined (optimization) criteria, while preserving the aforementioned advantages of RS.

Organization of This Paper. In Sect. 2 we review RS and present our extension RSP. In Sect. 3 we give a generalized analysis under which conditions RSP outperforms RS. We complement this by showing simulation results for exemplary cost functions in Sect. 4. Finally, in Sect. 5 and 6 we review related work and give a short conclusion.

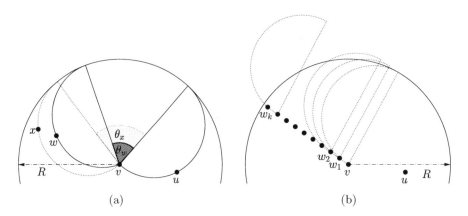

Fig. 1. Illustration of (a) the Sweep Circle delay function and (b) a constellation where the RS recovery path consists of arbitrarily many hops within v's transmission radius.

2   Descriptions of Algorithms and Correctness

Rotational Sweep Algorithm (RS). The currently best solution to the beaconless recovery problem is the contention-based *Rotational Sweep algorithm* (RS) [4]. For a given forwarder node v, execution of RS works as follows (see Fig. 1a for an illustration and [4, Sect. 3.1] for further details): Node v broadcasts a Request-To-Send (RTS) message containing its own as well as the position of previous hop u. All neighbors w of v which overhear the RTS start a delay timer proportional to angle θ_w which depends only on their own and the received positions. Neighbor w, whose timer expires first, immediately broadcasts a Clear-To-Send (CTS). This is overheard by the forwarder which, upon retrieval, immediately broadcasts the data packet to all neighbors. This terminates next-hop selection process and suppresses sending of other CTS answers.

Graphically speaking, the delay function evaluated individually by the nodes behaves like a sweep curve[1] hinged at and rotated around the forwarder v. Basically, a circle having diameter equal to the unit transmission radius and boundary touching v and previous hop u is rotated counter-clockwise around v until the boundary hits a node. This node will be the first to answer with a CTS.

The big advantage of RS is that it guarantees delivery while requiring only three messages (RTS-CTS-DATA) per hop, including data forwarding. The major drawback of this algorithm is that within transmission range of a single forwarding node, the recovery path may consist of arbitrarily many hops [5]. See Fig. 1b, where $\langle v, w_1, ..., w_k \rangle$ is part of the RS recovery path, v and w_k are neighbors in the network graph, yet, the data packet will be forwarded k times from v to w_k and k may depend on the number of network nodes n.

---

[1] For reasons of simplicity, we restrict ourselves in this work with the simpler Sweep Circle (SC) delay function. However, we claim similar results for the more sophisticated Twisting Triangle (TT) delay function [4, Sect. 3.3] and the extensions of RS introduced recently in [5].

Rotational Sweep Probing (RSP). We now introduce a simple, yet effective extension of RS in order to optimize its data forwarding decisions regarding a user defined optimization function.

Let v be a forwarding node on any RS recovery path. Furthermore, let $\langle v = w_0, w_1, w_2, ..., w_k \rangle$ be the longest consecutive subsequence of this path such that all nodes w_i, $1 \leq i \leq k$, are contained by v's transmission radius. Moreover, let f be some user defined utility function which is known to the nodes.

The RS-Probing algorithm proceeds as follows: v executes RS and acknowledges the CTS from w_1 by broadcasting a probing packet instead of forwarding the data. Generally, when receiving a probing packet from w_{i-1}, node w_i starts executing RS and awaits the CTS from w_{i+1}. Node w_i forwards the probing packet to w_{i+1} if the following conditions are satisfied: (1) w_{i+1} is contained in the forwarder's transmission radius. (2) Node w_{i+1} is not closer to the destination d compared to the distance between the initiator of the recovery process to the destination. (3) According to function f, the costs of transmitting the data packet from forwarder v to w_{i+1} are smaller or equal to the costs when transmitting the data from v to w_i. Otherwise, w_i announces via broadcast a *full stop* of the probing process, including its own position. When overhearing a full stop message by node w_i, neighbors of w_i stop their delay timers and forwarder v, which has initiated the corresponding probing process, acknowledges the full stop by immediate forwarding of the data to the node w_i having announced the full stop.

Similarly to the combined RS Routing algorithm in [4], a node switches a data packet back to greedy mode and uses Greedy routing if it determined an RS neighbor being strictly closer to the message's destination than the local minimum which has initiated the corresponding recovery process.

Extensions and Variations. In fact, in the version described above, the algorithm simply determines the first local optimum of function f for transmission of the data packet from a forwarder to its unit disk neighbors succeeding it on the RS recovery path.

Clearly, with minimal changes of the algorithm other variants can easily be obtained. For instance, if the probing process is continued, regardless of extrema, until the first node on the recovery path outside the transmission radius of the forwarder is detected, locally optimal routing decisions, in the aforementioned sense, can be taken.

Likewise, the algorithm can be changed in order to select the best choice for routing within the first k-hops, for some fixed constant k. This way, the message overhead produced for next-hop selection can be controlled and is guaranteed to be bounded by a constant. Constant k could also be adapted based on recent routing history.

Generally, f can be any cost function, however, if f requires information, which is neither part of the a priori node knowledge (such as the geographic position), nor part of the information attached to a control packet (e.g., distance to the data message's destination), then prior to the algorithm's execution an additional phase for data collection has to be added.[2]

Correctness and Message Complexity

Theorem 1. *Let $\Pi = \langle v_0, ...v_k \rangle$ be any recovery path produced by RS for the purpose of routing some message m to destination d. The execution of RSP with start node v_0 involves exactly all nodes $v_i \in \Pi$ and after its termination, m is held by v_k.*

Proof. During execution, every node v_i, $1 \leq i \leq k$, eventually receives a probing packet and determines its successor on the RS recovery path, using the same routine as in RS. In addition, only these nodes are considered for being forwarded a probing packet or the data message m, respectively. This holds in particular for v_k. When v_k receives a probing message from v_{k-1}, it determines the next node, say v_{k+1}, using RS. Because v_{k+1} is not an successor of v_k on Π it must hold that the Euclidean distance from v_{k+1} to destination d is strictly smaller compared to the Euclidean distance between v_0 and d. But then, v_k broadcasts a full stop and is being forwarded message m.   □

From the above Theorem and the fact that the combination of Greedy forwarding with RS guarantees message delivery in connected unit disk graphs [4, Thm. 1], the following Corollary immediately follows.

Corollary 2. *The combination of Greedy forwarding and RSP guarantees message delivery in connected unit disk graphs.*

Theorem 3. *In the worst-case, RSP requires at most $7/3$ times the number of packet transmissions required by RS, in order to route a message along an RS recovery path. In the best-case, the overhead is an additive constant of 4.*

Proof. Let $\Pi = \langle v_0, ..., v_k \rangle$ be an arbitrary RS recovery path. In order to route some message from v_0 to v_k, RS requires in total $3k$ messages. Let k' be the number of full stops RSP produces while routing the message along Π. Then, the overall number of messages sent by RSP is $3k + 3k' + k'$: $3k$ RTS-CTS-Probing steps, $3k'$ RTS-CTS-FullStop steps, and k' transmissions of the data packet. Clearly, the probing process ends at least once and at most k times, which reflects the cases where either all v_i, $1 \leq i \leq k$, are contained inside transmission radius of v_0, or where v_j, $i + 2 \leq j \leq k$, is never contained in the transmission radius of v_i, $0 \leq i \leq k - 2$. Thus, $1 \leq k' \leq k$. Therefore, the worst-case is given with $k = k'$, yielding a factor of $7/3$, whereas in the best case, $k' = 1$ and therefore, RSP requires only 4 more transmissions than RS.   □

3   Cost Analysis

When merely looking at the plain number of message transmissions, according to Thm. 3, RS outperforms RSP. However, assuming differences in message size

---

[2] Clearly, as RS and this extension take advantage from being beaconless, any preprocessing involving beaconing would contradict the main idea.

between simple control packets, such as RTS, CTS etc., and the actual data packets with payload, RSP starts paying off as RSP makes fewer data packet transmissions than RS in the average case. In order to evaluate the consequences of this difference, we introduce the following generalized cost function analysis.

Cost Functions. We consider any non-negative monotonically increasing cost function which maps a communication distance d to a cost value $f(d)$. Furthermore, we assume that this cost function is multiplicative with respect to the transmission duration, i.e., the cost $c(l, d)$ of a transmission over distance d with a duration l is given by $c(l, d) = l \cdot f(d)$.

We assume further that this cost function is additive, i.e., the cost for communicating along a communication path is equal to the sum of the cost of each individual communication step.

As an example consider the *total amount of bits*, when sending is done at a constant rate of b bits per time unit. The cost produced in one time unit is then independently of communication distance and is given by $f_h(d) = b$. The total cost of one communication step of duration l is then obviously $c(l, d) = l \cdot b$.

Another example is *total power consumption* per communication step under distance dependent required transmit power. A well established model to express power over communication distance d is given by $f_e(d) = d^\alpha + c$ (cf. [6]). The cost of one communication step over distance d and communication duration of l is then given by $c(l, d) = l \cdot (d^\alpha + c)$.

Cost Expressions for RS and RSP. For such cost functions we derive expressions c_{RS} and c_{RSP} of RS and RSP, respectively. The following analysis uses transmission times which are normalized to the time required for control messages. That means, the time required for control packets is assumed to be 1, while the time required for the data packet is assumed to be δ. In other words, δ expresses the relation between required transmission time for a data packet and a control packet. For the remainder, let R denote the maximum communication range of each node.

Analysis of RS. Assume that RS has produced a recovery path consisting of k hops. Let x_i be the communication distance covered in communication step i. RS first sends out an RTS at full signal strength since all nodes in the neighborhood have to be reached. Thus, RTS produces costs $f(R)$ in this (and any other) communication step. RTS is then answered by a CTS. Since the distance between sender and receiver is known upon CTS reply the cost for CTS reply is $f(x_i)$. Finally, a data packet is sent back at full signal strength. The cost for this amounts $\delta \cdot f(R)$. Full signal strength is required to assure that each neighbor node becomes aware of the data transmission. Summing all cost values together we get:

$$c_{\mathrm{RS}} = \sum_{i=1}^{k} f(R) + f(x_i) + \delta \cdot f(R) \ . \tag{1}$$

Analysis of RSP. Along each edge of the recovery path produced by RS our scheme sends out an RTS and a probing packet at full signal strength and a CTS at reduced signal strength since distance to the previous communication hop is known upon RTS reception. Thus, this three way handshake produces total cost

$$\sum_{i=1}^{k} 2f(R) + f(x_i) \ . \tag{2}$$

Let k' be the number of hops the data packet follows when using RSP instead of RS (i.e., k' also expresses the number of times a shortcut has been taken). Once RSP has found the next shortcut along the RS recovery path, it causes sending of another RTS at full signal strength and one CTS with signal strength adapted to the distance to the RTS sender. We estimate the cost of this CTS by $f(R)$ from above. Thus, the total cost of each last RTS/CTS handshake before the shortcut takes place is upper bounded by

$$\sum_{i=1}^{k'} 2f(R) \ . \tag{3}$$

Let x_i' be the distance covered in communication step i of the shortcut path the data packet is traveling along according to RSP. After the previously mentioned RTS/CTS handshake our scheme sends a *full stop* packet to the node where probing started at. This node then sends the data packet along the discovered shortcut. The stop packet has to be sent out at full signal strength in order to reach all neighbor nodes. When sending the data packet thereafter, the distance of the packet receiver is known and the signal strength can be adjusted accordingly. Thus, these two additional communication steps add the following costs:

$$\sum_{i=1}^{k'} f(R) + \delta \cdot f(x_i') \ . \tag{4}$$

Putting (2), (3), and (4) together yields:

$$c_{\text{RSP}} \leq \sum_{i=1}^{k} 2f(R) + f(x_i) + \sum_{i=1}^{k'} 3f(R) + \delta \cdot f(x_i') \ . \tag{5}$$

Deriving an Upper Bound for δ. In the following we are interested in δ, the relation between data and control packet lengths; since we assume the same constant bit rate for both data and control packets, value δ expresses both the relation in terms of transmission time and in terms of number of transmitted bits. We are interested at which relation δ the cost of the path produced by RSP is lower compared to the cost of the path produced by RS. That means we ask for δ satisfying $c_{\text{RS}} \geq c_{\text{RSP}}$. With (1) and (5) we get:

$$\sum_{i=1}^{k} (f(R) + f(x_i) + \delta \cdot f(R)) \geq \sum_{i=1}^{k} (2f(R) + f(x_i)) + \sum_{i=1}^{k'} (3f(R) + \delta \cdot f(x_i')) \ .$$

Solving the latter inequality for δ yields:

$$\delta \geq \frac{(k + 3k') \cdot f(R)}{k \cdot f(R) - \sum_{i=1}^{k'} f(x'_i)} \quad . \tag{6}$$

Estimating an Expression in Estimated Costs per Hop. The previous estimate on δ requires knowledge on the hop distances of the paths followed by RS and RSP, respectively. We now derive an estimate on δ which does not require knowledge on a concrete path but just the *maximum cost per hop*, the *average path cost per hop*, as well as the *average path cost savings*.

The *maximum cost per hop* is defined as before by

$$c_{\max} = f(R) \quad .$$

The *average path cost per hop* and the *average path cost savings per hop* is the statistical means of the cost of the total path averaged over the number of visited nodes. For the average path cost per hop we have to consider the cost of the path visited by RS

$$c_{\text{avg}} = \frac{\sum_{i=1}^{k} f(x_i)}{k} \quad .$$

For *average path cost savings*, we consider the difference between the cost of the path constructed by RS and the cost of the path used by RSP and average this over the total number of visited nodes. Note, both RS and RSP visit the same sequence of nodes; the cost savings are due to the data packet taking the shortcut path:

$$s_{\text{avg}} = \frac{\sum_{i=1}^{k} f(x_i) - \sum_{i=1}^{k'} f(x'_i)}{k} \quad .$$

Obviously, since RSP constructs shortcuts over the original RS path it trivially holds that $k' \leq k$. Thus, we can replace (6) by the following inequality, as any value for δ satisfying it, necessarily satisfies (6) as well

$$\delta \geq \frac{4k \cdot f(R)}{k \cdot f(R) - \sum_{i=1}^{k'} f(x'_i)} \quad .$$

Factoring out k in the denominator then yields

$$\delta \geq \frac{4f(R)}{f(R) - \frac{\sum_{i=1}^{k'} f(x'_i)}{k}} \quad ,$$

and extending the denominator with $\frac{\sum_{i=1}^{k} f(x_i)}{k} - \frac{\sum_{i=1}^{k} f(x_i)}{k}$ yields

$$\delta \geq \frac{4f(R)}{f(R) - \frac{\sum_{i=1}^{k} f(x_i)}{k} + \frac{\sum_{i=1}^{k} f(x_i) - \sum_{i=1}^{k'} f(x'_i)}{k}} \quad .$$

Finally we get:

$$\delta \geq \frac{4c_{\max}}{c_{\max} - c_{\text{avg}} + s_{\text{avg}}} \quad . \tag{7}$$

Hop Count Example. Consider the initially mentioned cost function $f(d) = b$ counting the total amount of bits when sending is done with a constant rate of b bits per time unit. In this case, (6) simplifies to

$$\delta \geq \frac{(k + 3k')b}{kb - k'b} = \frac{k + 3k'}{k - k'} \ . \tag{8}$$

This estimate requires, however, knowledge of the path lengths k and k'. The second estimate derived in (7), in contrast, requires only knowledge of the average cost savings per hop, which in this case amounts $1 - (k'/k)$. Furthermore, in this example maximum and average cost per hop in one time unit are both 1, since we normalized transmission length to the length of control packets. Thus, with (7) we get:

$$\delta \geq \frac{4}{1 - \frac{k'}{k}} \ . \tag{9}$$

That means, given an exact ratio k'/k or some rough estimate of it, for instance obtained via simulations, one can simply obtain an overestimate on the factor between control packet and data packet size required for RSP to outperform RS. Our simulations in Sect. 4 show that for any average node density in the interval $[5, 15]$, the ratio k'/k can be upper bounded by $7/10$. In combination with (9) we obtain a rough overestimate of $\delta \geq 13.\overline{3}$. When using (8) it can be obtained that $\delta \geq 10.\overline{3}$ is actually sufficient.

4   Simulation

We have investigated our scheme by use of simulations. We are interested in the two exemplary cost functions $f_h(d) = b$ and $f_e(d) = d^\alpha + c$ (see Sect 3).

Function f_h counts the total amount of bits when sending is done with a constant rate of b bits per time unit. It is used to find locally optimal hop paths for forwarding the data packet along an RS recovery path.

Function f_e counts normalized total power consumption under distance dependent required transmit power over a distance d. In [7] this cost function is used for local optimization of total power consumption in case of localized Greedy routing. In their *power progress* formula, costs $d^\alpha + c$ are divided by the progress made by a node towards the destination. We adopt this idea but use $(d^\alpha + c)/d$ for local optimization of total power consumption for the following reason: Considering recovery paths, there neither is necessarily a neighboring node on the recovery path being actually closer to the destination, nor is such a neighbor necessarily a suitable hop for reducing power consumption. We therefore suggest this simple heuristic, where the selected neighbor minimizes the power spent per unit of progress made in terms of advancing the recovery path.

Simulation Setup. We have implemented the *combined Rotational Sweep algorithm* [4] with recovery routines RS and RSP (using the Sweep Circle delay function), respectively, as described in the beginning of Sect. 2.

We performed simulations using the Java-based simulator for network algorithms *Sinalgo* [8] using a simplified MAC layer model and uniform transmission radii. Communication is performed in synchronous communication rounds. As we are interested in algorithmic average case performance, rather than considering problems arising with wireless communication, issues such as unreliable message transmission and interference are ignored.

The simulation area is a rectangle of size 500×500 with a unit disk radius of $R = 50$ units of length. We use a rectangular void region of 25% of the simulation area in the center, which makes recovery situations more likely as we choose the source/destination pair to be in the lower left and upper right corners, respectively. In each simulation run, nodes are placed uniformly at random in the remaining simulation area. The number of nodes being placed on the simulation area varies with different node densities (average number of neighbors of a node) ranging from 5 to 15.

For each integer value in $[5, 15]$ we generated random connected graphs until 500 were found satisfying the following requirements: There exists a path between the source and the destination node and this path has at least one local minimum node, i.e., recovery occurs at least once. Given such a graph, two data packets are routed from source to destination using the aforementioned algorithm, using RS and RSP for recovery, respectively.

Considering f_e, we chose path loss exponent $\alpha = 2$. Constant $c = 625$ was chosen such that for fixed R and α, the global minimum of f_e is always attained at distance $R/2$ (i.e., not outside the unit disk radius, in which case optimization of power consumption would degenerate into hop-count optimization).

For a single routing task, the following data has been collected: Overall number of control packets sent during greedy and during recovery mode as well as the overall number of hops of the data packet during greedy and during recovery mode. For simulations concerning power consumption optimization, in addition, data on power consumed for transmission of control packets and for data packets in greedy as well as in recovery mode, has been collected.

All plots are complemented by 95% confidence intervals.

Simulation Results. The number of control packets used by RSP for recovery always exceeds the number used by RS. Although the ratio of number of control packets used by RSP over RS is monotonically nonincreasing with increasing node density, from 2.26 at density 5 to 2.13 at density 15, this factor remains almost constant. The same holds when considering complete source-to-destination routing paths instead of recovery paths only. Thus, control packet overhead of RSP is roughly twice as much as with RS. This is contrasted by Fig. 2a showing the number of hops a data packet is being forwarded. Using RSP, in average a data packet in recovery mode is forwarded 32-40% less often compared to RS.

In the following, let δ denote the factor between control and data packet size. Figures 2b–2d show the normalized cost for recovery (including cost for data and control packet transmissions) for different values of δ. If $\delta = 1$ (Fig. 2b), these cost coincide with the total number of packet transmissions for recovery. Then, RS outperforms RSP due to fewer control packet transmissions. However,

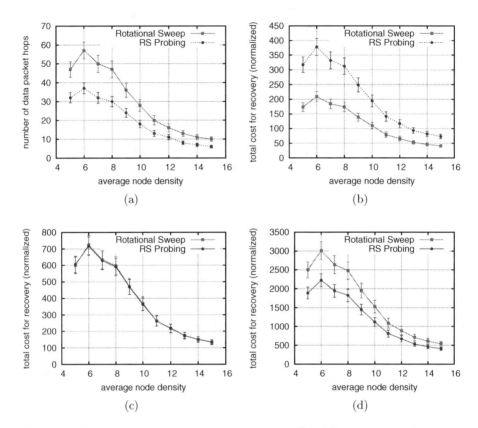

Fig. 2. (a) Number of data packet hops for recovery; (b)–(d) normalized total cost for recovery (including costs for control and data packet transmissions) with (b) $\delta = 1$, (c) $\delta = 10$, and (d) $\delta = 50$.

already for $\delta = 10$ (Fig. 2c), the algorithms perform almost identically well, as forecasted by the cost analysis in Sect. 3. When considering $\delta = 50$, RSP exceeds RS significantly.

Total power consumption for transmission of control packets in RSP is significantly higher than for RS (the factor varies between 2 and 3 for different node densities), which is due to RSP requiring sending of more control messages. This is contrasted by a significant reduction in normalized total power consumption for data packet transmissions (Fig. 3a). Independent of the node density, RS constantly consumes twice as much power for data transmission on recovery paths than RSP.

Figures 3b-3d show the aggregated total power consumption for recovery including costs for both control and data packet transmissions. For the case $\delta = 1$, i.e., where sending of a data packet over fixed distance produces the same cost as sending of a control packet over the same distance, RS consumes less power than RSP (Fig. 3b). This is due to the fact that RSP requires sending of more control

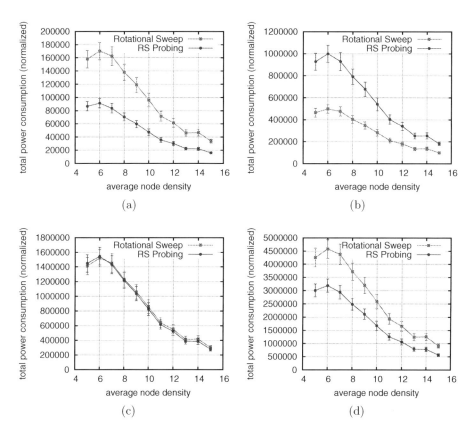

Fig. 3. (a) Normalized total power consumption for data packet transmission during recovery; (b)–(d) Normalized total power consumption for recovery (including costs for control and data packet transmissions) with (b) $\delta = 1$, (c) $\delta = 7$, and (d) $\delta = 25$.

packets. However, with increasing value of δ, RSP takes on greater significance. For $\delta = 7$, RSP is roughly as power consuming as RS (Fig. 3c), whereas for $\delta = 25$ (Fig. 3d) RSP is already significantly more efficient than RS.

5   Related Work

The beaconless recovery problem with guaranteed delivery can be solved with or without prior network planarization. In the following let u be a local minimum node w.r.t. destination d and let n be the number of nodes in a unit disk graph.

Node u can recover from local minimum situation by computing its incident edges in a connected planar subgraph of the network graph, i.e., a graph free of edge intersections, and subsequently use Face routing [1]. Repeated application of these steps then provably guarantees delivery [9]. There are two algorithms that allow for beaconless computation of the incident edges of a node in a connected and planar subgraph. *Beaconless Forwarder Planarization* [3] and *reactive PDT* [10]. However, focusing on reducing the message overhead, direct selection of the next forwarding edge on a recovery path without prior planarization is preferable.

The first algorithm to solve the beaconless recovery problem with guaranteed delivery using direct forwarding edge selection is *Angular Relaying* (AR) [3], the predecessor of RS. Instead of rotating a sweep curve around the forwarding node, a sweep line is used. This scheme has one major drawback: In order to select the next-hop, $\mathcal{O}(n)$ message transmissions are required in the worst-case and the edges selected belong to a planar non-spanner.

RS, in particular when using the Sweep Circle delay function, remedies the shortcomings of AR. RS requires only a constant number of transmissions per hop and edges selected by it have been proven to belong to PDT, a spanner with constant spanning ratio [11]. Simulations suggest that in the average case RS using the Twisting Triangle (TT) delay function outperforms RS with Sweep Circle (SC) delay function regarding overall recovery path length [4]. However, from a theoretical point, in contrast to RS using SC, it is not known if edges constructed via TT belong to a spanner with constant spanning ratio. For this reason, in [5] the beaconless algorithm *RS Shortcut* has been introduced. It aims for combining the positive aspects of RS paths using SC and TT, respectively. Basically, it assures that a data packet is always being forwarded along the SC path and cuts short with certain TT edges. For some scenarios this approach helps to save as much as $\Theta(n)$ message transmissions per hop. Yet, when considering scenarios like in Fig. 1b it is of no use and does not perform better than RS, in contrast to the approach introduced here.

6   Conclusion

In this work, we present a simple but powerful idea to optimize data routing decisions w.r.t. user defined cost functions on Rotational Sweep recovery paths. Generalized theoretical analysis complemented by simulations for exemplary cost functions clearly indicate that as of certain proportions between control packet and data packet sizes the approach presented can help to reduce routing costs compared to the foundational RS algorithm. For instance, if the size between data packets and control packets differs by a factor of 10, RS Probing already helps to reduce total power consumption for source-to-destination routing.

Acknowledgments. This work was supported by the German Research Foundation (DFG), grant "FR 2978/1-1".

References

1. Bose, P., Morin, P., Stojmenović, I., Urrutia, J.: Routing with Guaranteed Delivery in Ad Hoc Wireless Networks. Wireless Networks 7(6), 609–616 (2001)
2. Finn, G.: Routing and addressing problems in large metropolitan-scale internetworks. Tech. Rep. ISI/RR-87-180, University of Southern California (March 1987)

3. Rührup, S., Kalosha, H., Nayak, A., Stojmenović, I.: Message-Efficient Beaconless Georouting With Guaranteed Delivery in Wireless Sensor, Ad Hoc, and Actuator Networks. IEEE/ACM Trans. Netw. 18(1), 95–108 (2010)
4. Rührup, S., Stojmenović, I.: Optimizing Communication Overhead while Reducing Path Length in Beaconless Georouting with Guaranteed Delivery for Wireless Sensor Networks. IEEE Trans. Comput. 62(12), 2440–2453 (2013)
5. Neumann, F., Frey, H.: Path Properties and Improvements of Sweep Circle Traversals. In: Proc. of the IEEE 9th Intl. Conference on Mobile Ad-hoc and Sensor Networks (MSN), Dalian, China, pp. 101–108 (December 2013)
6. Rodoplu, V., Meng, T.H.: Minimum energy mobile wireless networks. IEEE J. Sel. Areas Commun. 17(8), 1333–1344 (1999)
7. Kuruvila, J., Nayak, A., Stojmenović, I.: Greedy localized routing for maximizing probability of delivery in wireless ad hoc networks with a realistic physical layer. Journal of Parallel and Distributed Computing 66(4), 499–506 (2006)
8. Distributed Computing Group at ETH Zurich, Switzerland: Sinalgo – Simulator for Network Algorithms. Online ressource, http://www.disco.ethz.ch/projects/sinalgo/ (accessed last: January 4, 2014)
9. Frey, H., Stojmenović, I.: On Delivery Guarantees and Worst-Case Forwarding Bounds of Elementary Face Routing Components in Ad Hoc and Sensor Networks. IEEE Trans. Comput. 59(9), 1224–1238 (2010)
10. Benter, M., Neumann, F., Frey, H.: Reactive Planar Spanner Construction in Wireless Ad Hoc and Sensor Networks. In: Proc. of the 32nd IEEE Intl. Conference on Computer Communications (INFOCOM), Turin, Italy, pp. 2193–2201 (April 2013)
11. Neumann, F., Frey, H.: On the Spanning Ratio of Partial Delaunay Triangulation. In: Proc. of the 9th IEEE Intl. Conference on Mobile Ad-hoc and Sensor Systems (MASS), Las Vegas, NV, USA, pp. 434–442 (October 2012)

Towards Bottleneck Identification in Cellular Networks via Passive TCP Monitoring

Mirko Schiavone[1], Peter Romirer-Maierhofer[1],
Fabio Ricciato[2], and Andrea Baiocchi[3]

[1] FTW Forschungszentrum Telekommunikation Wien
{schiavone,romirer}@ftw.at
[2] AIT Austrian Institute of Technology
fabio.ricciato@ait.ac.at
[3] University of Roma "La Sapienza"
andrea.baiocchi@uniroma1.it

Abstract. The bandwidth demand of today's mobile applications is permanently increasing. This requires more frequent upgrades of the mobile network capacity in the radio access as well as in the backhaul section. In such quickly evolving scenario, the risk of capacity bottleneck is increased, therefore network operators need tools to promptly detect capacity bottlenecks or, conversely, validate the current network state. To this end, we propose to exploit the passive observation of individual TCP connections. Being a closed loop protocol, the performances of every TCP connection depend on the status of the whole end-to-end path. Leveraging on this property, we propose a method to infer the presence of a capacity bottleneck along the path of an individual TCP connection by passively monitoring the DATA and ACK packets at a single monitoring point. We validate our approach with test traffic in a real 3G/4G operational network. The realized monitoring algorithm offers a powerful tool to network operators for on-line performance assessment and network troubleshooting.

1 Introduction

Due to the increasing popularity of bandwidth-hungry mobile applications, operators of mobile cellular networks need to continuously upgrade the network capacity in order to meet performance requirements and user expectations. Generally speaking, in a 3G/4G cellular network *capacity bottlenecks* could arise whenever the provisioned capacity at some network element (link or node) is not sufficient anymore to sustain the growing traffic demand. The severity of the bottleneck relates to the degree of mismatching between traffic demand and available capacity. In a growing traffic scenario, the bottleneck severity generally increases with time, until the bottleneck is removed by capacity upgrades. In other words, a bottleneck at an early stage may impact the network performance only slightly but, if neglected, it will lead to increasing performance degradation with time. In this context, network operators need to continuously monitor the state of the network to detect (and localize) capacity bottleneck as early as possible, before a serious performance degradation becomes noticeable by the customers. Accordingly, we present an approach to detect capacity limitations affecting individual

S. Guo et al. (Eds.): ADHOC-NOW 2014, LNCS 8487, pp. 72–85, 2014.

TCP connections, while the ultimate goal of our work is to correlate such information to enable the identification of aggregated, network-wide capacity bottlenecks. To this aim, we analyze the evolution of individual TCP flows by stateful tracking and association of DATA and ACK packets in both directions. For each TCP connection we evaluate basic performance metrics — throughput, semi-RTT and packet loss — and provide a set of indicators about the most likely limiting factor — e.g., downstream capacity restriction, rate limitation at the source, flow control by the destination — as inferred by a set of heuristic rules. The metrics and indicators extracted for individual TCP connections collectively form a set of *microscopic data* which can be aggregated in space and/or time in order to identify *macroscopic* resource limitations (i.e. bottlenecks) affecting multiple flows sharing a common network section.

The remainder of this paper is organized as follows: In §2 we discuss related work, while the methodology overview, including the deployed monitoring tool, is described in §3. In §4 we characterize the algorithms exploited to extract TCP based performance metrics and limitation indicators. After presenting the validation of our approach by means of controlled test traffic monitored within an operational mobile network in §5, we draw our conclusions and point to direction for future work in §6.

2   Related Work

The idea of inferring TCP characteristics from passive measurements has been explored already by a few previous works. For instance, the passive estimation of TCP out-of-sequence packets based on a single measurement point in a large IP backbone network in order to infer the health status of end-to-end TCP connections was already presented in [1]. The authors report about 5% of out-of-sequence packets pointing to network problems as, e.g., routing loops or in network re-ordering. As an extension of this work Jaiswal *et al.* present the passive inferring of additional TCP performance metrics, such as, e.g., the sender congestion window and the end-to-end Round Trip Time (RTT) in [2]. The authors report that sender throughput is typically limited by the lack of data to send, while the flavor of the TCP congestion control algorithm has only limited impact onto TCP throughput. While the approaches proposed in [1, 2] are conceptually similar to our methodology, they were developed for wired networks, while our focus is specifically on 3G/4G mobile networks. It should be remarked that there are considerable differences between the access layer characteristics of the two environments in terms of delay, jitter, bandwidth profiles, handovers, etc., and for that reason the TCP dynamics in mobile networks are deeply different from the wired scenario. Moreover, differently from [1, 2], our approach takes into account also the popular CUBIC TCP flavour and it is designed for *online* operation, not limited to the analysis of offline traces.

In [3] they present a set of techniques to identify the primary cause for TCP throughput limitations based on a passively captured packet trace. This work relies on the iterative identification of bulk transfer periods for estimating throughput only during such periods. Such iterative analysis approach complicates the throughput estimation in the online setting, which is the ultimate goal of our approach. In our work we address this issue by analyzing smaller chunks of individual TCP flows in a streaming fashion. Also, while [3] was developed for ADSL networks, our focus is on 3G/4G mobile networks.

The idea of utilizing passive traffic monitoring for optimization and troubleshooting in the context of 3G cellular networks was already presented in [4, 5, 6]. In particular, in [6] the authors present the passive identification of bottlenecks in an operational 3G network. This work focused on inferring bottlenecks based on the presence of spurious retransmissions and the qualitative shape of aggregated throughput distributions. Our work can be considered an extension, since we consider more metrics and additionally provide hints about the possible root cause of a detected bottleneck (i.e., network congestion versus server limitation).

In [7], Gerber *et al.* present the "Throughput Index" methodology to estimate achievable download speed using passively collected flow records in a 3G network. This approach is based on classifying specific services and characterizing their typical throughput behavior. That is, based on the throughput distribution, a specific service may be classified as rate-limited or not. In fact, as mentioned in [7], the maximum achievable download throughput can only be measured for non rate-limited services. In contrast, our approach does not rely on any service-based classification for the identification of rate-limited flows, but we classify greedy and non-greedy flows by directly analyzing low-level TCP dynamics. This allows for a more generic, application-independent TCP performance estimation.

3   Methodology

The instantaneous throughput of a generic TCP connection varies during the connection lifetime due to several factors, e.g., fluctuation of the radio channel, change in the cell occupancy, etc. In order to track short-term dynamics we split each TCP flow into *chunks* of shorter fixed duration. Choosing an appropriate chunk size is not trivial. In fact the non-stationary nature of the network makes shorter chunk duration appealing. However, in order to appreciate the closed-loop dynamics of TCP, the chunk duration must include multiple RTTs. Considering that in a 3G network the RTT can be up to half a second [8, 9], we selected a chunk size 10 of seconds.

In this work we focus prominently on the identification of performance limitations affecting downloads, i.e., data transfers towards the mobile devices. That is, we do not aim at detecting any capacity shortages for uploads originating from mobile terminals. The rationale for this choice is that download traffic is strongly prevalent in the network under study, while, in most of the cases, the capacity provisioned in the two directions is symmetric. Therefore, unless differently specified, in the rest of this paper the sender maps to a server in the Internet, while the receiver maps to a mobile device.

3.1   Methodology Overview

The mobile device may limit the data-rate of a download by using the TCP flow control mechanism (e.g., due to temporarily exhausted computational resources at the mobile device). This is achieved limiting the outstanding data, i.e., the amount of data packets that a server can send without waiting for an answer from the receiver, with the definition of the client "Advertised Window" (awnd). For more details about TCP refer to [10]. The server also controls the outstanding data, therefore the data-rate, by determining the "Congestion Window" (cwnd), which is the basis for the congestion control mechanism. In particular the server should adapt its `cwnd` accordingly to the state of congestion of the network, in order to exploit the maximum available bandwidth. On the other hand, a server might not fully utilize the network capacity along the path to its receiver. This might happen for several reasons, e.g.: i.) in case of downloads from video streaming services, the traffic rate is determined by the quality of the video rather than by the available bandwidth (refer to §5 for an illustrative example), ii.) the computational resources of a server may be temporarily exhausted, or iii.) the download rate provided by a certain server may be limited by certain quality of service settings. Our aim is to classify each TCP connection chunk in one of the two following categories:

- **Greedy:** A connection chunk whose throughput is limited by the network path, i.e., it is fully exploiting the available network capacity.
- **Non-greedy:** A connection chunk whose throughput limitation resides in one of the two end-points (server or mobile device), not in the network.

Based on the above mentioned capacity limitations, we propose to detect bottlenecks in the mobile network by the following two-tier approach:

- **Micro-analysis**: for each connection chunk we estimate the throughput and measure (or estimate, when direct measurement is not possible) TCP relevant parameters, e.g., `cwnd`, outstanding data, RTT etc. At the end of each chunk, all the parameters are combined (as explained in §4) for identify whether a performance limitation is induced by the mobile terminal, by the server, by the *downstream path* (between the monitoring point and the mobile device) or by the *upstream path* (between the server and the mobile device).
- **Macro-analysis**: the set of metrics and indicators established in the Micro-analysis for each flow are then stored in our DBStream database [11], which was specifically designed for continuous online analysis of large-scale traffic data. Recall that our ultimate goal is to aggregate the microscopic per-flow TCP metrics across flows that share a common network (sub)section in order to identify potential bottlenecks in that specific section (e.g. overloaded cells or congested network nodes).

Summarizing, our micro-analysis represents a basic requirement for building a large scale analysis tool (i.e. macro-analysis) that can be applied to all TCP connections crossing a given section of the mobile operator network, e.g., those between a given RNC and its SGSN (ref. to Fig. 1). Correlating throughput limitation causes of all connections (chunk by chunk) enables the understanding of general trends and the detection of possible network capacity shortages, e.g., by comparing observations over suitable large time scales (e.g., days, weeks).

In the remainder of this paper we report on the algorithms developed for the Micro-analysis, while the Macro-analysis methods are still under development. A remarkable challenge of this approach is computational scalability. This applies particularly to the Micro-analysis, since our final goal is to analyze in real-time *the state of every individual TCP connection* passing through the monitoring point on a multi-gigabit link. Therefore, we cannot maintain an extensive amount of per-connection states nor deploy too complex algorithms. This means that, when required, precision needs to be sacrificed for performance. However, given that we aggregate the output of the Micro-analysis to detect macroscopic bottlenecks, we may tolerate a certain level of inaccuracy

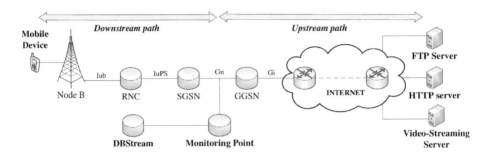

Fig. 1. Measurements setup: downstream v.s. upstream path

within the microscopic analysis of individual TCP chunks. That is, we assume that bottlenecks affecting larger sets of TCP flows are still identifiable when considering traffic aggregates, even if we misclassify a subset of individual chunks.

3.2   Measurement Setup

Our monitoring setup is depicted in Fig. 1: A controlled mobile device (smartphone) is connected via an operational mobile network to a set of servers (e.g. FTP or HTTP server). In our work we rely on both, controlled test servers as well as publicly accessible servers in the Internet. The mobile device deployed within our measurements supports connections via High Speed Downlink Packet Access (HSDPA) with a maximum download rate of 3.6 Mbit/s. For passively evaluating our test traffic on the Gn interface we rely on the monitoring system developed in the research project METAWIN — for more details refer to [12]. Packets originating from our mobile device pass subsequently through the mobile network and the Internet until the specific server is reached. Replies by the involved servers travel along the same route, but in reverse direction. Within our experiments we consider three different application scenarios: FTP downloads, HTTP downloads and video streaming. Our test traffic is captured on the so-called "Gn interface" links between the GGSN and SGSN (for more information about the 3G network structure refer to [13]). Moreover, in Fig. 1, the "Upstream path" refers to the path from the servers to our monitoring point, while the "Downstream path" refers to the path from the monitoring point to the mobile device.

3.3   TCP Connection-State Tracking

Fig. 2 depicts the high-level TCP transition-state-diagram used for tracking TCP connection states at the monitoring point (for more information about TCP state transitions refer e.g. to [10]). After successful completion of the "3-Way Handshake" initiated by the mobile device, the destined server starts sending DATA packets according to the "Slow Start" mechanism by increasing its cwnd exponentially. This phase ends whenever i.) the slow start threshold (ssthr) is reached or ii.) a packet loss is detected (as described in §4.2). The TCP protocol utilizes two mechanism for detecting possible

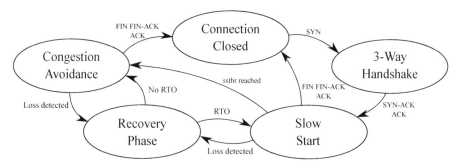

Fig. 2. TCP transition-state-diagram

packet losses in the network and triggering retransmissions: the "Retransmission Time-Out" (RTO) [14] and the "Duplicate acknowledgment" (Dup-Ack) [15]. In the first case a server enters a recovery state and retransmits all the packets that were in-flights when the loss was detected, while in case of Dup-Acks the server retransmits only the packet indicated by the Dup-Acks (fast-recovery) and waits for the acknowledgment. A "Recovery Phase" is over when an ACK for all the outstanding packets at the beginning of this phase is received at the monitoring point. The different loss detection strategies have also an impact on the cwnd. After an RTO the connection enters into the slow-start phase again, while a connection enters the "Congestion Avoidance" state after a "Dup-Ack" has been discovered. Within the congestion avoidance, the server adapts the cwnd according to its congestion avoidance algorithm. Finally, we consider a connection as closed whenever a mutual exchange of FIN-packets or a RST-packet is observed at the vantage point.

4   Extraction Algorithms for Performance Metric

In this section we describe the algorithms that we deploy in order to extract the basic performance metrics and the limitation indicators for each TCP connection passing through the monitoring point.

4.1   Round Trip Time

We consider RTT for the two separate paths of our setup (ref. Fig. 1): RTT upstream (RTT_U) and RTT downstream (RTT_D). RTT_U refers to the time elapsed between the observation of a device-originated ACK packet and the arrival of the subsequent server-originated DATA packet at our monitoring point. RTT_D refers to the time elapsed during the observation of a server-originated DATA packet and the arrival of the related device-originated ACK packet at our monitoring point. More specifically, RTT_D is computed as $RTT_{D_i} = t_{ack_i} - t_{data_i}$, where t_{ack_i} and t_{data_i} are the timestamps of the i-th TCP data segment sent by the server and of the associated acknowledgment sent by the mobile device.

An extensive RTT measurement campaign in the network under study, considering the RTT of TCP handshake packets, is reported in [8]. The authors report an 0.9-percentile of RTT_D that is around 400 ms (ref. [8, Fig. 3b]), while the 0.9-percentile of RTT_U is around 100 ms (ref. [8, Fig. 7a]). In case servers are located near the monitoring point, RTT_U exhibits an 0.9-percentile that is even below 4 ms (ref. [8, Fig. 7b]). According to the findings presented in [16], such low values of RTT_U are the dominant case in the network under study, as the Internet service providers largely rely on Content Distribution Networks for providing their content closely (and therefore with minimum RTT) to their end users. Based on the finding that RTT_D is typically much larger than RTT_U, in the following we rely solely on computing RTT_D. That is, we assume RTT_D provides a reasonable approximation a connection's overall RTT, instead of computing the total RTT as $RTT = RTT_U + RTT_D$. For sake of simplicity, we use "RTT" to refer to RTT_D in the following sections.

4.2   Retransmission Detection

In [1] Jaiswal *et al.* defined a set of rules to classify out-of-sequence packets observed at a monitoring point based on the different classes: retransmission, network duplication and in-network reordering. In case of retransmission, the authors also provide heuristics to detect both when a packet is lost in the upstream path (ref. Fig. 3a) or in the downstream path (ref. Fig.3b). Here, we adopt a similar approach. Based on the RTT estimation explained, we compute the RTO as specified in [14]. Finally, if the time since the observation of the last DATA packet is larger then the RTO, we assume a retransmission induced by an RTO expiration. In case we observe more than four ACK packets with the same acknowledgement number, we assume a retransmission that is induced by Dup-Ack mechanism. As depicted in Fig. 3 a retransmission triggered by packet loss in upstream (referred to as *upstream retransmission*) can be distinguished from a retransmission induced by packet loss in downstream (*downstream retransmission*) depending on whether the original data packet has been previously observed at the monitoring point or not. Finally, in order to detect packet duplicates and reordering, the IPID field of the IP datagram encapsulating the TCP segment is exploited. Since this value is unique and monotonically increasing during the lifetime of the TCP connection, we may assume in-network reordering if both the IPID and TCP sequence number of a packet are out of sequence. In case the IPID is monotonically increasing, but the TCP sequence number is out of sequence, we may assume a TCP-level retransmission. In contrary to the work in [1], we have access to the whole TCP header and all its options. Accordingly we may also rely on the optional 32-bit TSval field of the TCP-timestamp option [17], which provides the same functionality as the 16-bit IPID, but it is more robust against sequence wrapping problems due to its larger type range.

4.3   Outstanding Data, awnd and cwnd

Recall from §3 that the amount of outstanding data can be controlled by both the server and the mobile device with the cwnd and the awnd, respectively. In particular, in the whole lifetime of a TCP connection the outstanding data is limited as: $outstanding\_data < min(awnd, cwnd)$. We compute every sample of outstanding

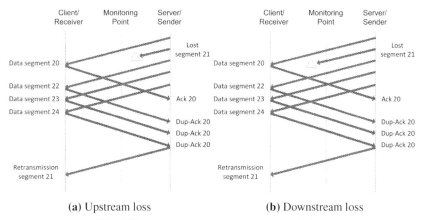

(a) Upstream loss   (b) Downstream loss

Fig. 3. Downstream v.s. upstream retransmissions

data as $outstanding\_data(n) = high\_seq(n) - high\_ack(n)$, where $high\_seq(n)$ is the highest sequence number sent from the server and $high\_ack(n)$ is the highest acknowledgment number sent from the mobile device when the n-th segment is observed at the monitoring point. However, this represents an underestimation of the outstanding data seen at the server side. In fact DATA packets might be in-flight between a server and our monitoring point and, additionally, we observe the device-originated ACKs before the destined server. This effect is more relevant the larger is the RTT_U between the monitoring point and the server.

While the awnd is explicitly notified by the mobile device to the server in every ACK and accordingly available at our monitoring point, a server's cwnd is never explicitly communicated to the mobile device and therefore not directly measurable at the monitoring point. In order to estimate it, we emulate the server behavior by increasing the cwnd every time an ACK from the mobile device is observer, and by decreasing its value whenever a retransmission is detected. This approach involves several challenges as discussed below. First, different servers deploy different congestion control algorithms (e.g., AIMD, Westwood, CUBIC, etc.). Based on active measurements in 2011 the author in [18] report that only 16-25% of the most popular web servers utilize the traditional AIMD [19] (additional increase, multiplicative decrease) algorithm, while 44% of web servers use the BIC [20] or the CUBIC [21] algorithms. Accordingly, in our analysis we estimate the cwnd following both: AIMD and CUBIC algorithms. Hence, we address the two most commonly used algorithms, while also other popular algorithms tend to behave in a similar manner (ref. [18, Fig. 3]). Second, ACKs of the mobile device are observed at the monitoring point before the server actually receives them. This results in an early increase, and thus in an overestimation of the cwnd. For mitigating this effect, together with the underestimation of the outstanding data, we rely on appropriate thresholding during the comparison between cwnd and outstanding data as discussed in §4.4. Third, servers typically deploy different ssthr for exiting the slow start phase during the initial phase of a flow. We consider a slow start phase as finished whenever: $i.$) a retransmission is detected or $ii.$) the amount of outstanding data increases less than one Maximum Segment Size (MSS) after one RTT interval.

Algorithm 1. Outstanding data limitation: initialization

1: $total\_sample \leftarrow 0$
2: awnd_cnt $\leftarrow 0$
3: $AIMD$_cwnd_cnt $\leftarrow 0$
4: $CUBIC$_cwnd_cnt $\leftarrow 0$
5: $final\_decision \leftarrow NULL$

Algorithm 2. Outstanding data limitation: main loop

1: $total\_sample \leftarrow total\_sample + 1$
2: **if** $outstanding\_data \geq awnd \cdot \gamma$ **then**
3:    $awnd\_cnt \leftarrow awnd\_cnt + 1$
4: **else**
5:    $AIMD\_diff = |outstanding\_data - AIMD\_cwnd|$
6:    $CUBIC\_diff = |outstanding\_data - CUBIC\_cwnd|$
7:    **if** $AIMD\_diff \leq CUBIC\_diff$ **then**
8:       **if** $outstanding\_data \geq AIMD\_cwnd \cdot \mu$ **then**
9:          $AIMD\_cwnd\_cnt \leftarrow AIMD\_cwnd\_cnt + 1$
10:      **end if**
11:   **else**
12:      **if** $outstanding\_data \geq CUBIC\_cwnd \cdot \mu$ **then**
13:         $CUBIC\_cwnd\_cnt \leftarrow CUBIC\_cwnd\_cnt + 1$
14:      **end if**
15:   **end if**
16: **end if**

4.4   Decision on the Limitation of Outstanding Data

We exploit the estimation of outstanding data, awnd, AIMD cwnd and CUBIC cwnd to decide whether a TCP connection chunk is "greedy" or "non-greedy" as follows. At the beginning of every TCP connection chunk we initialize our decision variables as shown in algorithm 1.. Every time a new sample of the outstanding data is available, and the TCP connection is in the "Congestion avoidance" state, the procedure depicted in algorithm 2. is invoked. First, the number of total samples ($total\_sample$) is increased. Then, the new sample of outstanding data is compared with awnd, and the closest of the two estimated cwnds, and our decision variables are increased accordingly. In order to take into account inaccuracy in the estimation of outstanding data, awnd and cwnd, we introduce the parameters: $0 < \mu, \gamma < 1$, refer to §5 for more details. Finally, at the end of every chunk, the algorithm 3. is exploited to identify the type of limitation. For conducting this identification all the decision variables are compared with the $total\_sample$ counter and a majority decision is taken. In order to decide for a specific limitation cause only in case of an absolute majority, we rely on the scaling factor $0.5 \leq \lambda < 1$. In case none of the above mentioned limiting condition is chosen, we classify the connection chunk as "not-greedy".

Algorithm 3. Outstanding data limitation: final decision

1: **if** $awnd\_cnt \geq total\_sample \cdot \lambda$ **then**
2:      $final\_decision \leftarrow awnd\_LIMITED$
3: **else**
4:      **if** $AIMD\_cwnd\_cnt \geq total\_sample \cdot \lambda$ **then**
5:          $final\_decision \leftarrow AIMD\_GREEDY$
6:      **else**
7:          **if** $CUBIC\_cwnd\_cnt \geq total\_sample \cdot \lambda$ **then**
8:              $final\_decision \leftarrow CUBIC\_GREEDY$
9:          **else**
10:             $final\_decision \leftarrow NOT\_GREEDY$
11:         **end if**
12:     **end if**
13: **end if**

4.5 Throughput Estimation

In our work we consider as throughput the average rate of data that has been ac-knowledged by the user within a connection chunk. More specifically, we rely on the formula $throughput(n) = \frac{acked\_end(n) - acked\_end(n-1)}{chunk\_duration}$, where $acked\_end(n)$ and $acked\_end(n-1)$ are the highest sequence numbers seen at the end of n-th and (n-1)-th chunk. As stated earlier $chunk\_duration$ refers to a time interval of 10 seconds. We are aware that this calculation might result in an underestimation of a connection's maximum achievable performance when data segments are sent only within a fraction, rather than during the whole duration of a chunk. To mitigate this, we do not consider the last chunk of a connection.

5 Validation in a Real Network

In this chapter we present the validation of our approach based on applying it in a real operational network. In the first part we present a validation based on involving controlled test servers, while in the second part we consider popular, publicly accessible servers in the Internet. Within our experiments we considered the following values for the thresholds introduced in §4.4: $\mu = 0.8$, $\gamma = 0.9$. In particular, since the awnd is explicitly communicated to the server by the mobile device, the scaling factor γ has to counterbalance only the underestimation of the outstanding data at the monitoring point. On the contrary, the threshold μ is used for the comparison between outstanding data and the two estimated cwnd, therefore this has to take into account also possible overestimation of the cwnd estimation. Finally, we set the threshold for the majority decision to $\lambda = 0.6$.

5.1 Analysis of TCP Connections to Controlled Test Servers

We start with validating our approach by comparing the amount of outstanding data ob-served at a controlled server with the value estimated at the monitoring point. As shown

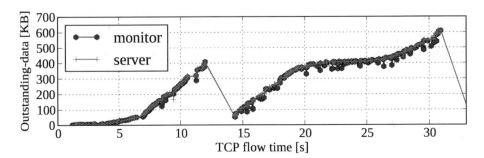

Fig. 4. Outstanding data: monitoring point v.s. server side

in Fig. 4 both follow the same qualitative trend. However, the amount of outstanding data inferred at the monitoring point is slightly below the one at the server (indicated by the slight notches visible in Fig. 4). Next, we test our module by deploying the CUBIC algorithm in our FTP server. The time-series of outstanding data, CUBIC cwnd, AIMD cwnd and the retransmissions during the whole TCP connection are shown in Fig. 5a. As expected, the outstanding data follows the CUBIC cwnd trend. After the first ramp up of the outstanding data, we observe an upstream retransmission (indicated by ▲) pointing to congestion in the upstream path. For all following increases of outstanding data we observe downstream retransmissions (depicted as ●) indicating bandwidth saturation in the downstream path. Additionally, we observe that both above mentioned upstream and downstream retransmissions triggered the expiration of the RTO and the subsequent retransmissions (refer to the ■ in Fig. 5a). As a consequence, the server reacts by drastically decreasing the size of cwnd. In contrast, Fig. 5b depicts the case of a non-greedy connection. Here we enforce a data rate limitation at the application layer on the server side. In the first chunk, the server increases its cwnd not yet reaching the introduced data rate limit. Due to the presence of a slow start phase our algorithm does not classify connection limitation for the first chunk. In the subsequent chunks, none of the estimated cwnds matches the outstanding data, accordingly our module correctly classifies the connection as non-greedy. Note that we also considered the case of a connection limitation enforced by limiting the awnd at the mobile device within our study. There, we were able to classify all connection chunks accordingly as $awnd\_limited$ at the monitoring point (not shown herein for the sake of brevity).

Finally, our experiments collectively show that a chunk size of 10 seconds adequately reflects the short-term dynamics observed in Figures 5a and 5b, and enables us to correctly classify the limitation types for all the evaluated chunks.

5.2   Analysis of TCP Connections to Internet Servers

In this section we present the results considering publicly accessible servers in the Internet. For instance, Fig. 6a depicts the results derived from downloading large objects via the HTTP protocol. We report that CUBIC cwnd follows properly the amount of outstanding data clearly pointing to a greedy connection. However, since the server observably utilizes a more aggressive way to increase the cwnd than considered by our

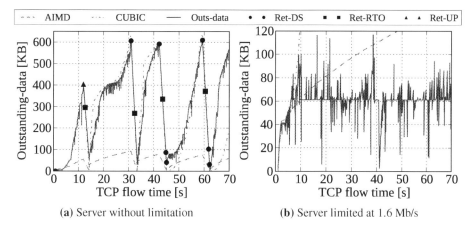

(a) Server without limitation   (b) Server limited at 1.6 Mb/s

Fig. 5. Controlled tests: greedy v.s. non-greedy connections

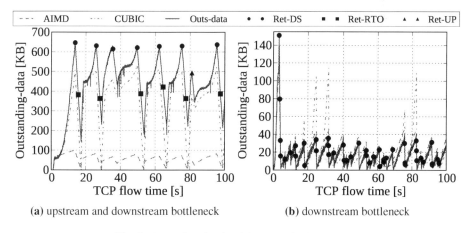

(a) upstream and downstream bottleneck   (b) downstream bottleneck

Fig. 6. HTTP downloads with servers in the Internet

module operated at the monitoring point. This explains why we obtain values of outstanding data that are higher than the estimated CUBIC `cwnd`. Moreover, we observe several retransmissions in Fig. 6a indicating congestions in the two different paths of the network. In fact, shortly after 80 seconds a packet loss in the upstream path is detected, while all the remaining retransmissions are caused by losses in the downstream path. Additionally, several times the server RTO expires triggering a slow start phase at the server side. In absence of an RTO expiration (observed at 40 and 80 seconds), the congestion avoidance phase is entered resulting in a decrease of outstanding data. Another case where a TCP flow experiences abnormally high packet loss in the downstream path — indicating severe congestion in this path — is shown in Fig. 6b. Such loss events are particularly relevant, since they may point to a bottleneck in the mobile network, specifically if more than a user exhibit the same loss pattern simultaneously.

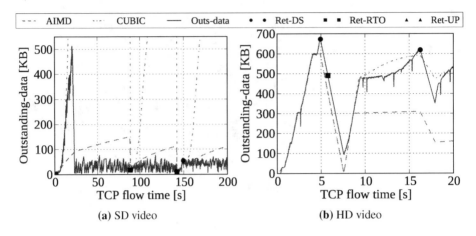

Fig. 7. YouTube: standard definition v.s. high definition video

Finally, we characterize the TCP chunk performance limitation for downloads from the popular video-streaming service: "YouTube". In particular, Fig. 7a depicts the results obtained while downloading standard-definition (SD) video. In the initial phase (i.e. first 20 seconds) the connection is greedy as the server sends bursts of data in order to fill in the application buffer at the mobile device. After this first phase, the data-rate at the server side is limited by the rate of the deployed video encoding. This is reflected by the outstanding data that is limited even without any packet loss. In fact, its value remains stable throughout the whole remaining TCP connection. The decision algorithm classifies all the chunks, but the first two (i.e. slow start phase), as "non-greedy". In contrast, Fig. 7b reports our results gained by downloading a High-definition (HD) video. In this case the server pushes the data to the mobile device as fast as possible resulting in a correctly classified, greedy connection. We may speculate that the fast pushing of HD content is required for avoiding any video stalling effects at the mobile device (ref. to [22] for more details about YouTube stalling events in the network under study).

6 Conclusions and Ongoing Work

In this paper we have presented a method to infer rate limitations and performance metrics of individual TCP connections in a mobile network. We have implemented the proposed method in a software module that is able to track and process all TCP traffic on a multi-gigabit link with a commercial high-end computers. The algorithm was successfully validated in a real operational network where controlled flows towards a test mobile device were embedded in the production traffic. We have considered both controlled test servers and uncontrolled Internet servers. The test results indicate a good inference power of the proposed algorithm at the level of individual TCP connections (microscopic analysis). In the progress of the work, we are currently developing robust aggregation and analysis methods to perform macroscopic analysis on the basis of the microscopic data produced by our module.

Acknowledgments. The Telecommunications Research Center Vienna (FTW) is supported by the Austrian Government and the City of Vienna within the competence center program COMET.

References

[1] Jaiswal, S., et al.: Measurement and classification of out-of-sequence packets in a tier-1 IP backbone. IEEE/ACM Trans. Netw. 15(1), 54–66 (2007)
[2] Jaiswal, S., et al.: Inferring TCP Connection Characteristics Through Passive Measurements. In: INFOCOM (2004)
[3] Siekkinen, M., et al.: A root cause analysis toolkit for TCP. Computer Networks 52(9), 1846–1858 (2008)
[4] Ricciato, F., Hasenleithner, E., Romirer-Maierhofer, P.: Traffic analysis at short timescales: an empirical case study from a 3G cellular network. IEEE Trans. Netw. Service Manag. 31(8), 1484–1496
[5] Ricciato, F.: Traffic monitoring and analysis for the optimization of a 3G network. IEEE Wireless Commun. (2006)
[6] Ricciato, F., Vacirca, F., Svoboda, P.: Diagnosis of capacity bottlenecks via passive monitoring in 3G networks: An empirical analysis. Computer Networks 51(4) (2007)
[7] Gerber, A., et al.: Speed testing without speed tests: estimating achievable download speed from passive measurements. In: Internet Measurement Conference. ACM (2010)
[8] Romirer-Maierhofer, P., Ricciato, F., D'Alconzo, A., Franzan, R., Karner, W.: Network-Wide Measurements of TCP RTT in 3G. In: Papadopouli, M., Owezarski, P., Pras, A. (eds.) TMA 2009. LNCS, vol. 5537, pp. 17–25. Springer, Heidelberg (2009)
[9] Laner, M., et al.: A comparison between one-way delays in operating HSPA and LTE networks. In: WINMEE 2012, 8th Int'l Workshop on Wireless Network Measurements, Paderborn, Germany (May 18, 2012)
[10] Stevens, W.R.: TCP/IP Illustrated, vol. 1. Addison Wesley, Reading (1994)
[11] Bär, et al.: DBStream: an online aggregation, filtering and processing system for network traffic monitoring. In: TRAC (2014)
[12] Ricciato, F., et al.: Traffic monitoring and analysis in 3G networks: lessons learned from the METAWIN project. Elektrotechnik und Informationstechnik 123(7-8), 288–296 (2006)
[13] Bannister, J., Mather, P., Coope, S.: Convergence Technologies for 3G Networks: IP, UMTS, EGPRS and ATM. Wiley (2004)
[14] Paxson, V., Allman, M.: Computing TCP's Retransmission Timer. RFC 2988 (2000)
[15] Allman, M., Paxson, V., Stevens, W.: TCP Congestion Control. RFC 2581 (April 1999)
[16] Casas, P., Fiadino, P., Bär, A.: IP mining: Extracting knowledge from the dynamics of the internet addressing space. In: International Teletraffic Congress, pp. 1–9. IEEE (2013)
[17] Jacobson, V., Braden, R., Borman, D.: RFC 1323: TCP Extensions for High Performance (May 1992)
[18] Yang, P., et al.: TCP Congestion Avoidance Algorithm Identification. In: Proceedings of IEEE ICDCS, Minneapolis, MN (June 2011)
[19] Jacobson, V.: Congestion Avoidance and Control. In: ACM SIGCOMM, pp. 273–288 (1988)
[20] Xu, L., Harfoush, K., Rhee, I.: Binary Increase Congestion Control (BIC) for Fast Long-Distance Networks. In: INFOCOM (2004)
[21] Ha, S., Rhee, I., Xu, L.: CUBIC: a new TCP-friendly high-speed TCP variant. ACM SIGOPS Operating Systems Review 42(5), 64–74 (2008)
[22] Schatz, R., Hoßfeld, H., Casas, P.: Passive youtube qoE monitoring for ISPs. In: IMIS, Palermo, Italy, July 4-6. IEEE (2012)

Connectivity-Driven Attachment
in Mobile Cellular Ad Hoc Networks

Julien Boite and Jérémie Leguay

Thales Communications & Security, Gennevilliers, France
{julien.boite,jeremie.leguay}@thalesgroup.com

Abstract. Cellular wireless technologies (e.g. LTE) can be used to build *cellular ad hoc networks*. In this new class of ad hoc networks, nodes are equipped with two radio interfaces: one being a terminal, the other one being an access point. The nodes can initiate only one outgoing attachment towards a neighbor using their terminal interface, while they can receive multiple incoming attachments through their access point. In this context, attachment decisions based on traditional criteria (e.g. signal quality) may lead to network partitions or suboptimal path lengths, thus making access point selection critical to ensure efficient network connectivity. This paper proposes a distributed greedy attachment strategy to reach near optimal network connectivity. Through simulations, we show that our strategy leads to network connectivity almost as good as in pure ad hoc networks, with small impact on path length. We also study the impact of thresholds to avoid low value attachments.

Keywords: Cellular Technologies, Ad Hoc Networks, Attachment Strategy, Network Connectivity.

1   Introduction

Most ad hoc networks have been conceived and deployed with wireless equipments that use a dedicated ad hoc radio mode. This is the case for WiFi community networks such as FunkFeuer in Austria [1] or Guifi [2] in Spain, as well as for public safety [3] and tactical networks [4]. This paper considers a particular class of ad hoc networks, that we call *cellular ad hoc networks*, where wireless technologies in cellular mode such as WiFi (infrastructure) or LTE are used in a specific way to build an ad hoc network. This kind of setup has been shown to be promising [5] as cellular modes offer higher data rates than ad hoc modes, when available, in off-the-shelf products. This paper proposes a distributed attachment strategy to maximize network connectivity for such networks.

In *cellular ad hoc networks*, nodes are composed of two radio interfaces as presented in Fig. 1: one acting as a terminal, the other one acting as an access point. This combination allows each node (i) to attach to one of its neighbors using the terminal interface, and (ii) to receive connections from neighboring nodes thanks to the access point radio interface. The network links are thus built

S. Guo et al. (Eds.): ADHOC-NOW 2014, LNCS 8487, pp. 86–99, 2014.
© Springer International Publishing Switzerland 2014

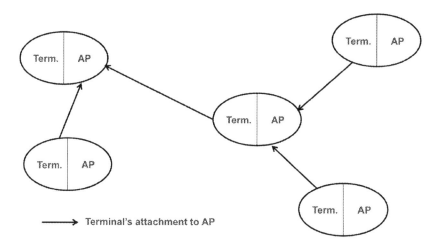

Fig. 1. Cellular ad hoc network

from the attachments that nodes initiate with one another using their terminal interface. A routing protocol makes multi-hop communications possible between connected nodes.

By *mobile cellular ad hoc networks*, we mean that the above-described nodes can move. Due to nodes mobility, attachments may change over time, leading to a dynamic network topology that has to be carefully maintained to ensure that the highest number of nodes can communicate. Attachment decisions from one node to another are the unitary network events that build the network topology. However, while neighbors in the radio range can communicate all together using the traditional ad hoc mode, the cellular mode introduces a particular cons-traint: the terminal interface of a node can attach to only one neighbor node. Therefore, the attachment decisions have to be made with caution since they directly impact the network topology and, subsequently, the global network con-nectivity. This paper proposes an attachment strategy to maximize the global network connectivity in this context.

Traditionally, off-the-shelf terminal interfaces apply an attachment strategy which is driven by pre-configured preferences and signal strength maximization. They select the access point to which they attach or towards which they should start a handover based on signal quality, user preferences or historical data. Because these local decisions do not consider connectivity at the network scale, they can create problematic connectivity conditions as depicted in Fig. 2. As shown in the figure, the network may (i) be disconnected while the radio range would allow full connectivity or may (ii) suffer from long routes between nodes that could be closer in the network space. To avoid such situations, we propose to drive attachment decisions by the objective of maximizing connectivity from a global network viewpoint. With this approach, ad hoc nodes dynamically attach or handover to a neighboring node so that connectivity is maximized, i.e. so that the number of node pairs that could exchange data is the highest possible.

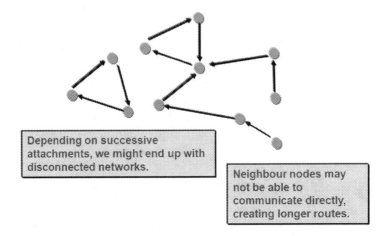

Fig. 2. Connectivity issues may arise with signal-based attachment strategies

Other issues should be addressed when deploying a cellular ad hoc network such as the allocation of non-overlapping channels to the different access point interfaces of nodes. Distributed spectrum allocation techniques may be used to solve this issue [6] in case that the number of channels is lower than the number of nodes. In this paper, we assume that this problem has been solved and focus on making good attachment decisions with regards to network connectivity.

We propose a distributed attachment mechanism that significantly improves network connectivity. It adopts a greedy approach by attaching smaller network partitions to larger ones. We show that this strategy can be implemented on top of existing technologies such as WiFi or LTE, and requires only one piece of information to be shared between nodes using existing control channels. We present a performance evaluation in simulation and compare our distributed mechanism with the optimal solution (pure ad hoc) and the state of the art solution (signal strength based attachment strategy). Two main criteria have been considered: (i) the percentage of connected pairs of nodes, and (ii) the average path length between connected pairs. Our mechanism outperforms traditional attachment strategies and offers close to optimal network connectivity while the average path length is just a bit longer. As frequent changes in attachments may engender a high number of handovers and thus cause disruptions to ongoing flows, we introduce thresholds to avoid handovers that do not bring significant gain on network connectivity or would impair too many flows. With identified threshold values, we show that near-optimal results still hold when only the most beneficial handovers are triggered.

The remainder of the paper is structured as follows. Section 2 presents the related work on cellular ad hoc networks. The greedy strategy that we propose and its implementation are presented in Section 3. Section 4 shows the performance evaluation. Finally, Section 5 concludes this paper and gives some perspectives for future work.

2   Related Work

The combination of access point and terminal functions to build a cellular ad hoc networks has been first proposed in 2007 [7]. However, this setup has only received an interest from the research community rather recently. In [5], authors have evaluated this approach using WiFi nodes. They have introduced preference information so that nodes do not oscillate when they have several neighbors, as the signal strength may vary back and forth between one neighbor and another. They have also introduced the possibility to turn off the access point part on nodes in case of high density. Indeed, when all nodes are in range, a good solution can be that everybody attach to the same node. In [8], authors have considered virtualization functionalities on a single WiFi card to allow a node to act as an access point and a terminal at the same time. The paper proposed a tree-based logical structure to perform routing that could be maintained following the attachments.

A number of techniques have been proposed to modify the access point selection procedure in wireless networks. In [9], the authors propose to run a battery of performance tests once being attached. If the connection is not satisfying enough, another access points is evaluated. One of the key feature of our proposition is that the decision can be achieved pro-actively from the information retrieved in the discovery phase. To the best of our knowledge, there is no related work that addresses the problem of making efficient attachments with regards to network connectivity in cellular ad hoc networks.

3   A Greedy Attachment Strategy

The objective of the mechanism we propose is primarily to maximize the number of nodes that can communicate. For this, we adopt a greedy attachment strategy that connects nodes to their best connected neighbors. The decisions are made locally by each node, using two sources of information: (i) the network topology that nodes maintain, limited to the network part they belong to, and (ii) a simple property of the network part behind neighboring nodes. Section 3.2 explains how this property, that we call *connectivity information*, can be exchanged with no overhead. As a secondary purpose, the algorithm tries (i) to minimize the number of flows that may be impaired by reattachment decisions, and (ii) to improve the connectivity of already connected areas by minimizing the average path length. Overall, the attachment decisions trade off between improving connectivity (increasing the number of connected pairs of nodes, reducing the average path length) and reducing flow impairments.

3.1   Architecture

The distributed attachment mechanism we propose relies on a few system modules embedded on each node. Fig. 3 presents the main interactions between the attachment module, that executes the attachment algorithm detailed in section 3.3, and other system components from both the network side (routing, flow management) and the radio side (wireless interfaces). These interactions and their purpose are described below.

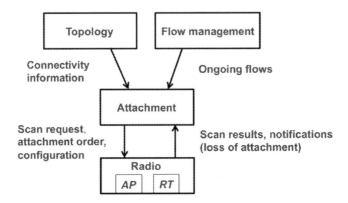

Fig. 3. Main interactions between the attachment module and other system modules

Topology module: the attachment module needs topology information to make decisions. It relies on the topology module to maintain such information, built from the routing protocol exchanges. This information represents the network topology around the node, i.e. a graph with nodes and links between them within the node's connected network area. We call this network area composed of reachable nodes the node's *connected component* in the rest of this paper. From this topology view, a node can thus verify if a neighboring node is already in its connected component and thus determine if it is worth attaching to it (the more new nodes a neighboring node allows to reach, the more interesting the attachment to this node). The attachment module can also compute properties of the node's connected component such as its size (number of nodes) and the average length of paths between nodes. These performance indicators from a connectivity viewpoint are key information used in the attachment decision process.

Radio module: depending on the implementation, this module sits either on top of two radio interfaces (represented in Fig. 3 as the access point AP and the relay terminal RT) or on top of one interface having both functionalities. To discover neighboring nodes, the attachment module can ask the radio module (more precisely the RT part) for neighbors discoveries. Most of cellular technologies support this action which is generally called a scan. Nodes can thus retrieve, for each neighboring node, the frequency used, the signal strength, and other specific information which depends on the technology (e.g. SSID for WiFi). In addition, when an attachment decision is made, the attachment module sends an order to the radio module (RT) so that the relay terminal performs the attachment. The attachment module can also ask for radio reconfigurations such as frequency change or meta-data updates (e.g. the SSID announced by the AP). This allows to take benefit of existing communication channels, used by access points to announce meta-data, for exchanging *connectivity information* between nodes during neighbors discoveries (cf. next section). Finally, the radio module notifies the attachment module of connectivity events, especially when the terminal losses its attachment.

Flow management module: an attachment decision made while the node is already attached can impact ongoing flows. The number of these flows and their relative importance are thus parameters that the attachment module should take into account. The flow management module maintains a list of flows relayed by the node. The attachment module can gather this list and thus evaluate the impact of a reattachment, e.g. according to QoS or criticity policies.

Note that the solution presented so far can be extended without losing its general principles. For instance, using several terminal interfaces per node is one possible extension.

3.2   Connectivity Information Exchange

In order to make good attachment decisions, a node needs to compare the properties of its connected component with those of candidate neighbors. One node can easily compute parameters relative to its connected component from the graph maintained by the Topology module. However, when a candidate neighbor is not in the connected component of the node, the required information is not available since this neighbor in not in the local graph. For that reason, connectivity information must be exchanged between nodes during a neighbor discovery process.

In the attachment process we present in Section 3.3, the only required connectivity information is the size of each neighbor's connected component. Exchanging this small piece of information can be implemented using several means. It can be achieved using a dedicated communication channel. However, off-the shelf cellular technologies do not support this explicitly. Existing fields can be used instead, such as SSID in WiFi or the PLMN (Public Land Mobile Network) identifier in LTE. To take benefit from these native communication channels, the attachment module in the architecture described above can configure the radio module (AP) so that these text fields are dynamically updated. A node can thus update its SSID AP_ID_X with X the connectivity information announced to other nodes, basically the number of nodes in its connected component.

3.3   Attachment Decision

The attachment procedure evaluates the interest of attaching to a neighbor with regards to the global network connectivity. It considers the following factors:

1. Opportunity to increase the size (number of nodes) of the node's connected component. The attachment strategy firstly looks for a neighbor that would allow to increase the number of nodes that can be reached. The number of nodes and the presence of a neighbor node in the current connected component is an information retrieved from the Topology module. The size of the

connected components to which neighboring nodes belong is obtained during the neighbor discovery process. Since the decision to modify an attachment can disrupt or impact ongoing flows (packet losses, increased path length and delays), the attachment strategy estimates this impact and finally decides whether or not to change the attachment. Information on ongoing flows are retrieved from the Flow management module.

2. Opportunity to reduce paths length inside the node's connected component. When the size of the connected component cannot be increased, the attachment strategy looks for a neighbor that would substantially reduce paths

---

Algorithm 1. Pseudocode for the attachment process.

---

function UPDATEATTACHMENT
      ▷ Retrieve list of neighbors and their respective connectivity information
   triggerNeighborsDiscovery()
▷ Find neighbor with highest Connected Component (CC) that is not in node's CC
  bestNeighbor ← *findBestNeighborOutsideCc*()
  if (! node.*isAttached*()) **then**   ▷ The node is not attached
    if (bestNeighbor != NULL) **then** ▷ A candidate neighbor outside CC exists
      attachTo(bestNeighbor)
    else   ▷ No candidate neighbor outside CC
      optimizePathLengthInsideCc()
    end if
  else   ▷ The node is already attached
    if (bestNeighbor != NULL)   ▷ A candidate neighbor out of CC exists
&& (bestNeighbor.*getCcSize*() ≥ node.*getCcSize*()) **then**   ▷ and has a larger CC
                    ▷ Simulate handover to compute flow disruption rate
      flowDisruptionRate ← *computeFlowDisruptionRate*()
      if (flowDisruptionRate < thresholdFlowDisruptions) **then**
        attachTo(bestNeighbor)
      end if
    else   ▷ No candidate neighbor outside CC
      optimizePathLengthInsideCc()
    end if
  end if
end function

function OPTIMIZEPATHLENGTHINSIDECC()
        ▷ Find neighbor that minimizes average path length inside CC
  bestNeighbor ← *findBestNeighborInsideCc*()
  if (bestNeighbor != NULL) **then**   ▷ A candidate neighbor inside CC exists
                ▷ Simulate attachment to compute average path length gain
    pathLengthGain ← *computePathLengthGain*()
    if (pathLengthGain > thresholdPathLengthGain) **then**
      attachTo(bestNeighbor)
    end if
  end if
end function

---

length in the connected component to optimize network connectivity. The average paths length before and after simulated attachments to candidate neighbors can be computed from the connectivity graph maintained by the Topology module.

The attachment procedure is greedy as it permanently tries to increase the size of nodes' connected component. Algorithm 1 details this procedure that applies in several cases: (i) periodically when nodes are not attached to any neighbor, (ii) immediately after a node loses its attachment, and (iii) periodically when nodes are already attached with the aim to further increase connectivity, reduce path length, or update the topology according to the mobility of nodes.

This attachment strategy makes use of two thresholds to control reattachments:

- *thresholdFlowDisruptions* prevents reattachments if too many flows are impacted (weights can be associated to the importance of flows),
- *thresholdPathLengthGain* is defined to avoid changes of attachment if the gain generated on the average path length is not high enough (it filters low value attachment decisions and increases network stability).

The combination of these two thresholds allows to trade off between the improvement of connectivity and the negative impact that frequent changes of attachments may have.

4   Performance Evaluation

Our distributed attachment mechanism has been evaluated in the simulator shown on Fig 4. This tool that we have developed generates reproducible mobility

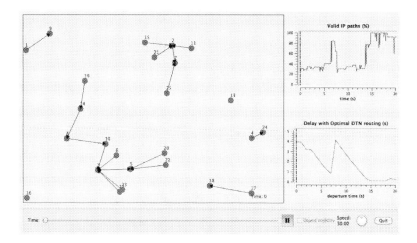

Fig. 4. Simulator GUI

scenarios using the classical random waypoint model and simulates the progressive attachments between nodes. We performed several simulations to analyze the network connectivity obtained using the following attachment strategies:

- *Signal*: a strategy that selects the neighbor with the highest signal quality
- *Optimal*: a strategy that provides the best ad hoc connectivity. This strategy simulates the traditional ad hoc interface where nodes can exchange data as soon as they are in radio range (i.e. without the constraint of the cellular mode, where one node can only attach to one of its neighbors)
- *Greedy*: the greedy strategy that we propose.

To evaluate the performance of these different strategies, we have considered the following measures during and at the end of each simulation:

- Percentage of connected pairs of nodes (i.e. for which an IP path exists)
- Average path length between connected nodes, in number of radio hops
- Average minimum Delay Tolerant Network [10] (DTN) delay, i.e. the average minimum time that a message, or *bundle*, would need to reach its destination using store, carry and forward transfers. This indicator reflects the quality of network connectivity since messages propagate from one node to another thanks to contacts between mobile nodes.
- Number of handovers (reattachments).
- Percentage of flow disruptions, i.e. the portion of flows that are rerouted or interrupted by handovers. We have considered that each node maintains a flow with every nodes in its connected component.

We have considered a scenario with 25 nodes and performed simulations to (i) analyze the performance results and compare our greedy attachment strategy against the others, and to (ii) focus on our greedy strategy to study the impact of thresholds we can set to keep only the most beneficial reattachments (maximum percentage of impacted flows, minimum gain in average path length). The table shown in Fig. 5 gives the simulation parameters that are common to all the simulations. Nodes are moving according to a random waypoint model with no pause time and a new speed selected randomly at each direction change.

# nodes	25
Width × Heigth	600m x 600m
Radio range	120m
Mobility speed	[0:30] m/s
Simulation duration	1 minute

Fig. 5. Global simulation parameters

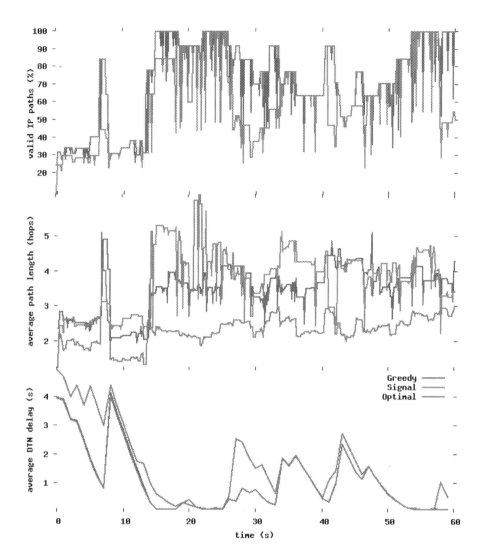

Fig. 6. Percentage of connected pairs of nodes (top), average path length (middle), and average DTN delay (bottom) over time

4.1   Global Performance Comparison

We have first compared the different strategies regarding their global performance results, without setting any of the thresholds defined above for the greedy strategy. Fig. 6 presents several simulation results. From top to bottom, these plots show the evolution over time of (top) the percentage of connected pairs of nodes, (middle) the average path length, and (bottom) the average DTN delay. We can observe that:

- the percentages of connected pairs of nodes for Greedy and Optimal are very close. We analyzed that the average percentage of connected pairs of nodes (over the whole simulation) for Greedy, Signal and Optimal is respectively 70.4%, 59.8% and 73.2%. The greedy strategy is thus very close to optimal while Signal leads to less than 60% of pairs of nodes connected in average. This confirms the network connectivity problem that occurs with traditional attachment strategies, and demonstrates the efficiency of our greedy strategy to alleviate this problem: the network connectivity is near-optimal almost all the time.
- the average path length for Greedy is higher than for Optimal, but most of the time lower than for Signal. We analyzed that the average path length is 1.16 hop longer for Greedy than Optimal for which it is 2.33 hops in average. This path elongation is due to the connectivity constraint (one node can attach to only one other node) that does not exist in pure ad hoc mode. Greedy does not reach Optimal due to this constraint, but it still performs better than Signal regarding average path length.
- the plots of the average DTN delay for Greedy and Optimal overlap, which means that Greedy offers the same performance than Optimal for delay-tolerant message delivery. In average, the time that is necessary to deliver all messages sent from one node to any other node for Greedy, Signal and Optimal is respectively 1.09, 1.56 and 1.07 seconds. Greedy and Optimal plots very similar performances, as for the first result on connected pairs of nodes. This confirms that the network connectivity obtained by Greedy and Optimal is very close.

These results show that our greedy attachment strategy outperforms traditional access point selection schemes in terms of connectivity between network nodes. It provides connectivity performances very near to those of a traditional ad hoc network. The single attachment constraint results in slightly longer average path length compared to the optimal, but this does not significantly impact the delivery of data packets.

4.2   Impact of Thresholds

We also studied the impact of the threshold values we can set in order to control reattachments. These thresholds define (i) the relative gain on average path length above which the attachment change is considered valuable, and (ii) the maximum portion of ongoing flows that can be impacted (considering one bi-directional flow between each node pair). We performed several simulations with different values for each of these thresholds to evaluate their impact on the number of handovers, the average percentage of connected pairs of nodes, and the mean of average DTN delay, over a whole simulation. Fig. 7 presents these results.

In top figures we show, for several threshold values set to the minimum gain in the average path length, the impact on (left) the number of connectivity events, (middle) the average percentage of valid IP paths between nodes, and (right) the

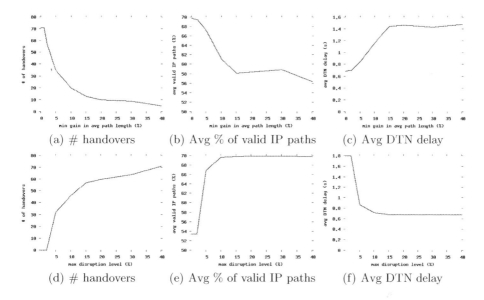

Fig. 7. Impact of the path length gain threshold (top figures) and the flow disruption threshold (bottom figures) on handovers, network connectivity and DTN delay

average DTN delay. We observe a fast decrease of the number of handovers when the threshold on expected gain in average path length increases (cf. Fig. 7(a)). For instance, it decreases by 50% for a threshold set to only 5%. On the other side (cf. Fig. 7(b) and Fig. 7(c)), such a low threshold value does not severely impact the connectivity performance since the average percentage of connected pairs of nodes only decreases from 70% to 67%, and the average DTN delay sligthly increases of 0.1 second. However, setting higher threshold values has a more significant impact on connectivity and DTN delay.

We can conclude that setting this threshold to a low value (such as 5%) allows to avoid many reattachments that do not generate significant gains in average path length, while it has a limited impact on network connectivity and average DTN delay.

In bottom figures we show, for several threshold values set to the maximum portion of ongoing flows a reattachment can disturb, the impact on (left) the number of connectivity events, (middle) the average percentage of valid path between nodes, and (right) the average DTN delay. As expected, we see from Fig. 7(d) an important increase of the number of handovers when the threshold value increases. Indeed, many reattachments do not occur with low threshold values. For instance, the number of connectivity events is divided by 2 for a threshold around 7%. For such a low threshold value, we observe from Fig. 7(e) that the average portion of valid IP paths between nodes is not so much affected, while important losses of connectivity occur for a threshold under 5%. The average percentage of connected pairs of nodes rapidly converges towards a maximum of 78% for a threshold set to 10% and above. Combining these two results, we can say that many handovers that would disturb ongoing flows do not significantly improve the network connectivity. This confirms the relevance of limiting such reattachments. We also note from Fig. 7(f) that setting the threshold under 5% can lead to higher DTN delay.

We can conclude that setting this threshold to a low value (7 to 10%) allows to avoid many reattachments that do not improve connectivity so much while they would generate flow disruptions, with limited impact on the DTN delay.

These two thresholds are finally useful to trade off between achieving the best connectivity and reducing the negative impact of frequent changes of attachments. To highlight this, we ran one more simulation to evaluate our greedy strategy with the threshold on flow disruptions set to 10%, that we call Greedy.10%. Fig. 8 compares the results to those obtained with Optimal and Greedy. Using this threshold preserves a near-optimal network connectivity. At the same time, we analyzed that the number of handovers decreased by 54%. Using only a low value for the threshold on flow disruption thus allows to filter many handovers that are not useful while preserving a near-optimal network connectivity.

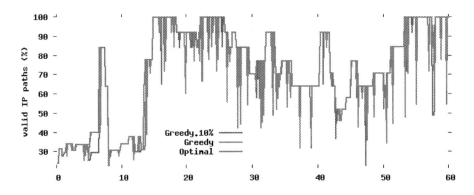

Fig. 8. Influence of the flow disruption threshold on the percentage of connected pairs of nodes

5 Conclusions and Perspectives

This paper has presented a greedy attachment strategy in order to improve the global network connectivity in mobile ad hoc networks built with cellular wireless technologies. Our strategy is driven by the objective of maximizing the number of nodes that can communicate so that the network connectivity is the best possible. We have presented a practical way of implementation that takes benefit from existing control channels of WiFi and LTE technologies to exchange connectivity information between nodes with no overhead. Through simulations, we show that our strategy leads to a network connectivity almost as good as in pure ad hoc networks, while the average path length is a bit longer. We also study the impact of thresholds we fix on the gain in average path length and on flow disruption to maximize the utility of reattachments. The evaluation we made has shown that this strategy gives results close to the optimal and outperforms the traditional signal strength based strategy. The thresholds that can be configured to control reattachments allow to trade off between best connectivity and the impact of those reattachments.

The attachment decision strategy presented in this work may be included in a large multi-criteria algorithm, taking for instance into account signal strength, the social proximity of nodes as it influences traffic patterns, or the number of terminals already attached to a node. We plan to study such complex decisions in future work. We also plan to work on prototyping activities. We conjecture that the path elongation that we observed between cellular and pure ad hoc networks should be compensated by the more efficient MAC layer (infrastructure mode) used in cellular ad hoc networks.

References

1. FunkFeuer: FunkFeuer: free net, `http://www.funkfeuer.at`
2. Guifi: Guifi network, `http://guifi.net/`
3. Knopp, R., Nikaein, N., Bonnet, C., Aiache, H., Conan, V., Masson, S., Guibe, G., Martret, C.: Overview of the widens architecture, a wireless ad hoc network for public safety. In: Proc. of SECON (2004)
4. Iovine, R., Bouis, J.: The flexnet-waveform in the international sdr arena. In: Military Communications Conference. In: IEEE MILCOM (2009)
5. Wirtz, H., Heer, T., Backhaus, R., Wehrle, K.: Establishing mobile ad-hoc networks in 802.11 infrastructure mode. In: ACM Chants (2011)
6. Cao, L., Zheng, H.: Distributed spectrum allocation via local bargaining. In: Proc. of SECON, pp. 475–486 (2005)
7. Bereski, P., Soulie, A.: Flexible radio network (patent), `http://patent.ipexl.com/EP/EP1936872.html`
8. Sarshar, M.H.: Poo K.H., Abdurrazaq, I.A.: NodesJoints: A Framework for Tree-Based MANET in IEEE 802.11 Infrastructure Mode. In: IEEE Symposium on Computers and Informatics (2013)
9. Nicholson, A.J., Chawathe, Y., Chen, M.Y., Noble, B.D., Wetherall, D.: Improved access point selection. In: Proc. of ACM MobiSys (2006)
10. Fall, K.: A delay-tolerant network architecture for challenged internets. In: Proc. of ACM SIGCOMM (2003)

Hybrid Model for LTE Network-Assisted D2D Communications

Thouraya Toukabri Gunes[1], Steve Tsang Kwong U[1], and Hossam Afifi[2]

[1] Orange Labs, Issy-les-Moulineaux, France
{thouraya.toukabrigunes,steve.tsangkwongu}@orange.com
[2] Telecom SudParis, Evry, France
hossam.afifi@telecom-sudparis.eu

Abstract. In the evolution path towards the "Always-Connected" era and the trend for even more context-aware services, Device-to-Device communications (D2D) promise to be a key feature of the next-generation mobile networks. Despite remaining technical issues and uncertain business strategies, D2D-based services represent a new market opportunity for mobile operators that would manage to smoothly integrate these new technologies as a complement or even an efficient alternative to cellular communications. Existing research efforts on the integration of D2D technologies in cellular networks have mostly failed to meet user expectations for service simplicity and reliability along with operator requirements regarding lightweight deployment, control and manageability. This paper proposes a hybrid model for D2D communications assisted by mobile operators through the LTE network: it includes a lightweight D2D direct discovery phase and an optimized data communication establishment for proximity services. The proposed hybrid model is appraised against the existing solutions in literature and the current standardization effort on Proximity Services (ProSe) within the 3GPP.

Keywords: D2D communications, LTE networks, Proximity Services.

1   Introduction

Today's mobile networking world facts are: the mobile industry is shipping more smartphones and tablets than PCs; success stories of social networking services like Facebook has become a social trend from which mobile users developed the need to be connected anywhere to their surroundings; statistics in [4] envision an exploding number of more than thousand billion wireless connections around the world in 2020. Meanwhile, revenues of mobile services have been growing at a much slower rate than the growth of mobile connections since 2011[1]. The challenge for mobile operators, who struggle each quarter to turn a profit on voice and SMS services, is yet to face the threat of Over-The-Top (OTT) providers who have put their foot down at the mobile market with apps that supply instant messaging, multimedia services like photo sharing and video conferencing and other popular services for free. In this context,

---

[1] http://www.wwpi.com/

S. Guo et al. (Eds.): ADHOC-NOW 2014, LNCS 8487, pp. 100–113, 2014.
© Springer International Publishing Switzerland 2014

Device-to-Device communications have become the new driver in wireless networking and mobile market.

Defined as a short range direct communication between devices without the involvement of the network infrastructure, D2D communications have been proposed as an underlay to cellular networks. Such a solution will evolve cellular networks toward a layered topology in which multiple network layers (femto-network, D2D-network, Wifi-network…) would coexist with a main macro-cell layer.

With these new types of communications mainly based on context and proximity information, a new generation of user-centric mobile services will rise, offering at the same time the opportunity for operators to extend their mobile networks' capacities and to alleviate the traffic in their core networks; for instance, smart cities services, real-time social discovery of nearby persons, targeted and personalized hyper-local services (advertising, couponing/ticketing, restaurant/hotels booking, content download, etc.). Besides, when including group communications and relay mechanisms, D2D communications could be a relevant fallback alternative for the public safety services (police, firefighters, emergency services, etc…) in disaster situations (earthquake, Fire, etc.): using a specific D2D-enabled Public safety device, an officer/agent can exchange data and transmit information to other devices through a D2D group communication. Moreover, in poor radio coverage areas, relay-based D2D mechanisms could be an efficient way to extend network connectivity.

Surfing on the wave of the successful worldwide launch of 4G LTE (Long Term Evolution) mobile networks, a new short range technology based on LTE (LTE Direct) has been developed by Qualcomm[2]. Envisioned to be the next trendy D2D technology that best meets the requirements of the above mentioned types of services and successfully tested in a recent research work[3], 3GPP (3rd Generation Partnership Project) has then initiated a standardization effort on the integration of LTE Direct in mobile networks [12]. If this effort first addresses Radio Access Network (RAN) requirements and technical issues for the support of D2D-based services, it comes also along with a feasibility study [5] and a technical specification on the architecture enhancements for the support of Proximity Services (ProSe) [6].

In literature, many research works have been done on D2D communications and their integration within LTE networks. The earliest ones addressed mainly the radio aspects such as D2D radio interference management with cellular communications, power control, radio resources allocation/sharing methods and spectrum regulatory aspects (use of a licensed or unlicensed band for D2D). Studies have also been made on D2D discovery and communication mechanisms. However, the few proposed solutions in these fields are still immature and don't answer basic user concerns for simplicity, reliability and QoS when using D2D-based services. Otherwise, if these requirements have been answered by the recently standardized solutions proposed in 3GPP [6], the current specification is globally lacking from a more extensible and evolutionary vision of the D2D integration in current and next generation networks.

---

[2] http://www.qualcomm.com/solutions/wireless-
networks/technologies/lte/lte-direct
[3] http://english.etnews.com/internet/2909211_1299.html

This paper proposes a novel LTE-based D2D discovery and communication mechanism. The solution aims to integrate D2D as native features in the current LTE architecture in order to promote the new paradigm of decentralized and locally-scoped communications. It mainly includes a reliable Direct Discovery phase and an efficient and optimized communication establishment phase. In the following sections, we give an overview on D2D use cases classification and Proximity Services (ProSe) as well as a discussion on existing D2D mechanisms. Then, we describe our D2D hybrid approach for discovery and communication and compare it to the current 3GPP standardized solution.

2   Overview on D2D Use Cases

Similarly to location-based communications, a D2D communication benefits from the proximity of devices in order to establish a direct link between them for a local data exchange. Basically, devices could be any device equipped with a D2D technology suitable for short range communication such as smart phones, tablets, laptops, network printers, cameras, or even connected vehicles.

When dealing with "devices proximity", the definition of "proximity" becomes questionable and different proximity levels could be defined (i.e. geographical proximity, network topology proximity (e.g. devices within the same subnet), radio range, etc.). For mobile operators with already a rich amount of valuable user data, context and proximity information are assets that need to be empowered to offer new value-added services to users and apace mainstream the adoption of D2D-based services.

Table 1. D2D use cases classification

Use case category	Applications
Commercial and Social Proximity Services: An evolution of LBS services through hyper-local and dynamic proximity data.	• Discovery-centric services: Context-aware applications, Social networking applications, location enhancement applications, Social gaming, and smart cities services… • Communication-centric services: content and video sharing services
Public safety services based on group and relay communications: Secure services used on specific D2D enabled devices and deployed on a dedicated non-public network	• Direct communication between public safety agents in or outside network coverage: push-to-talk, group communication, priority handling… • Dedicated network access sharing for out of coverage devices through peer-to-peer connections to nearby in-coverage devices.
Services for network capabilities enhancement	• Offloading services: offload of local data traffic or video/voice call traffic. • Multi-hop access services: Internet connection sharing through devices acting as relays, Connectivity extension to heterogeneous networks (UE acting as a gateway to Sensor network (MTC), UE in a vehicle acting as a cooperative relay to an ITS network infrastructure, etc.)

2.1   D2D Use Cases Classification

D2D-based services is a business opportunity for operators and present multiple at-tractive use cases going from public/commercial services to more specific fields like public safety and military services. Generally, D2D applications aim at proposing and improving local services for users while optimizing the throughput over the radio network, reducing the overall core network load and enhancing connection delays. Multiple use case classifications were made in [2], [5] and [6]. Generally, D2D-based services can be classified into three main categories like described in table 1.

2.2   Operator Role in D2D

D2D communications are definitely a business opportunity for mobile operators. Op-erators' strengths consist in having a powerful network infrastructure that will allow the deployment of new D2D-based services while assuming their original contractual duty on keeping data security and users' privacy. Despite the set of technical issues and challenges that operators may face with the integration of D2D in their LTE net-works, they still have rich valuable assets that meet user expectations toward new D2D services. Operators' important assets include:

— Service security and QoS through secure and uninterrupted connections.
— Identity management, authentication and Privacy when using a D2D service.
— Context information exposure for more attractive services and better QoE.
— Devices management: user's cellular and non-cellular devices are associated to his user profile and included into the operator's subscriber database and automatically associated with the owner's cellular devices.
— Ensure the consistency of the user experience including reachability and mobility aspects (seamless offload, seamless handover, etc.).

3   D2D Peer Discovery and Communication Establishment

A D2D communication has mainly two phases: A discovery phase in which devices aware of their location detect surrounding devices/services, and a Communication phase in which D2D peers exchange or share data on a D2D link.

Two communication modes were identified in [1], [2], [3] and [6]: the D2D Direct mode or the D2D network-assisted mode. As both D2D discovery and communication phases could be relatively independent, different models can be derived from each D2D mode according to the role of the operator in D2D phases. In a network-assisted D2D mode, two schemes could be derived: a fully-controlled scheme in which the operator have a full control on discovery and communication phases, and a loosely-controlled scheme in which discovery and communication phases are partially assist-ed by the operator, for instance with authentication mechanisms and radio resource allocation.

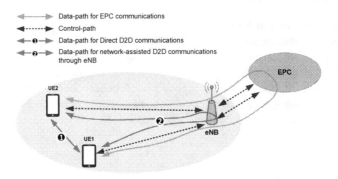

Fig. 1. Data path in direct and network-assisted D2D communication

3.1   D2D Peer Discovery Approaches

D2D peers need to discover each other before initiating a D2D communication. This discovery phase could be done directly between devices in an ad-hoc manner or with the support of an operator network that can either control the entire discovery phase by detecting D2D candidates at the core network level or only assist it as a trusted third party. Two main discovery approaches were identified [1] [2] [3]:

- Direct discovery approach:

According to 3GPP, D2D direct discovery is defined as the process of detecting and identifying devices in proximity using E-UTRA[4] direct radio signals [6]. Two discovery models were identified in ProSe: the "I'm Here" model (A) and the "who is there? / are you there?" model (B). In model A, a UE (User Equipment) could be an Announcing UE that broadcasts some information to its surrounding at pre-defined discovery intervals, or a monitoring UE that monitors certain information of interest from devices in proximity. In model B, the discovery model is more accurate about what is exactly needed to discover and thus, defines two roles for the UE: a "discoverer UE" which sends certain information about what it is interested to discover and a "discoveree UE" which replies with some information to the discoverer. The Direct discovery approach has the advantage of flexibility and scalability as it can adapt to an increasing number of D2D connections and allow in this way the offload of core network traffic to local D2D communications. However, such a method may have an impact on the UE complexity as it needs to support power management mechanisms to avoid battery consumption issues caused by an "always-on" discovery. Besides, specific radio resource scheduling and allocation mechanisms need to be defined in order to overcome interference issues with cellular communications. Moreover, this distributed approach may suffer from the lack of security and authentication mechanisms that need to be deployed in order to meet users' concerns about privacy.

---

[4] E-UTRA: Evolved UMTS Terrestrial Radio Access.

- Centralized discovery approach:

This approach involves at least one or more network entities in the discovery procedure. As the operator network has a wider vision on the overall traffic and on UE mobility context, centralized discovery approaches aim at exploiting mobile operator core network assets about devices micro/macro mobility in order to provide a more accurate and efficient discovery information. The control of the discovery phase at the core network ensures the delivery of a better QoS and a more reliable service, thanks to authentication and privacy mechanisms offered natively by the operator. Centralized discovery is defined as EPC-level discovery in the 3GPP [5]. However, this approach could be less scalable than the direct approach as it could have performance and overload impacts on the core network with the additional D2D traffic. Efficient load-balancing and lightweight signaling mechanisms should be implemented at the core network in order to make D2D profitable for the operator and the user.

3.2 D2D Communication Establishment

After the discovery phase, D2D peers establish a communication link for data exchange. As shown in figure 1, the exchange of data could be done either directly between D2D devices, or using an optimized path through the eNB.

Basically, the setup of an EPC-based communication [8] [11] [13] consists on the establishment of an EPS (Evolved Packet System) bearer composed of a radio bearer (E-UTRAN bearer between UE-eNB), an EPC bearer and packet filters. There are two types of EPS bearers: the default bearer and dedicated bearers. At the initial network attach of the UE, an EPS default bearer is setup by the establishment of a PDN (Packet Data Network) connection with the PDN Gateway and the allocation of an IP address to the UE. In the case of a D2D communication, a dedicated D2D bearer with specific radio resources allocated by the eNB is setup. However, there is no need to setup an EPC bearer as the data path is optimized for D2D communications. Since the D2D UE has already an IP address associated with its default bearer, it is used for all bearers within the same PDN connection including the D2D dedicated bearer. Based on the concept of QoS-based flow aggregates, multiple D2D flows corresponding to simultaneous D2D connections could be carried within the same D2D data radio bearer.

According to the current 3GPP standard on ProSe [5] [6], direct D2D communications are allowed only when devices are out of coverage and only for Public safety services and bearer mechanisms are not yet defined for these types of communications. In [8], a D2D dedicated bearer mechanism is proposed for a D2D offloading service. Generally, existing D2D literature lacks from detailed mechanisms for D2D bearer establishment for D2D-based services other than public safety and offloading.

3.3 Existing Solutions

Several solutions were proposed in D2D literature for D2D discovery and communication establishment mechanisms. They are summarized as follows:

— Solution 1 [3]: an EPC-based D2D discovery through MME using the Session Initiation Protocol (SIP). A SIP dedicated handler is proposed as an extension the MME in order to process D2D SIP packets. D2D SIP packets are encapsulated in NAS packets. The main limitation is the protocol layer violation between the User plane and the Control plane when processing SIP messages in the MME.
— Solution 2 [3]: an EPC-based D2D discovery through gateways for offloading. A D2D filtering is implemented in the gateways to discover D2D candidates. All IP flows associated to the same tunnel endpoints are filtered and marked as D2D. Radio measurements are made by eNB to decide to offload or not the traffic. The deployment of such a solution would have an impact on EPC entities performance (congestion, resource consuming).
— Solution 3 [8]: an EPC-based D2D discovery through a ProSe function implemented at gateways for offloading. A D2D dedicated bearer control mechanism is proposed. The limitation of this solution is that the dedicated bearer setup procedure is done in EPC while the data exchange is done directly between devices. Also, the additional IP address allocated for the dedicated EPC bearer would break the session at the handover from EPC to D2D link.
— Solution 4 [6]: a Direct D2D discovery through a dedicated ProSe server. A ProSe authorization mechanism is proposed before the direct discovery. The Direct discovery is made over IP through a ProSe protocol that is to be defined in [12]. D2D bearer control procedures are not yet defined and ProSe communication mechanisms are specified for one-to-many schemes only.

Table 2. Existing D2D solutions comparison

	Solution 1	Solution 2	Solution 3	Solution 4
Direct or EPC-based discovery?	EPC-based discovery	EPC-level discovery for offloading	EPC-level discovery for offloading	Direct discovery
D2D bearer mechanism?	No	No	Yes	No
EPC impacted entities	MME	PGW, SGW, eNB	SGW, PGW, MME, eNB	ProSe, HSS, MME
Support for D2D Authorization?	No	No	Yes	Yes
Support for session continuity at handover?	No	No	No	No

As described in table 2, most of the existing solutions propose EPC-based discovery mechanisms with weak contributions on the communication establishment mechanisms. D2D authentication and authorization are not seriously considered except at the 3GPP ProSe solution. Besides, IP session continuity at handover from EPC-based communications to D2D is not yet discussed in any of the above-mentioned solutions. Generally, the proposed LTE improvements to support D2D communications are lacking from a long term vision to the viability of the EPS architecture: adding complexity to a system without considering its evolution in the future and regardless to economic and scientific advances may lead to its fast obsolescence. In next sections, we expose a long term vision of the deployment of D2D communications through a hybrid LTE network-assisted D2D solution for Discovery and communication.

4   A Hybrid Model for D2D Discovery and Communication

4.1   Motivations and Proposal

The main motivation for D2D-based services was originally the offload of local communications through an optimized data path (D2D path) in order to alleviate the traffic overload in the core network. With the expansion of context aware applications market, the motivation has evolved toward a new type of mobile services based on proximity information. In the vision of future cellular networks [9], studies have demonstrated the tendency to move toward more flat architectures in which core network functions are virtualized in order to enhance the performance and reduce operational costs of core network extensions for operators. In this context, Cloud-Radio Access Network (C-RAN) has been proposed as an enhancement to the LTE RAN. It consists mainly on the optimization of eNB deployments through the virtualization of some of its functions. Composed of a BBU (Baseband Unit) and a RRH (Remote Radio Head), the eNB BBU function is virtualized and centralized [10] in order to reduce deployment costs in urban areas, efficiently use processing resources, limit inter-eNB interference issues and improve scalability on the RAN through collaborative radio mechanisms. In the following, we introduce a beyond-LTE vision of the D2D integration within the current EPS architecture.

4.2   Enhanced eNB Role for Local D2D Services

From a network design point of view, the closest network entity to UEs is the eNB. Being the anchor point between the RAN and the EPC, the eNB disposes of location information of each UE in the cell (for devices in the active mode). As such, we propose to integrate main D2D functions in the eNB. The solution is based on a hybrid approach that combines a lightweight D2D Direct discovery mechanism assisted by the operator for the Authentication and Service Authorization aspects, and an optimized and QoS-enabled D2D communication through the eNB. Besides, we choose to minimize the involvement of EPC entities (MME and HSS) in order to avoid too much overhead in EPC. We also propose some D2D extensions to the LTE protocol stack and to implement the following functions as an evolution to the current eNB implementation:

— Locating and checking D2D peers positions in cells (for devices being in the Active mode).
— Identification of an active D2D communication.

- Authorization and allocation of dedicated D2D discovery resources for the D2D pairs.
- Mode selection by the means of specific radio measurements on the cell (global congestion check on the cell) and with the UEs.
- Dedicated D2D Radio bearer establishment with D2D peers and resource allocation for the D2D communication.

4.3   A Reliable Direct Discovery Mechanism

Before initiating the D2D procedure, we assume that D2D enabled UEs have registered to a specific D2D service through a service platform. A D2D service registration phase consists basically on granting access credentials for a specific D2D service from an Application Server (AS) using a client application embedded on devices. From an LTE point of view, the service authentication could be done through the IMS (IP Multimedia Subsystem) if the service is IMS-based or directly between the UE and the AS through some authentication services offered by the operator (i.e. GBA[5]) if the service is IMS-independent. The proposed solution in this paper is agnostic to whether the service is IMS-based or not and the D2D service registration mechanism is out of this paper's scope. As depicted in figure 2, our solution is based on the following steps:

- Initial NAS Attach, D2D Authentication & Authorization:

In order to start a D2D Discovery, D2D peers need to exchange their identities and to be authenticated by the operator network as trusted D2D enabled devices. In order to relieve identity generation and management, we propose to use the Single Sign-On Authentication concept (SSO)[6]: each D2D enabled UE has a unique D2D identity (D2D_id) stored locally in the UE. Such unique identifier is also mapped to the user IMSI and is stored among the subscriber profile information at the HSS. Basically, at power on, a UE initiate a Network Attach procedure in order to beneficiate from the NAS-level services, after which a UE context is setup at the MME, a default bearer is established between the UE and the PDN-GW and a UE IP address is allocated. As a pre-discovery phase and in order to secure the discovery between D2D enabled UEs, we propose to include a D2D authorization procedure during the network attach: the IMSI[7] is sent into the **NAS_attach_req()** message to the MME with an indication to check the D2D_id during the authentication phase with the HSS. An **Update_location_req()** message is then sent by the MME to the HSS in order to get the subscriber profile information [7]. The D2D_id is then retrieved from the HSS user profile by mapping with the IMSI together with the list of D2D authorized services for that specific UE. This D2D authorization information is then stored at the eNB on a per UE D2D_id basis and is refreshed at the expiration of a validity timer.

---

[5] http://en.wikipedia.org/wiki/Generic_Bootstrapping_Architecture
[6] http://en.wikipedia.org/wiki/Single_sign-on
[7] IMSI: International Mobile Subscriber Identity.

- PDN connection & default bearer establishment:

Once the D2D UE is authenticated and authorized, the default bearer is created through the PDN connection and an IP address is allocated to the UE. This bearer is maintained during the D2D communication in order to ease the handover to the EPC path if the D2D connection is broken. Moreover, as the UE does not change its PDN connection while initiating a D2D communication, the same IP address allocated at the initial NAS attach is used by the D2D dedicated bearer described hereafter.

- Resource allocation for direct discovery:

Assuming that a D2D application is used at the UE to access a specific D2D service, a specific D2D request message is sent from the UE to the eNB. The **D2D_direct_discovery_request (UE D2D_id, D2D_app_id)** is an extented RRC message that is sent to the eNB to request the authorization to use direct discovery resources. The packet is processed at the eNB which checks the authorization with the MME for the specified D2D_id in order to allocate the resources for the direct discovery over the E-UTRA. Once the authorization for direct discovery resource allocation is granted, the eNB generates a temporary D2D_id to the requesting UE for the direct discovery purposes and sends back an acceptance notification to the UE with the temp_D2D_id, as well as the needed configuration information for the authorized and allocated discovery resources (uplink/downlink, transmission power, etc.). The D2D generated temporary identifier is used to identify an active D2D connection at the eNB and expires with a validity timer.

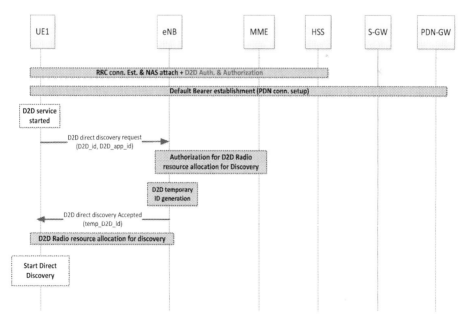

Fig. 2. D2D control signaling for a direct discovery over E-UTRA

- D2D direct discovery over E-UTRA:

After granting the authorization and the temp_D2D_id for discovery, the UE sets up through the D2D application its discovery mode according to the discovery models described in section (3.1). Proximity discovery messages are then sent over the air to discover devices of interest.

4.4   An Optimized and QoS-Enabled Communication through eNB

After the direct discovery, a data path should be setup. In the specific case of a D2D communication, the communication is optimized and the data path could be established whether directly between devices or through the eNB which plays the role of a network anchor for the communication. In both schemes, the establishment of an additional EPC bearer is not needed as the data are exchanged locally between the UEs.

In our proposal, we choose to have a controlled communication between D2D peers by using an optimized data path through the eNB. For the operator, charging and legal interception rules would be facilitated when the data passes through the eNB; this avoids implementing additional complex mechanisms on the UE to support such functions. In addition, the control of the D2D bearer establishment through the eNB allows adapting QoS parameters according to the D2D service type. It also maintains the session continuity; even if one of the D2D UEs is not anymore in the D2D range, the session is not broken as long as the D2D communicating UEs are in the coverage range of the eNB. Instead, a seamless handover would be done in order to bring back the D2D data path to normal EPC communication path. The concept of D2D dedicated bearer is explained in figure 3.

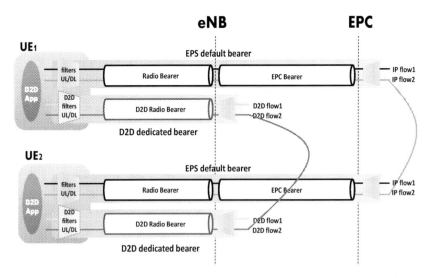

Fig. 3. D2D dedicated bearer concept for D2D communications meaning an optimized data path through the eNB

In figure 4, we propose a signaling call-flow for the establishment of a D2D communication after a discovery phase. Both discovered UEs send a D2D direct communication request to the eNB with their respective D2D_id. At the reception of these requests, the eNB will ask the MME for the establishment of a dedicated D2D radio bearer with each UE. The role of the MME here is to maintain a context for active D2D bearers in order to ease the handover to normal EPC connection when a D2D connection is broken. Each D2D dedicated bearer is identified at the MME by a D2D_bearer_id. A D2D bearer context is a D2D bearer in which an IP flow is routed. As explained before, there is no need to allocate a new IP address for the D2D dedicated bearer: using the same initially allocated IP address, the IP flow is routed on the D2D dedicated bearer through the eNB using a routing rule setup on the UE: the D2D packet filter is mapped with the correspondent radio bearer setup between the UE and eNB. At the eNB side, D2D flow coming from UE1 on the radio bearer RB1 are linked with the D2D flow coming from UE2 on the radio bearer RB2. Thus, the eNB maintain a layer 2 routing table in which D2D flows are associated with their correspondent radio bearers.

After the setup of the dedicated bearer between the eNB and each UE, a notification of acceptance message is sent back to the UEs to inform them about the radio resources allocated for data transmission through the eNB.

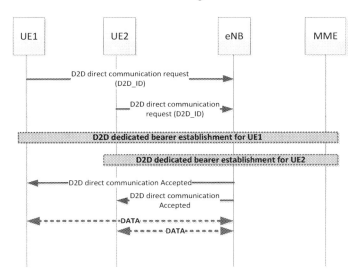

Fig. 4. eNB assisted D2D communication establishment

5   Discussion

The hybrid mechanism proposed above brings new features compared to existing D2D solutions proposed in literature. First, the direct discovery approach is combined with a network-assisted communication establishment approach. The reason of this combination is that operator networks have an important role to play in the deployment of

D2D-based services as trusted parties which can satisfy users' concerns about security and privacy. It is then proposed a control plane signaling for D2D authentication and authorization during the initial attach to the network: this SSO based mechanism reduces the number of exchanged messages between the UE and the EPC network. From a long term architecture point of view, we proposed to evolve the current eNB functionalities to support D2D functions: including D2D mechanisms at the EPC level would be contradictory to the basic concept of D2D communications, which is offloading D2D traffic to local communications. One of the main objectives of our approach is to take down the D2D functionality to the lowest network entity, i.e. the closest entity to UEs. Moreover, we also propose a dedicated bearer establishment for a D2D communication using an optimized path through the eNB. As such, a service-adapted QoS could be applied through the proposed dedicated bearer establishment mechanism.

6   Conclusion

This paper proposes a long term operator-assisted D2D communication mechanism through a trusted procedure for a Direct D2D discovery over E-UTRA. The operator role in assisting the D2D authentication and authorization of the D2D users is highlighted. Then, a dedicated and optimized signaling for D2D communication establishment was proposed with a specific mechanism to apply a per-service QoS. In a more global view, the solution envisions a more active role of the eNB in the EPS architecture in order to meet emerging local and proximity services deployment requirements. Ongoing simulation and modelization work are carried out in order to evaluate and compare the performance of the proposed solution in terms of delays and estimate the impact of the D2D load on RAN and EPC entities.

References

1. Fodor, G., Dahlman, E., Mildh, G., Parkvall, S., Reider, N., Miklos, G., Turanyi, Z.: Design Aspects of Network Assisted Device-to-Device Communications. IEEE Communications Magazine 50, 170–177 (2012)
2. Lei, L., Zhong, Z., Lin, C., Shen, X.: Operator Controlled Device-to-Device Communications in LTE-Advanced Networks. IEEE Wireless Communications 19, 96–104 (2012)
3. Doppler, K., Rinne, M., Wijting, C., Ribeiro, C., Hugl, K.: Device-to-Device Communication as an Underlay to LTE-Advanced Networks. IEEE Communications Magazine 47, 42–49 (2009)
4. David, K., Dixit, S., Jefferies, N.: 2020 Vision The Wireless World Research Forum Looks to the Future. IEEE Vehicular Technology Magazine 5(3), 22–29 (2010)
5. 3GPP TR 22.803, Feasibility study for Proximity Services (ProSe) (2013)
6. 3GPP TS 23.303, Architecture enhancements to Support Proximity Services (ProSe) (2014)
7. Srinivasa Rao, V., Gajula, R.: Protocol signaling procedures in LTE. Radysis white paper (September 2011)

8. Yang, M.J., Lim, S.Y., Park, H.J., Park, N.H.: Solving the data overload: device-to-device bearer control architecture for cellular data offloading. IEEE Vehicular Technology Magazine (March 2013)

9. Cavalcanti, D., Agrawal, D., Cordeiro, C., Xie, B., Kumar, A.: Issues in integrating cellular networks, WLANs and MANETs: A futuristic heterogeneous Wireless Network. IEEE Wireless Communications (June 2005)

10. Sundaresan, K., Arslan, M.Y., Singh, S., Rangarajan, S., Krishnamurthy, S.V.: FluidNet: A Flexible Cloud-based Radio Access Network for Small Cells. In: MobiCom 2013 (September 2013)

11. 3GPP TS 23.401, General Packet Radio Service (GPRS) enhancements for Evolved Universal Terrestrial Radio Access Network (E-UTRAN) access (2014)

12. 3GPP TR 36.843, Feasibility Study on LTE Device to Device Proximity Services - Radio Aspects (2014)

13. 3GPP TS 36.300, Evolved Universal Terrestrial Radio Access (E-UTRA) and Evolved Universal Terrestrial Radio Access Network (E-UTRAN); Overall description; Stage 2 (2014)

On the Problem of Optimal Cell Selection and Uplink Power Control in Open Access Multi-service Two-Tier Femtocell Networks

Eirini Eleni Tsiropoulou, Georgios K. Katsinis, Alexandros Filios,
and Symeon Papavassiliou

Network Management & Optimal Design Laboratory (NETMODE),
School of Electrical & Computer Engineering,
National Technical University of Athens (NTUA),
9 Iroon Polytechniou str. Zografou 15773, Athens, Greece
{eetsirop,gkatsinis}@netmode.ntua.gr,
{el09039,papavass}@mail.ntua.gr

Abstract. In this paper the problem of optimal cell selection and power control in the uplink of multi-service open access two-tier femtocell networks is addressed via a game theoretic approach. Each user is associated with a properly defined QoS-aware utility function, which depends on the two tier architecture and the class of requested service, i.e. real or non-real time service. An optimal cell selection game is formulated and is proven to be a potential game. Its solution constitutes a feasible pure strategy Nash equilibrium (NE), which guarantees that the users select the most appropriate cell to connect in order to maximize their QoS-aware performance. Subsequently, an optimal power allocation is adopted, towards achieving an energy efficient resource allocation, given users' assignment to the cells. A learning algorithm is proposed towards determining the cell selection game's NE, combined with the power allocation mechanism. The effectiveness of the proposed approach is evaluated through modeling and simulation.

Keywords: Cell Selection, Power Control, Potential Game, Two-Tier Femtocell Networks, Nash Equlibrium.

1 Introduction

Femtocells have arisen as a promising solution in order to confront users' Quality of Service (QoS) prerequisites and decongest the macrocell network from the increasing number of mobile users and corresponding traffic. According to the statistical data in [1], nearly *90%* of data services and *60%* of phone calls nowadays are taken place in indoor environments. Thus, femtocells, which usually reside in a home/office environment and consist of an Access Point (AP) with short range, i.e. *10-50m*, connected to service provider's internet network [2], have gained great attention, as they can provide large indoor coverage and capacity due to the short communication distance to users. Furthermore, femtocells are characterized by low cost installation over an existing macrocell network. The entire overlaid network is referred as two-tier

S. Guo et al. (Eds.): ADHOC-NOW 2014, LNCS 8487, pp. 114–127, 2014.

femtocell network and consists of femtocells and the conventional macrocell network. Moreover, there are two main access methods, which have been defined by the Third Generation Partnership Project (3GPP) as follows: (a) closed access, where only a subset of users who have already registered can connect to the femtocell, and (b) open access, where all users are permitted to access femtocell [2].

In the area of two-tier femtocell networks, the overlapping of the coverage areas of femtocells and macrocells offers the ability to the users to connect either to the macrocell or to one of the neighboring femtocells according to their quality of service criteria. Thus, the cell selection problem is an essential step in order to assign the users to the cells and subsequently (or jointly) implement an optimal resource allocation in the two-tier femtocell network.

Towards confronting the cell selection problem, a cell selection game in two-tier femtocell networks is proposed in [3] and formulated via adopting the concept of primary and secondary users in the cognitive radio, in order to resolve the prioritized access issue. Furthermore, in [4] an evolutionary game model is presented to describe the dynamics of the cell selection process and authors consider the evolutionary equilibrium as the solution. Moreover, in [5] the authors extend the capacity based cell selection mechanism to account for both uplink and downlink transmissions in a co-channel operation. Additionally, a cell selection scheme is proposed in [6] that adopts the downlink capacity of the two-tier femtocell network to the traffic load while guaranteeing QoS constraints. Moreover, considerable research efforts have been devoted to the individual problem of utility-based resource allocation, either via adopting different utility function [12], [13] among macro-users (MUs) and femto-users (FUs), or the same utility function [14], [15].

In this paper, we address the problem of optimal cell selection and uplink power control in open access two-tier femtocell networks, supporting multiple services (i.e. both real-time and non-real time) with various and often diverse QoS prerequisites via a game theoretic framework. Initially, a novel well designed multi-domain utility function is proposed, which reflects user's service QoS aware performance efficiency according to both the tier that the user is connected to, as well as the type of service he requests (Section 2). Then, the cell selection problem is formulated as a potential game aiming at maximizing each user's utility. The game's potential function is determined and the existence of a cell selection Nash equilibrium for all users residing in the two-tier femtocell network is proven (Section 3). Furthermore, since it is difficult or impractical to let all users in the two-tier femtocell network have full knowledge of the complete information, a non-cooperative cell selection algorithm based on learning and optimal power control has been proposed as well (Section 4). To demonstrate the effectiveness and efficiency of the proposed approach detailed numerical results are provided in Section 5, while Section 6 concludes the paper.

2   System Model and Background Information

We consider a two-tier femtocell cellular network that consists of a central macrocell base-station (MBS) B_0 serving a particular service area $\mathfrak{R}$, providing a cellular coverage of radius R_0. Within the service area $\mathfrak{R}$, F co-channel femtocells B_j ($j = 1,...,F$) reside in the second tier of the overall network architecture.

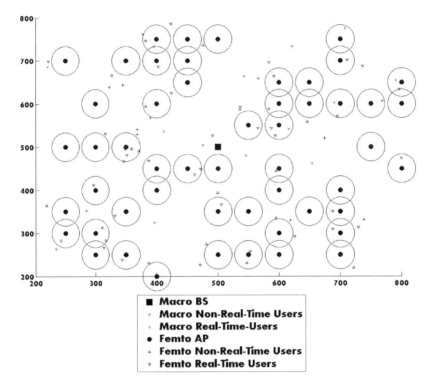

Fig. 1. Two-tier network topology

Each femtocell supports a region $\mathbb{C} \subset \mathfrak{R}$ and consists of a disk of radius $R_c \ll R_0$. The entire two-tier femtocell network topology is presented in Fig. 1.

In our proposed framework, we consider open access femtocells, i.e. registration in not mandatory for mobile users to be served by femto Access Points (AP). Thus, each AP and consequently the MBS, is not aware of the specific number of users that reside within its coverage area. Therefore, cell selection mechanism is one of the key challenges for achieving on one hand load balancing in order to avoid network congestion and performance degradation, while on the other hand achieve the maximization of users' perceived satisfaction due to the fulfillment of their QoS prerequisites.

Moreover, at each time slot t, the two-tier network supports $M(t)$ continuously backlogged users, where $\mathfrak{M}(t)$ denotes their corresponding set. The set of users $\mathfrak{M}(t)$ consists of macro-users (MUs) and femto-users (FUs) requesting either real-time (RT) services with strict short-term QoS constraints (i.e. $\mathfrak{M}_{MRT}(t)$ and $\mathfrak{M}_{FRT}(t)$) or non-real time (NRT) services (i.e. $\mathfrak{M}_{MNRT}(t)$ and $\mathfrak{M}_{FNRT}(t)$). Furthermore, let $M_{MRT}(t)$, $M_{FRT}(t)$, $M_{MNRT}(t)$ and $M_{FNRT}(t)$ denote the number of MUs and FUs, requesting real-time and non-real time service, respectively. It is noted that throughout the rest of the analysis in this paper it is omitted the notation of the time slot t for simplicity of the presentation.

At the beginning of each time slot a cell selection and power allocation mechanism is responsible to assign the M users to the j, $j=0,1,...,F$ cells (i.e. macrocell and femtocells) and correspondingly determine users' optimal uplink transmission power, towards maximizing their satisfaction received by system's optimal resource allocation. Given that a user $i \in \mathfrak{M}$ is connected to the MBS B_0 or to an AP B_j ($j=1,...,F$), then his uplink transmission power is denoted by P_i and his path channel gain from the transmitter of user i to the receiver of user $i' \neq i$ by $G_{ii'}$. The channel gains are represented via adopting the simplified path loss model in the IMT-2000 specification [8].

Moreover, let σ^2 be the variance of Additive White Gaussian Noise (AWGN) at B_j ($j=1,...,F$). Thus, the received SINR γ_i of user i at B_j is given as:

$$\gamma_i = \frac{WG_{ii}P_i}{R\left(\sigma^2 + \sum_{i' \neq i} G_{ii'}P_{i'}\right)} \tag{1}$$

where W [Hz] denotes the available spread spectrum and R [bit/sec] is the bit rate.

A. Utility Functions and Multi-service QoS Prerequisites

Aiming at aligning real-time and non-real time user's various services flow characteristics under a common optimization framework, each mobile user is associated with a properly defined QoS aware utility function, which represents its degree of satisfaction to the cell selection strategy B_j ($j=1,...,F$) and the tradeoff between his utility-based actual uplink throughput performance and the corresponding energy consumption per time slot. Moreover, the properties of the utility function depend on the tier that the user belongs to (i.e. two-tier architecture) and user's requested service, i.e. real-time and non-real time services. Therefore, the proposed utility function is formulated as a multi-domain function as follows.

$$U_i(B_{ij}, P_i, \boldsymbol{P}_{-i}) = \begin{cases} \left. \dfrac{R_F f(\gamma_i)}{P_i} \right|_{B_{ij}} &, i \in \mathfrak{M}_{MRT} \\[2ex] \left. \dfrac{\log(1 + D \cdot f(\gamma_i))}{P_i} \right|_{B_{ij}} &, i \in \mathfrak{M}_{MNRT} \\[2ex] \left. \dfrac{R_F f(\gamma_i)}{P_i} - c\left(e^{P_i} - 1\right) \right|_{B_{ij}} &, i \in \mathfrak{M}_{FRT} \\[2ex] \left. \dfrac{\log(1 + D \cdot f(\gamma_i))}{P_i} - c\left(e^{P_i} - 1\right) \right|_{B_{ij}} &, i \in \mathfrak{M}_{FNRT} \end{cases} \tag{2}$$

for all $B_{ij} \in \boldsymbol{B}_j$ ($j = 1,...,F$), where R_F, D are positive constants and $f(\gamma_i) = \left(1 - e^{-A\gamma_i}\right)^M$ denotes user's efficiency function, which represents the successful packet transmission at fixed data rates depending on the modulation and coding schemes that are being used [8].

Considering the formulated utility function, the FUs are penalized via a convex pricing policy with respect to their uplink transmission power in order to mitigate the caused interference to the MUs, as well as to give higher priority to the MUs which in general are characterized by worse channel conditions compared to FUs, which appear proximity to their corresponding AP. Furthermore, considering non-real time users, their utility-based uplink throughput performance is formulated via adopting Shannon's formula in order to present users' greedy behavior to achieve high transmission rate. On the other hand, considering real time users, we adopt the sigmoidal utility function $R_F f(\gamma_i)$ with respect to the bit energy to interference density ratio γ_i to present the achievable data rate. At the inflection point γ_i^{Infl} of the proposed sigmoidal function, $R_F f(\gamma_i)$, we map the ideal value of user's transmission rate (i.e. voice service $R_F f(\gamma_i) = 64Kbps$) at which his service QoS prerequisites are fulfilled. This formulation enables real time users to request system resources and express their strict and short-term QoS prerequisites.

3   Optimal Cell Selection: A Game Theoretic Approach

In this section, we propose a potential game model to formulate and address the cell selection problem. Since each user $i \in \mathfrak{M} = \mathfrak{M}_{MRT} \cup \mathfrak{M}_{FRT} \cup \mathfrak{M}_{MNRT} \cup \mathfrak{M}_{FNRT}$ in the two-tier femtocell network aims at the maximization of his utility U_i, and due to the absence of coordination between the two tiers and the users, all the users are assumed to be willing to participate to a non-cooperative cell selection game.

Let $G_N = \left\{ \mathfrak{M}, \{B_{ij}\}_{\substack{i \in \mathfrak{M} \\ j=0,1,...,F}}, \{U_i\}_{i \in \mathfrak{M}} \right\}$ denote the NOn-cooperative CEll Selection game (NOCES game). The rational mobile users represent the players of the game consisting the set $\mathfrak{M}$, the strategies $\{B_{ij}\}_{\substack{i \in \mathfrak{M} \\ j=0,1,...,F}}$ are users' choice of different cell selection strategy and the payoff $\{U_i\}_{i \in \mathfrak{M}}$ is the utility obtained by each user.

Formally, the NOn-cooperative CEll Selection game (NOCES game) can be represented as follows.

(*NOCES game*)

$$\max_{\boldsymbol{B}_j(j=0,1,...F)} U_i(\boldsymbol{B}_j, \boldsymbol{P}_i)\big|_{B_{ij}} \; for \; all \; i \in \mathfrak{M} \tag{3}$$

where B_{ij} denotes the specific strategy of cell selection of user i from the feasible set of strategies $B_j \left(j = 0,1,...,F \right)$. As presented in NOCES game (3), each user aims at selfishly maximizing his utility, via selecting a cell selection strategy and an optimal value of his uplink transmission power. Thus, NOCES game is a non-cooperative and distributed game, where the decision intelligence lies at user's part.

In the following section, the Nash equilibrium approach is adopted towards seeking analytically the solution of the NOCES game, which is most widely used for game theoretic approaches. The Nash equilibrium point is a set of cell selection strategies, such that no user in the two-tier femtocell network has the incentive to change his cell selection strategy, since his utility cannot be further improved by making any individual changes on his strategy, given the choices of cell selection strategies of the rest of the users.

Definition 1: The cell selection strategy vector $B_{ij}^* = \left\{ B_{1j}^*, B_{2j}^*, ..., B_{Mj}^* \right\}$, $j \in B_j \left(j = 0,1,...,F \right)$, $\forall i \in \mathfrak{M}$ is a Nash equilibrium, if for every $i \in \mathfrak{M}$ holds true that $U_i \left(B_{ij}^*, B_{-ij}^* \right) \geq U_i \left(B_{ij}, B_{-ij}^* \right)$ for all $B_{ij} \in B_j \left(j = 0,1,...,F \right)$.

A. NOCES Potential Game

Following the previous discussions and analysis, the potential game model is proposed towards proving the existence of Nash equilibrium of NOCES game. A potential game is a special normal form game, which is characterized by a potential function $P(B_j)$, where $B_j \left(j = 0,1,...,F \right)$ is the set of cell selection strategies. The fundamental property of a potential game is that when a deviation occurs, the change in P, i.e. ΔP, is reflected in the change in value seen by the unilaterally deviating user i, $\forall i \in \mathfrak{M}$, ΔU_i [10].

Definition 2: A strategic game $G_N = \left\{ \mathfrak{M}, \{B_{ij}\}_{\substack{i \in \mathfrak{M} \\ j=0,1,...,F}}, \{U_i\}_{i \in \mathfrak{M}} \right\}$ is called exact potential game, if there exists an appropriate potential function $P(B_j)$, such that:

$$U_i \left(B_{ij}, B_{-ij} \right) - U_i \left(B_{ij}', B_{-ij} \right) = P \left(B_{ij}, B_{-ij} \right) - P \left(B_{ij}', B_{-ij} \right) \tag{4}$$

for every $\forall i \in \mathfrak{M}$ and for every B_{-ij} and B_{ij}, $B_{ij}' \in B_j \left(j = 0,1,...,F \right)$.

Analytically, potential games are the means towards presenting in a unified mode, via the potential function, the deviations in the actions of any user in the two-tier femtocell wireless networks. Furthermore, it is noted by the above definition that it is possible for many potential functions to exist for the same game. However, the difference between any two potential functions for the same game must be a constant [11].

Theorem 1: NOCES game $G_N = \left\{ \mathfrak{M}, \left\{ B_{ij} \right\}_{\substack{i \in \mathfrak{M} \\ j=0,1,..,F}}, \left\{ U_i \right\}_{i \in \mathfrak{M}} \right\}$ is an exact potential

game with potential function

$$P(B_j) = \sum_{\substack{i=1 \\ j=(0,1,...,F)}}^{M} U_i\left(B_{ij}\right) = \sum_{\substack{k=1 \\ k \neq i \\ j=(0,1,...,F)}}^{M} U_k\left(B_{kj}\right) + U_i\left(B_{ij}\right)\Big|_{j=(0,1,...,F)} \tag{5}$$

Proof: Based on definition 2 in order to prove that NOCES game is an exact potential game, it should be concluded that expression (5) holds true.

Assume that $B_{ij} \in B_j$ $(j=0,1,...,F)$ is an arbitrary strategy of user $i \in \mathfrak{M}$ and the strategies of the rest of the users remain unchanged. From (2) and (4), we have $\forall B_{ij}, B_{ij}'$

$$U_i\left(B_{ij}, B_{-ij}\right) - U_i\left(B_{ij}', B_{-ij}\right) =$$

$$= \sum_{\substack{k=1 \\ k \neq i \\ j=(0,1,...,F)}}^{M} U_k\left(B_{kj}\right) + U_i\left(B_{ij}\right)\Big|_{j=(0,1,...,F)} - \sum_{\substack{k=1 \\ k \neq i \\ j=(0,1,...,F)}}^{M} U_k\left(B_{kj}\right) + U_i\left(B_{ij}'\right)\Big|_{j=(0,1,...,F)} =$$

$$= P\left(B_{ij}, B_{-ij}\right) - P\left(B_{ij}', B_{-ij}\right)$$

So that the game $G_N = \left\{ \mathfrak{M}, \left\{ B_{ij} \right\}_{\substack{i \in \mathfrak{M} \\ j=0,1,..,F}}, \left\{ U_i \right\}_{i \in \mathfrak{M}} \right\}$ satisfies potential game

definition [10] and G_N is an exact potential game, where $P(B_j)$ is the potential function of G_N .   ∎

B. Towards a Nash Equilibrium for the Optimal Cell Selection Game

Based on definition 2 of the exact potential game, it is immediately implied that the

Nash equilibria of NOCES game $G_N = \left\{ \mathfrak{M}, \left\{ B_{ij} \right\}_{\substack{i \in \mathfrak{M} \\ j=0,1,..,F}}, \left\{ U_i \right\}_{i \in \mathfrak{M}} \right\}$ and

$\left\{ \mathfrak{M}, \left\{ B_{ij} \right\}_{\substack{i \in \mathfrak{M} \\ j=0,1,..,F}}, \left\{ P \right\}_{i \in \mathfrak{M}} \right\}$ coincide [11]. Moreover, a valuable property of a

potential game is the existence of at least one Nash equilibrium given that B_j $(j=0,1,...,F)$ is finite.

Proposition 1: NOCES game $G_N = \left\{ \mathfrak{M}, \left\{ B_{ij} \right\}_{\substack{i \in \mathfrak{M} \\ j=0,1,..,F}}, \left\{ U_i \right\}_{i \in \mathfrak{M}} \right\}$ with payoff function

$U_i(B_{ij}, P)$, $B_{ij} \in B_j$ $(j=0,1,...,F)$ and compact strategy set B_j has a non-empty set of Nash equilibrium.

It should be mentioned that the feasibility of Nash equilibrium has already been proven. However, in order to implement the determination of Nash equilibrium, it is obvious that each user is not informed of the strategies of the rest of the users, when he chooses to which cell to connect to. Therefore, in the next section aiming at implementing the optimal cell selection mechanism with user's personal information only, we propose a distributed cell selection learning algorithm. It should be clarified that alternatively a centralized controller could be used to solve the problem via adopting an exhaustive algorithm. However, a centralized controller is not available in the practical two-tier femtocell networks and would conclude to high computational analysis and high convergence time of the cell selection problem. Thus, the adoption of a learning-based algorithm of user's past personal information arises as a promising solution, which results in low-complexity approach that determines NOCES game's pure cell selection strategy Nash equilibrium.

4 Non-cooperative Cell Selection Algorithm Based on Learning and Power Control

The knowledge of users' path channel gain is perfectly available at the base station and the Access Points, but it is unknown at each user. Based on this observation, the proposed non-cooperative cell selection algorithm based on learning method can help users to learn via utilizing their own past knowledge and choose cell selection strategies without the complete information of other users.

Therefore, the cell selection game G_N is extended to a mixed strategy form in order to support the learning algorithm. The probability vector of user i with which he selects the j-th $(j=0,1,...,F)$ cell to access is denoted by $Pr_i = \{\Pr_{i0}, \Pr_{i1}, ..., \Pr_{iF}\}$ and it consists the mixed cell selection strategy of user $i, i \in \mathfrak{M}$.

The mixed strategy cell selection game is played iteratively per time slot t, in order to determine the optimal users' cell selection. Each user is represented by a learning player and the actions of each learning player are the feasible cell selection strategies of the corresponding user. Thus, the mixed strategy at each iteration r of NOCES algorithm $Pr_i(r) = \{\Pr_{i0}(r), \Pr_{i1}(r), ..., \Pr_{iF}(r)\}$ is the strategy probability distribution of the i-th learning player at the iteration r of the learning cell selection algorithm.

We define the reaction $r_i(r)$ of learning player's $i, i \in \mathfrak{M}$ as the normalized payoff of user $i, i \in \mathfrak{M}$. Let $r_i(r) = c \cdot U_i(r)$ represent the normalized payoff $U_i(r)$, where $U_i(r)$ is the payoff function of NOCES game G_N. The constant c, $0 \prec c \prec 1$, contributes to guarantee that learning player's reaction $r_i(r)$ lies in the interval $[0,1)$. Furthermore, the adaptive normalized parameter mechanism is adopted, due to the fact that the learning players have no prior knowledge of their utilities. Thus, at NOCES algorithm's iteration r, if $\forall i \in \mathfrak{M}$, $U_i(r) \succ \dfrac{1}{c}$, we define

$c = \dfrac{1}{U_i(r)+\tau}$, where τ is a positive constant, otherwise c remains unchanged.

Finally, the game is played iteratively in order the learning players to gain knowledge about the Nash equilibrium.

Furthermore, the NOCES algorithm is guaranteed that it converges to a pure strategy Nash equilibrium of NOCES game, as it was shown in Proposition 1 from theorem 4.1 in [9]. On the other hand, considering power control problem, we adopt as an independent component the approach that has been proposed in our previous work [8]. It should be clarified that the NOCES game concludes to users' assignment to the cells. Thus, the specific distribution of users per cell will be the output of cell selection mechanism, which will act as input to the utility-based uplink power control algorithm in multi-service two-tier femtocell networks (UPC-MSF algorithm), as it has been proposed in [8].

A. NOCES Algorithm

Step 1: At the beginning of each time slot t, set the initial cell selection probability

vector $\boldsymbol{Pr}_i(r=0)$ as $\mathrm{Pr}_{ij}(r=0) = \dfrac{1}{\displaystyle\sum_{j=0}^{F} x_{ij}}$, $\forall i \in \mathfrak{M}$, $(j=0,1,...,F)$, where

$x_{ij} = 1$, if user i is feasible to connect to cell j (i.e. he resides within j-*th* femtocell coverage area), otherwise $x_{ij} = 0$. Afterwards, each user chooses a cell selection strategy according to his cell selection probability vector $\boldsymbol{Pr}_i(r=0)$.

Step 2: At every iteration $r \succ 0$, each user chooses a cell selection strategy B_{ij}, according to his cell selection probability vector $\boldsymbol{Pr}_i(r)$.

Step 3 (Optimal Power Allocation): Given that all users have chosen the cell selection strategy, the macro base station and each Access Point knows the specific number of users that reside within their coverage area, implement UPC-MSF algorithm as it has been proposed in our previous work [8] to determine users' optimal power allocation.

Step 4: Given the optimal power allocation, each user i, $\forall i \in \mathfrak{M}$ each user calculates his payoff $U_i(r)$ and correspondingly his reaction $r_i(r) = c \cdot U_i(r)$.

Step 5: Each user updates his action probability vector via the following rule, where $0 \prec b \prec 1$ is a step size parameter.

$$\mathrm{Pr}_{ij}(r+1) = \mathrm{Pr}_{ij}(r) - b \cdot r_i(r) \cdot \mathrm{Pr}_{ij}(r), \ if \ B_{ij} \neq B_{ij}^{ite}$$

$$\mathrm{Pr}_{ij}(r+1) = \mathrm{Pr}_{ij}(r) + b \cdot r_i(r) \cdot \left(1 - \mathrm{Pr}_{ij}(r)\right), \ if \ B_{ij} = B_{ij}^{ite}$$

Step 6: If each user i, $\forall i \in \mathfrak{M}$, there exists a component of $\mathrm{Pr}_{ij}(r)$ which is larger than a value approaching one (e.g. *0.999*), which means that user i is willing to choose cell j, then stop and assign user i to cell j. Otherwise, return to step 2.

It should be clarified that NOCES algorithm can be characterized as a low complexity algorithm, due to the simplicity of the calculations that it performs. Moreover, the NOCES algorithm combined with UPC-MSF algorithm determine jointly per time slot t user's optimal cell selection strategy, as well as users' optimal uplink transmission power.

5   Numerical Results and Discussions

In this section, some indicative numerical results are provided to illustrate the operation and features of the proposed framework, as well as demonstrate the effectiveness of the proposed cell selection and power allocation framework, i.e. NOCES algorithm.

Throughout our study, we consider a single macro-cell CDMA system, where $F=15$ femtocells reside within the macrocell. The total number of users is set to $M=80$ (thus we consider a congested two-tier femtocell network), where a portion of them requests real-time service (e.g. voice service) and the rest of them request non-real time service (e.g. data transfer). The users are distributed randomly within the two-tier femtocell network, where the macrocell has a radius $R_0=400m$ and each femtocell has a radius $R_c=50m$. The maximum uplink transmission power is set to $P_i^{Max} = 2Watts$, while $W=10^6 Hz$ and $\sigma^2 = 5*10^{-15}W$. The efficiency function that is adopted has a sigmoidal form, i.e. $f(\gamma_i) = \left(1-e^{-A\gamma_i}\right)^M$, where A, M are positive constants, which determine the slope of the efficiency function. Moreover, each user is able to choose a feasible cell selection strategy, if he resides within cell's coverage area.

In Fig. 2, the evolution of the choice probabilities of user $i, i \in \mathfrak{M}$, is presented under various step size parameters, i.e. $b=0.2$, $b=0.5$ and $b=0.75$. Fig. 2 shows the fast convergence of the cell selection part of NOCES algorithm. In the scenario presented in Fig. 2, we have chosen user $i=1$, $i \in \mathfrak{M}$, who finally is connected to the macrocell, as NOCES algorithm indicates. The specific scenario is simulated for various step size parameters for the same channel realization and it is observed that user *1*, selects macrocell base station to be connected to, given that all users positions within the two-tier femtocell network has not been changed. Moreover, it is observed that when $b=0.2$, user *1* converges to a pure strategy Nash equilibrium B_{10}^* after about *250* iterations, when $b=0.5$ he converges to B_{10}^* after about *180* iterations and when $b=0.75$ he converges after about *30* iterations. Thus, it is concluded that NOCES algorithm converges to a pure strategy Nash equilibrium and the convergence speed is high when the step size parameter b is large, as it is better shown in Fig. 3.

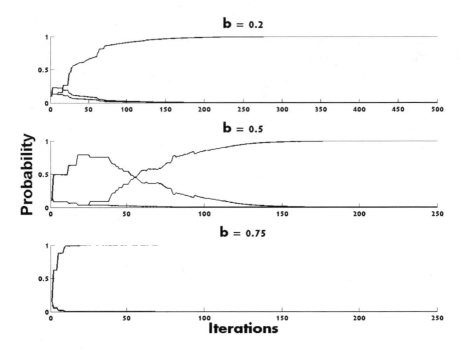

Fig. 2. Cell selection probability convergence under various step size parameters b of user $i=1$

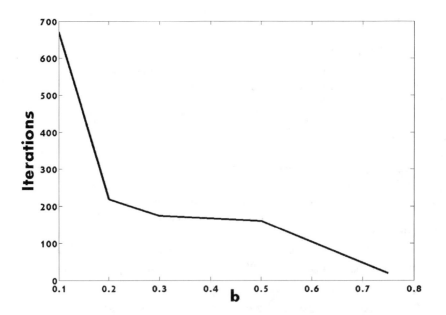

Fig. 3. Iterations of NOCES algorithm with respect to the step size parameter b

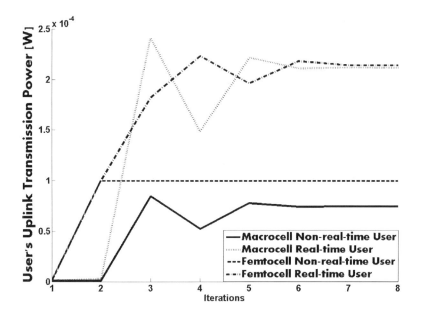

Fig. 4. Convergence of users' uplink transmission powers

Moreover, in Fig. 4 we present the convergence of users' uplink transmission power, considering the optimal power allocation part of NOCES algorithm. Specifically, we consider four different type of users, i.e. two users requesting real and non-real time services respectively and being served by the macrocell, and two users requesting these two different type of services, but being served by a femtocell. In Fig. 4, the NOCES algorithm has converged to a pure cell selection strategy Nash equilibrium and as it is observed, the convergence time of the power allocation part of NOCES algorithm is high, due to the fact that only *8* iterations are required in order the NOCES algorithm to determine users' optimal power allocation. Moreover, it is observed that the values of users optimal uplink transmission power are low, due to the fact that the users select to connect to the cell that satisfies their QoS prerequisites, while in parallel maximizing their utility function, which concludes to power saving.

6   Concluding Remarks

In this paper, the problem of optimal cell selection and power control in the uplink of multi-service open access two-tier femtocell networks is addressed via a game theoretic approach. Aiming at simultaneously treating multiple services with diverse QoS prerequisites, as well as two-tier femtocell network special characteristics under a common framework, a generic utility-based framework has been introduced and analyzed. Specifically, a non-cooperative cell selection game (NOCES game) was formulated and it was proven that it is a potential game. Moreover, the existence of a pure cell selection strategy Nash equilibrium was proven and a distributed low-complexity algorithm was proposed, in order to determine it, based on a learning approach of users' previous cell selection strategies. It should be noted that our approach facilitates the creation of a general framework, where the control intelligence and decision making process lie at the users considering both the cell selection, as well as the resource allocation. Such a vision is very well aligned with the current efforts for the realization of users' self-optimization functionalities, envisioned by 3GPP LTE and LTE Advanced efforts and specifications.

Acknowledgement. This research is co-financed by the European Union (European Social Fund) and Hellenic national funds through the Operational Program 'Education and Lifelong Learning' (NSRF 2007-2013).

References

[1] Mansfield, G.: Femtocells in the US Market - Business Drivers and Consumer Propositions. In: Femtocells Europe 2008, London, UK (June 2008)

[2] Chandrasekhar, V., Gatherer, A., Andrews, J.G.: Femtocell networks: a survey. IEEE Com. Mag. 46(9), 59–67 (2008)

[3] Lin, J.-S., Feng, K.-T.: Game Theoretical Model and Existence of Win-Win Situation for Femtocell Networks. In: 2011 IEEE International Conference on Communications (ICC), June 5-9, pp. 1–5 (2011)

[4] Ziqiang, F., Lingyang, S., Zhu, H., Niyato, D., Xiaowu, Z.: Cell selection in two-tier femtocell networks with open/closed access using evolutionary game. In: 2013 IEEE Wireless Communications and Networking Conference (WCNC), April 7-10, pp. 860–865 (2013)

[5] Tarasak, P., Adachi, K., Sumei, S.: Cell selection for TDD two-tier cellular networks based on uplink-downlink capacity. In: 2013 IEEE Wireless Communications and Networking Conference (WCNC), April 7-10, pp. 2016–2021 (2013)

[6] De Domenico, A., Strinati, E.C., Duda, A.: An energy efficient cell selection scheme for Open Access femtocell networks. In: 2012 IEEE 23rd International Symposium on Personal Indoor and Mobile Radio Communications (PIMRC), September 9-12, pp. 436–441 (2012)

[7] Guidelines for evaluation of radio transmission technologies for IMT-2000. ITU Recommendation M.1225 (1997)

[8] Tsiropoulou, E.E., Katsinis, G.K., Vamvakas, P., Papavassiliou, S.: Efficient uplink power control in multi-service two-tier femtocell networks via a game theoretic approach. In: 2013 IEEE 18th International Workshop on Computer Aided Modeling and Design of Communication Links and Networks (CAMAD), September 25-27, pp. 104–108 (2013)

[9] Sastry, P.S., et al.: Decentralized learning of Nash equilibria in multiperson stochastic games with incomplete information. IEEE Trans. Syst., Man, Cybern. 24, 769–777 (1994)

[10] Monderer, D., Shapley, L.S.: Potential games. Games and Economic Behavior 14, 124–143 (1996)

[11] Voorneveld, M.: Potential Games and Interactive Decisions with Multiple Criteria. PhD Dissertation, Tilburg University (September 1999)

[12] Chandrasekhar, V., Andrews, J.G., Muharemovic, T., Shen, Z., Gatherer, A.: Power control in two-tier femtocell networks. IEEE Trans. on Wireless Comm. 8(8), 4316–4328 (2009)

[13] Su, T., Zheng, W., Li, W., Ling, D., Wen, X.: Energy-efficient power optimization with Pareto improvement in two-tier femtocell networks. In: IEEE 23rd International Symposium on Personal, Indoor and Mobile Radio Communications, pp. 2512–2517 (2012)

[14] Lu, Z., Sun, Y., Wen, X., Su, T., Ling, D.: An energy – efficient power control algorithm in femtocell networks. In: 7th International Conference on Computer Science & Education (ICCSE), pp. 395–400 (2012)

[15] Zheng, W., Su, T., Li, W., Lu, Z., Wen, X.: Distributed Energy – Efficient power optimization in two – tier femtocell networks. In: 2012 IEEE International Conference on Communications (ICC), pp. 5767–5771 (2012)

A Smart Bluetooth-Based Ad Hoc Management System for Appliances in Home Environments

Sandra Sendra[1], Antonio Laborda[2], Juan R. Díaz[1], and Jaime Lloret[1]

[1] Instituto de Investigación para la Gestión Integrada de zonas Costeras
Universidad Politécnica de Valencia
46730 Grao de Gandia, Valencia, Spain
sansenco@posgrado.upv.es, {juadasan,lloret}@dcom.upv.es
[2] Universidad SANJORGE
A-23 Zaragoza-Huesca Km. 299.
50830 Villanueva de Gállego, Zaragoza, Spain
alu.23324@usj.es

Abstract. The number of home devices integrating new technologies is continuously increasing. These advances allow us to improve our daily routines. In addition, the improvement in network infrastructure and the development of smart phones and mobile devices allow us access from any place to any of our systems over the Internet. Bearing in mind this idea, we have developed a low-cost ad hoc protocol based on Bluetooth technology that allows us to control all our home appliances and monitor the power consumption of our homes. Our proposal is based on an Android application installed on a mobile device which acts as server. The application allows users to program the various appliances. It is also able to check the status of the appliance, as well as controlling the power consumption of the house and its cost. The system is equipped with a smart algorithm able to manage all appliances and decide which ones should work as a function of various criteria such as time of day or power consumption. Finally, the system is able to detect faults in water and electricity supply for acting accordingly. All data received and sent by the server are stored in a database which the system can check and compare to make their own decisions.

Keywords: Electrical household appliances, Bluetooth, Android applications, Smart Algorithm, Ad-hoc network, Internet of Things.

1 Introduction

The advances of new technologies in regards to electronic, communication infrastructures [1] and the Internet [2] in the last 10 years led us to think today would be living in smart homes [3]. However, automation is not yet implemented in most of our homes. Perhaps, we have only observed significant progress in management appliances, making home life similar to that of years ago.

Currently, the smart appliances are eco-friendly, innovative and have very modern designs. This type of appliance can be managed using mobile phone applications.

S. Guo et al. (Eds.): ADHOC-NOW 2014, LNCS 8487, pp. 128–141, 2014.

Thus, our smart phones/tablets become the control center from which we can remotely monitor our home.

The biggest technology brands like Panasonic, Samsung and LG, have available smart appliances suitable for domestic use and affordable for any user. These improvements allow a real time monitoring [4] of these smart appliances [5]. An example is The Smart Kitchen project [6]. It focuses on the usage of many small and inexpensive devices. Many small devices can be networked over a fieldbus in a domotic application. The advantage of fieldbus technology over other networking methods such as Ethernet is the high optimization level to the application.

There are several wireless technologies that our current mobile devices offer such as IrDA, Bluetooth, WiFi or NFC, among others. Bluetooth is the technology which presents the simplest mechanism for device discovering and their configuration. This technology also allows the implementing of ad-hoc networks and presents low power consumption. The prices of bluetooth devices are low. Its data transfer rate is enough for data exchange between smartphones and tablets since the expected traffic is small [7].

In addition, operating systems (OS) providers as Google has presented the new versions of its OSs (Jelly Bean of Android 4.3) which supports the new version of Bluetooth known as Bluetooth Smart (i.e. Bluetooth 4.0 on a dual chip). This technology minimizes the power consumption during the measurement process and data transmission. Therefore, devices such as tablets or smartphones using this version of Bluetooth present a more efficient use of battery. This new technology was already available in Apple from model iPhone 4S.

In this paper, we are going to present a new system to manage and control smartly our appliances at home. Our proposal includes, the mobile application developed using Android, the communication protocol used between our appliances and mobile devices and the smart systems used to intelligently decide which appliances should work in each moments as well as controlling the energy consumption at home.

The rest of the paper is structured as follows. Section 2 shows the background of the Smart application for controlling devices in home environments. Section 3 describes our proposed architecture for controlling our appliances and the applications developed to be installed in the different devices. The database used for storing all data is explained in Section 4, meanwhile Section 5 shows the communication protocol and the structure of each packet. Section 6 present the Android application developed for our system and the network tests. Finally, Section 7 shows the conclusion and future work.

2 Related Work

Because the great evolution of new technologies and advances in the field of Internet of Things (IoT), we can find several applications and systems developed for home environments. The main developed systems are assistive environment applications and body area networks [8][9], healthcare [10], behavior monitoring systems [11], passive localization and people tracking and power metering [12], among others.

Other applications, in regards to smart homes, are the intelligent management of appliances in home environments. In this section, we are going to present some smart applications developed for controlling different electronic devices in home environments.

J. Lloret et al. presented in [13] a low-cost communication architecture that increases the safety and quality of life at home. The proposed architecture allows sensors, which detects physical parameter changes in household appliances to be managed via Twitter. Authors proposed a smart system able to gather data from sensors. The information from sensors is fused in order to notify the user by messages when some event has happened in order to take decisions.

A. Kamilaris [14] presented an application framework with concurrent, multi-user support that facilitates the development of advanced ubiquitous applications by habitants without programming experience to automate his house. Authors resume the main features and requirements that future Smart Homes should present. These systems should be open and accessible for simultaneous users for interacting directly with sensors and actuators, and they should allow continuous monitoring. They also mplemented a small number of RESTful Web Services in TinyOS and performed several test to define the application framework performance and the device discovery timing.

Finally, A. Kamilaris et al. also presented [15] a system that includes a 6LoWPAN-based wireless sensor network inside the home environment. Author compensate the overhead embedding the IPv6 stack on sensor devices by using HTTP caching to reduce the mean response time to access sensor data. Finally, authors developed a graphical interface to abstract home automation procedure for typical home residents. Through the different tests and technical evaluation, the system show several benefits enabling embedded sensors in terms of performance and energy conservation.

As we can see, there are lots of possibilities and developments to improve our daily live. Our proposal also tries to make easier our routines and chores. This system allows us the monitoring of our entire house including the activity of the different appliances and energy consumption monitoring in real-time.

3   Proposed Architecture and Control Applications

In this section, we are going to present our proposed architecture for controlling our appliances. Our system is based on a distributed communication architecture which allows us the fast communication and execution of actions on each appliance. We will also show the operation diagram for client and server and the smart system which will inform us about the power energy consumption at home and the current price as a function of our service provider and the kind of service.

The proposed architecture is based on a server (smartphone or tablet) and other smart appliances that are indoors. All of them are equipped with Bluetooth cards. Actions on each device are performed in a distributed manner.

Our network presents a decentralized topology where the server is located on the center of network and the appliances are the intermediate nodes. The server has embedded a database (generated using MySQL) which keeps information of the different elements to be managed.

We can find two different scenarios. On the one hand, all appliances and server are within the house. So, the server performs a local access to different devices (see Figure 1). The server collects the status of each appliance including the electric power consumption. The server queries the price of KWh as a function of the type of services and the smartphone shows in its screen the current value (in Euros) of the consumed energy. In this case, the server and appliances should be in the coverage area in order to communicate all nodes. If we would like to spread signal along the house, we should implement a scatternet. A scatternet is a number of interconnected piconets that supports communication between more than 8 devices [16].

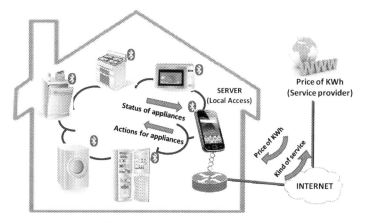

Fig. 1. Architecture for local access

As Figure 2 shows, the second situation places the server outside the home. It means that access to appliances will be remote. When the network does not detect the presence of the smartphone, a computer will act as local alternative server. This local server will act as a gateway between Bluetooth home network and the Internet. It is also able to store all events that happens in the house. When smartphone is in the Bluetooth network, the local alternative server will send this information to the smartphone.

The goal of eliminating the alternate server in the presence of the smart phone is to reduce the power consumption at home. The logic that manages the communication between the server and the various devices is performed by a mobile application (App) developed on Android. In addition, the specific software to interact with Bluetooth card devices is developed in Java.

Because we are using Bluetooth technology, the possible problem of nodes synchronization is solved by the characteristics of the standard where the system is synchronized using a global clock and a specific pattern of jumps, both unique. In our case, our master device provides the synchronization reference from its internal clock and other devices operate as slaves.

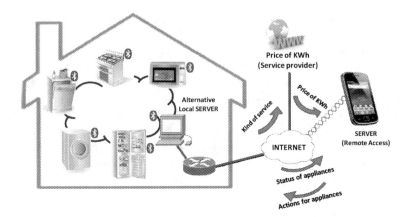

Fig. 2. Architecture for remote access

3.1   Control Application for Server

Our control application is able to communicate with any device connected to it. There exist different types of orders and actions to be sent to the devices. Thus, the user can interact with a single device by a specific order or send a global command to turn off all appliances, if there is any fault or serious incident.

The first step of control application is to register the time and check that there is no problem with water and energy supply. If an appliance such as washing machine is running and the water supply has been cut off, the device will probably break. If there is an outage of light, the control application will store the program that had the appliance and the appliance state where it left off.

For normal operation where there are no problems in energy/water supply, the application will register which appliances are scheduled to work during the day. Depending on the time of day, the system will decide which appliances should work.

These decisions can be configured by the user according to different criteria. The first one can depend on the energy consumption. The user can, e.g., specify the system that the dishwasher and the washing machine perform their functions at night to use the reduced prices rates.

Another criterion may be due to user preferences such as oven cook food in the hours near to lunch time instead of doing laundry. Thus, the user could maintain constant energy consumption in their homes.

After the training phase, the algorithm will analyze each received data which will be compared with the data from the database. After determining the priority of appliances and a possible alarm, the system will perform, on the one hand, a tagging and storing of the appliances priority and alarm, on the other hand, the communication to the device. Figure 3 shows the flow diagram for control application.

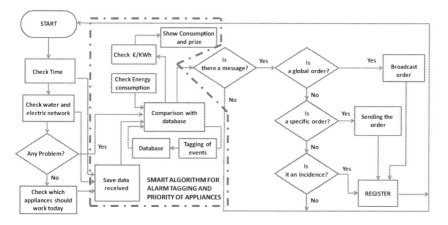

Fig. 3. Flow diagram of smart algorithm for control application

3.2   Control Application for Client

For each device, an application for actions managing is implemented. Figure 4 depicts the flowchart of the client application. The client application installed on each device analyzes what type of order has received (global or specific) and the origin of this order (from user or from the device itself).

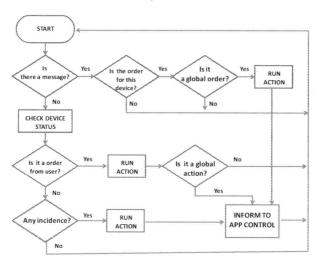

Fig. 4. Flow diagram for client app.

In this way, each Bluetooth card will act as its own master. But the control application will take the most important role. It concentrates the intelligence of system and it will be responsible for analyzing the environment, time and appliances that must work today. After the learning period where the user has made a regular use of their appliances, the system will be able to decide what appliances should work depending on the time of day. However, each device will be able to make their own decisions about how to improve their situation.

Decentralizing part of decisions on the actions of the device allows us to have a more reliable fault-tolerant system. If an error occurs in the control application or in the communications between the control application and the different appliances, each device may act separately, considering a series of actions based on the last recorded state. In addition, the information flow will be distributed in whole system.

4   Database Structure

This section explains the database structure used to manage and store the information.

In order to manage the flow of information between the device and the App control, it is needed a repository where performing queries. This repository or database can be easily developed using SQL statements. Our database will store information about data from each device, their programming and programs, their faults, the different commands sent to the devices as well as the status of programming and solution for a given fault.

The information of each device is structured and stored in tables and the relationships between each field are shown in Figure 5.

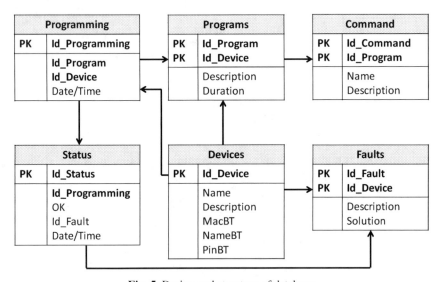

Fig. 5. Design and structure of database

The names given to each field and tables are representative of its content. Table "Programs" stores the description and duration of a particular program for a device. Table "Faults" stores a description a fault and a possible solution for a device. Table "Programming" stores for a device and its program, the date and time of the different programs that we have made. Table "Status" stores for a determined Programming, the status of device within its program, the status of a possible solution, and the date and time of each event. The "Id_Fault" field in "Status" table stores a fault code generated for an appliance. The system tries to find a solution as a function of this fault code. The server sends a set of orders to the appliance accordingly to the available information within the field "Solution". The Date field displays the last programming updated in the status field of that device. Table "Devices" store a full description of each device. Finally, Table "Command" stored the set of commands that correspond to a determined program.

5   Communication Protocol

In order to communicate all appliances and control application, it is needed to design a communication protocol. Our proposal is an open protocol. It is very basic and simple but very effective. This section shows the different kinds of packets we have. Each packet identifies the different fields it has.

When we need to design a communication protocol for applications monitoring, it can be use closed protocols or owners of a particular brand. Sometimes these protocols may be variants of standard protocols. However, this fact will limit us to use devices from a particular brand. We may also use open protocols defined between several companies in order to unify criteria. Usually, there are not patents related on this protocol. So, any manufacturer can develop applications and products which carry this communication protocol. Some examples of the most widely used open protocols are KNX [17], Lonworks [18], X10 [19], among others.

For our application, we have designed an easy protocol that defines three types of packets. The first one shows information about settings that server can send to the appliances. Figure 6 shows the structure of programming packet. The second packet (See Figure 7) is in charge of reporting the status of program for a given appliance, from the appliance to the server. Finally the third packet allows us managing any faults that may occur on a specific appliance. The way you manage faults is very simple. If a fault code is received, the server will query the database for possible solutions and after that, it will send the code of solution to the appliance (see Figure 8).

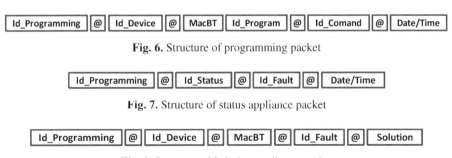

Fig. 6. Structure of programming packet

Fig. 7. Structure of status appliance packet

Fig. 8. Structure of Solution appliance packet

If server receives a fault code for which there is no solution, the management application will mark the state of the failed programming (OK = False). With this action, it avoids to rerun programming that generates a fault. The application will inform the user that he/she should contact to the customer support as soon as possible.

6   Test Bench and Operation System

This section presents the graphic user interface developed for managing all appliances. We have also performed a network performance to analyze the data traffic in Bluetooth network.

6.1   Graphic User Interface

To manage, view and send the necessary orders to the different appliances, a graphic user interface has been developed using android. Figure 9 shows the main screens of our application. The first one allows users controlling programs of each appliance, as well as turns on/off and put it on hold. The second and third screens show a general view of house status. On the one hand, the second one shows the power consumption in watts and economic cost depending on the type of service contract and the current price set by the state or company. Finally, the third screen inform the user about the overall state of the system and if there is a problem with the supply of water or electricity. If a problem is detected, the system will perform various verification tasks in order to generate the correct orders.

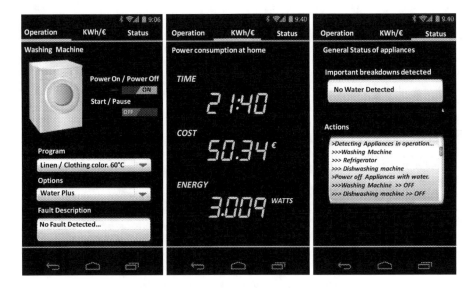

Fig. 9. Main Screens of control application

With the selected information, the database is updated and the different schedules are sent to the appropriate device. The communication between the application and the different devices is done through sockets.

The server only works with requests from one device although it may be simultaneously processing information from other devices.

6.2   Test Bench

The last step to ensure the correct operation of our system is to check the connectivity and operation of our network. In this subsection, we are going to perform a test bench with an appliance and our server. To perform our test, we have used a washing machine provided with a Bluetooth card and a mobile phone as a server. Figure 10 shows the scenario used in this test.

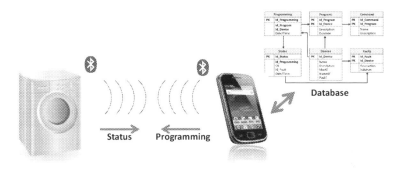

Fig. 10. Scenario used in the test bench

The appliance periodically connects to the server. This communication permits the updating of status and programming in the database. The modifications on the database are done through SQL statements. Figure 11 shows the flow diagram of messages between an appliance, the server and the database.

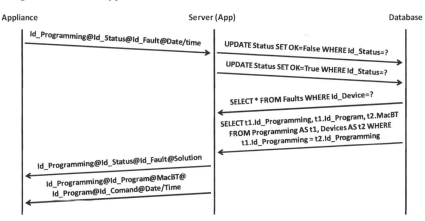

Fig. 11. Message diagram between the database, server App and appliance

To perform our test, we have established a connection between the device and the mobile phone. The mobile phone simultaneously runs the control application and a network sniffer in order to capture the packets traffic generated. Figure 12 shows the total packets per second recorded in the communication between both devices.

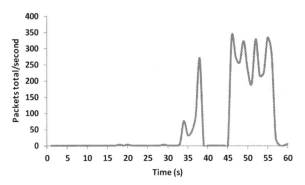

Fig. 12. Packets Total/second

Figure 13 shows the total bytes per second recorded in the communication between both devices.

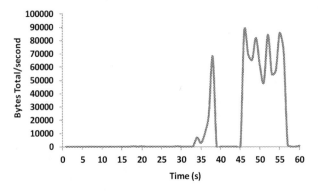

Fig. 13. Bytes Total/second

Figure 14 shows the bytes read per second recorded in the communication by server. As we can see the maximum volume of registered data is around 820 bytes.

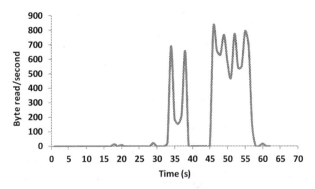

Fig. 14. Bytes read/second

Figure 15 shows the bytes written per second recorded in the communication by server. In this case, the maximum volume of registered data is around 9000 bytes.

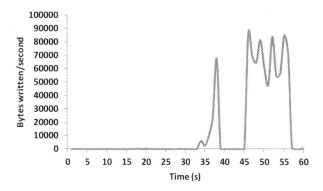

Fig. 15. Bytes written/second

The difference between bytes written and bytes read is because in this communication, the server is sending the different programming to the appliance and the appliance only sends information of its status.

Figures 12, 13, 14 and 15 show two zones with more data volume. The first of these zones is located between the 33rd and 40th second. In this period, the connection between devices is performed. The second period with big data volume is located between 45th and 55th second. During this interval, devices perform the transmission of information.

7   Conclusion

In this paper, we have presented a smart application for controlling appliances in home environments. Our application allows us monitoring and controlling, in real time, all of our appliances placed at home. The application has been implemented using Android, but we could develop this system for other mobile platforms.

The main advantage of this system is its versatility with any type of device, because we only need to provide each device with a Bluetooth cards. We can do it in different ways. First, there already have appliances with Bluetooth systems that can be used to connect with them through the server. However, for those devices that do not have this technology, we can use economic hardware such as Arduino and its complements that will allow us integrating this wireless technology with our appliances. The user interface is very simple and through it, it is very easy to control the power consumption at home. In addition, the smart algorithm is able to decide about which appliances should work as a function of the time and the home power consumption. This is possible due to the previous training period that system performs.

Our system allows us maintaining a stable energy and water consumption in our house. The system can also make use of reduced rates of light provider companies that will generate a considerable economic savings.

As future works, we want to develop a secure system to integrate it to our programs in order to avoid any kind of threat [20] and vulnerability that can endanger our system and the stored data.

References

1. Garcia, M., Sendra, S., Lloret, J., Canovas, A.: Saving energy and improving communications using cooperative group-based Wireless Sensor Networks. Telecommunication Systems 52(4), 2489–2502 (2013)
2. Liu, Y., Zhou, G.: Technologies and Applications of Internet of Things. In: Proceedings of 2012 Fifth International Conference on Intelligent Computation Technology and Automation (ICICTA), Zhangjiajie, China, January 12-14, pp. 197–200 (2012)
3. Aiello, M.: The Role of Web Services at Home. In: Proceedings of the Advanced International Conference on Telecommunications and International Conference on Internet and Web Applications and Services (AICT-ICIW 2006), Guadeloupe, France, February 23-25 (2006)
4. Mowafi, M.Y., Awad, F.H., Al-Batati, M.A.: Opportunistic Network Coding for Real-Time Transmission over Wireless Networks. Network Protocols and Algorithms 5(1), 1–19 (2013)
5. Gangadhar, G., Nayak, S., Puttamadappa, C.: Intelligent Refrigerator with monitoring capability through internet. International Journal of Computer Applications. Special Issue on "Wireless Information Networks & Business Information System 2(7), 65–68 (2011)
6. Soucek, S., Russ, G., Tamarit, C.: The Smart Kitchen Project—An Application of Fieldbus Technology to Domotics. In: Proceedings of 2nd International Workshop on Networked Appliances (IWNA 2000), New Brunswick, NJ, USA, November 30-December 1 (2000)
7. Zhang, W., Tan, G.-Z., Ding, N.: Traffic Information Detection Based on Scattered Sensor Data: Model and Algorithms. Adhoc & Sensor Wireless Networks 18(3-4), 225–240 (2013)
8. Ranjit, J.S., Shin, S.: A Modified IEEE 802.15. 4 Superframe Structure for Guaranteed Emergency Handling in Wireless Body Area Network. Network Protocols & Algorithms 5(2), 1–15 (2013)
9. Braeken, A., Singelee, D.: Efficient and Location-Private Communication Protocols for WBSNs. Adhoc & Sensor Wireless Networks 19(3-4), 305–326 (2013)
10. Augusto, J.C., McCullagh, P., McClelland, V., Walkden, J.A.: Enhanced healthcare provision through assisted decision-making in a smart home environment. In: Proceedings of the Second Workshop on Artificial Intelligence Techniques for Ambient Intelligence (AITAmI 2007), Hyderabad, India, January 6-7 (2007)
11. Zhang, L., Zhao, Z., Li, D., Liu, Q., Cui, L.: Wildlife Monitoring Using Heterogeneous Wireless Communication Network. Adhoc & Sensor Wireless Networks 18(3-4), 159–179 (2013)
12. Viani, F., Robol, F., Polo, A., Rocca, P., Oliveri, G., Massa, A.: Wireless Architectures for Heterogeneous Sensing in Smart Home Applications: Concepts and Real Implementation. Proceedings of the IEEE 101(11), 2381–2396 (2013)
13. Lloret, J., Macías, E., Suárez, A., Lacuesta, R.: Ubiquitous Monitoring of Electrical Household Appliances. Sensors 12(11), 15159–15191 (2012)

14. Kamilaris, A., Trifa, V., Pitsillides, A.: The smart home meets the Web of Things. International Journal of Ad Hoc and Ubiquitous Computing 7(3), 145–154 (2011)
15. Kamilaris, A., Trifa, V., Pitsillides, A.: An Application Framework for Web-Based Smart Homes. In: Proceedings of the 18th International Conference on Telecommunications, ICT 2011, Ayia Napa, Cyprus, May 8-11, pp. 134–139 (2011)
16. IEEE Std 802.15.1-2002 – IEEE Standard for Information technology – Telecommunications and information exchange between systems – Local and metropolitan area networks – Specific requirements Part 15.1: Wireless Medium Access Control (MAC) and Physical Layer (PHY) Specifications for Wireless Personal Area Networks (WPANs)
17. KNX international Site, http://www.knx.org/knx-en/index.php (last access: February 1, 2014)
18. LonWorks Technology. In: ECHELON web site, http://www.echelon.com/technology/lonworks/ (last access: February 1, 2014)
19. X10 protocol. In: X10 web site, http://x10-lang.org/ (last access: February 1, 2014)
20. Rohini Basak, R., Sardar, B.: Security in Network Mobility (NEMO): Issues, Solutions, Classification, Evaluation, and Future Research Directions. Network Protocols and Algorithms 5(2), 87–111 (2013)

A Distributed Time-Domain Approach to Mitigating the Impact of Periodic Interference

Nicholas M. Boers and Brett McKay

MacEwan University, Edmonton, AB, Canada
boersn@macewan.ca, mckayb24@mymacewan.ca

Abstract. Low-powered wireless transmissions, such as those of a wireless sensor network (WSN), are particularly susceptible to radio-frequency (RF) interference. When the interference exhibits regularities amounting to perceptible patterns, e.g., regularly-spaced short-duration impulses that correlate among multiple network nodes, opportunities exist for nodes to avoid impulses and consequently mitigate their negative impact on the packet reception rate. Rather than adopt special hardware for classification and mitigation, which is often done with cognitive radios, our research explores techniques that can enhance the medium access control schemes of the traditional off-the-shelf RF modules typically found in low-cost WSN nodes. This paper describes a distributed time-domain approach for identifying the periodicity of impulses and scheduling transmissions around them. The approach is evaluated using a simulator in terms of packet reception rates and latency, and the results show that it can significantly reduce packet losses.

1 Introduction

Within a wireless sensor network (WSN), contention-based medium access control (MAC) protocols typically assess the communication channel prior to each transmission. If an assessment shows significant radio-frequency (RF) activity, a node reacts by delaying its transmission until it senses an idle channel. In a network of nodes that follow the same protocol, this deferring of transmissions allows nodes to transmit practically uninterrupted after acquiring the channel.

Interruptions can occur, however, from RF interference originating outside of the network. Sources can vary significantly, and [13] classifies them as (a) *incidental* when they are designed to neither generate nor emit RF energy, e.g., a spark plug or electric motor; (b) *unintentional* when they intentionally generate, but unintentionally emit, RF energy, e.g., a microwave oven [20]; and (c) *intentional* when they both intentionally generate and emit RF energy, e.g., a cordless phone or WSN node. Given that external RF sources have an inability (a & b) or limited ability (c) to detect WSN transmissions, there exists a real possibility that their emissions will interrupt on-going communication. WSNs are increasingly subjected to high levels of man-made RF interference [2], and our work further explores the development of proactive, rather than strictly reactive, MAC protocols.

S. Guo et al. (Eds.): ADHOC-NOW 2014, LNCS 8487, pp. 142–155, 2014.

This work on developing a proactive (or interference-aware) MAC focuses on interference that exhibits regularities amounting to perceptible patterns, e.g., regularly-spaced short-duration impulses that correlate among multiple network nodes (Fig. 1). This pattern has been described within the literature (Sect. 2), and our own nodes have encountered it within a prototype WSN.

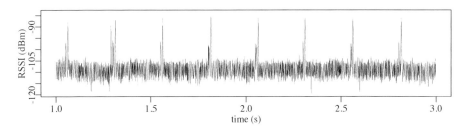

Fig. 1. Periodic impulsive interference encountered within a dense urban environment. A CC1100 transceiver, operating within the 900 MHz ISM band, collected received signal strength indicator (RSSI) samples at approximately 5000 Hz.

While this work focuses on periodic impulsive interference, we have no delusions about the pattern's prevalence. When reported in the literature, researchers have observed such interference within specific environments while they were performing experiments. Its presence certainly changes over time, since man-made interference is highly dynamic and unpredictable. To the best of our knowledge, no survey has yet specifically and comprehensively investigated perceptible patterns within RF interference. That said, the fact that this pattern has been observed motivates our work to develop techniques to avoid it.

This paper first presents a novel time-domain approach for determining the periodicity of an impulsive interference source within the constraints of resource-limited WSN-class nodes. The developed technique is suitable for devices with just kilobytes of memory and a few megahertz of processing power. It then describes a preliminary protocol for devices to exchange their classifications with their neighbours to build a more complete representation of any present interference. After describing how nodes discover multiple active interferers, the paper describes a proactive MAC whereby WSN-class nodes can synchronize with multiple interferers and identify uninterrupted transmission windows.

The paper is organized as follows. Section 2 provides a brief overview of related work. In Sects. 3 and 4, the proposed approaches for classification and synchronization are described, respectively. Section 5 evaluates these approaches within a simulator and compares the proposals against a traditional listen-before-transmit (LBT) MAC protocol. Finally, Sect. 6 concludes the paper with a brief review of its contributions.

2   Related Work

A number of researchers have investigated the interference generated by specific source devices. Chandra investigated specific incidental RF sources within a three-story building [7]. Using a spectrum analyzer, he looked at the noise generated by electronic equipment in a workshop, a photocopier, an elevator, and fluorescent tubes. Musăloiu-E. and Terzis investigated estimators and distributed algorithms that allow 802.15.4 (ZigBee) nodes to dynamically switch frequencies upon detecting 802.11 (WLAN) interference [12]. Srinivasan et al. presented a comprehensive study on the interaction among 802.11, 802.15.1 (Bluetooth), and 802.15.4 devices [18]. Rather than focus on specific sources or protocols, this work focuses on observed interference regardless of its source.

To observe interference, researchers have looked to the transceiver's received signal strength indicator (RSSI). In reasonably modern transceivers, such as the Texas Instruments CC2420, researchers have observed that the RSSI output is a promising indicator for use in link estimation [16]. Others, e.g., [12], have gone further and used (a) the RSSI of commercial off-the-shelf hardware to detect channel usage and (b) protocols to avoid noisy channels rather than coexist with them.

Within RSSI traces, a number of researchers have observed regularities that amount to perceptible patterns. Srinivasan et al. observed strong, spatially-correlated impulses up to -35 dBm or higher in their traces from sensor platforms, and they concluded that the impulses originated external to the nodes [17]. Lee et. al. and Rusak and Levis also observed impulsive interference sometimes as strong as 40 dB above the noise floor, and they noted that many of the impulses were periodic and that the noise patterns changed over time [9, 15]. In one of our prototype WSNs, nodes encountered 4 Hz impulsive interference that significantly impacted packet reception rates [3]. Srinivasan et al. observed 802.11b interference as high as 45 dB above the noise floor when sampling with a 802.15.4 transceiver, and in their time series plots, the interference appeared as periodic impulses at roughly 36 Hz [18]. More generally, Mitra and Lampe noted that temporal dependence within interference is imminent [11].

Our earlier work focused on classifying interference through analysis in the frequency-domain before reacting to it [6]. The classification was accomplished by simplifying the Lomb periodogram method of least-squares spectral analysis [10, 14] for use within resource-constrained WSN nodes. In both simulations and a hardware implementation, WSN nodes were capable of detecting the periodic interference, scheduling their transmissions around it, and reducing losses and retransmissions. While the approach performed well, its two greatest weaknesses were its (a) classification time, which lasted for over eighteen seconds during which nodes could not communicate and (b) memory usage, which scaled with the number of analyzed frequencies and required 10 B of storage per frequency. As an alternative to the earlier approach, this paper describes and evaluates a comprehensive distributed *time-based* approach to mitigate the impact of periodic impulsive interference, and the new approach has a number of advantages that are described in the following sections.

3   Distributed Classification

The proposed distributed classification approach uses a collection of nodes to classify multiple sources of periodic impulsive interference. For tractability, each individual node classifies only the strongest observed interferer and then shares its classification with its neighbours. When sharing, individual nodes use data binning for the received classifications, sort the bins by count, and use the bins with the highest counts to parameterize their MAC. This section describes these steps, and the next describes how nodes subsequently synchronize with the identified interferers.

The procedure described in this section should only occur when necessary, e.g., when a node joins the network or detects a change in the interference. It involves the collection of RSSI samples, which requires that the node enable the transceiver's receive mode. Although the receive mode requires less energy than the transmit mode, the cost is still nontrivial.

3.1   Identification of Strongest Interferer

Given nodes and interferers distributed in an environment, the concept of *strongest interferer* is relative to the observing node. Nodes determine an appropriate RSSI cutoff threshold that distinguishes the strongest interferer through a data binning technique. They create a new bin on-demand whenever a new sample falls outside the limits of all the existing bins. Each bin records the sum and count of the RSSI samples placed within it, and a predefined (constant) threshold around each bin's calculable mean determines its limits. After populating the bins, nodes identify the bin with the highest RSSI that meets a minimum count and adopt that bin's minimum limit as the strongest interferer threshold.

During this data collection phase, received signals belonging to packet transmissions should not be considered interference, and nodes must prevent them from impacting the threshold. Transceivers such as the Texas Instruments CC1100 support multiple clear channel assessment (CCA) options [19]; the most appropriate for our purposes assumes that the channel is clear when the device is not receiving a packet. With the transceiver configured to output the CCA to a digital input/output pin, the microcontroller can efficiently poll the CCA status.

The transceiver can only detect a packet transmission following the start symbol, so the preceding preamble complicates matters. To overcome this complication, our approach (a) samples the RSSI at a given frequency, e.g., once per millisecond, (b) queues each sample in a FIFO queue, (c) empties the queue whenever the CCA indicates that the node is receiving a packet, and (d) dequeues a sample only after the queue reaches a length that coincides with the duration of a packet's preamble. Using this approach, nodes collect RSSI samples that predominately represent interference.

3.2   Classification

To detect the period of a periodic impulsive interference source, a node effectively creates a histogram of the time elapsed between RSSI samples from the strongest interferer. It produces RSSI samples at 1000 Hz and obtains each sample from a FIFO queue to reduce the likelihood of erroneously interpreting a bona fide transmission as interference. During packet reception, the queue remains empty and RSSI samples do not impact the histogram.

In practise, considering only the elapsed time between impulses is insufficient because random impulses may appear at any point in time. To reduce the impact of random impulses, a node maintains a circular buffer of impulse time-stamps that exceed the threshold determined in Sect. 3.1. Upon observing an impulse above the threshold, a node computes the difference between the impulse's time-stamp and each buffered time-stamp. For each difference, it increments a count within the histogram and adds the new time-stamp to the circular buffer. Finally, the most frequent time difference in the histogram corresponds to the period of the interference.

To illustrate this point, please consider Fig. 2. In this simplified example, the top plot shows the RSSI values observed at a node, where periodic impulses occur at 0, 10, 20, and 30. At 7, the node receives a random impulse below the threshold (horizontal dashed line). At 17 and 24, the node receives random impulses over the threshold; these impulses are distractors to classification. In this example, the circular buffer has 3 slots, and over time, the node develops the histogram. At $t = 0$, an impulse occurs, and since the window is currently empty, the histogram remains unchanged and the node buffers the time-stamp 0. When the next impulse arrives at $t = 10$, the node computes the difference between 10 and the buffered time-stamp 0; the difference, 10, then contributes to the histogram, and the node buffers the time-stamp 10. When the next impulse arrives at $t = 17$, two time-stamps exist in the buffer: 0 and 10. The two differences, $17 - 0 = 17$ and $17 - 10 = 7$, are then incorporated into the histogram prior to buffering the 17. The procedure then continues for the remaining impulses, and at the end, the most common difference is 10, which corresponds with the period of the interference.

3.3   Distribution

After a node classifies the strongest interferer within its neighbourhood, it begins the distribution phase by announcing itself to the network. This announcement contains the node's identifier and classification, along with the maximum number of hops that the classification should propagate within the network. Nodes propagate classifications for several hops because even relatively weak interference can corrupt packet receptions. Neighbours do not rebroadcast this announcement; instead, they later include its content within their own packets.

When a node receives an announcement, it caches the identifier, classification, and decremented hop count. After receiving new information from a number of neighbouring nodes, it aggregates a random subset of the cached 3-tuples to produce a new driver packet for transmission. A random subset is used to include a variety of classifications within aggregate packets. As neighbours receive aggregate packets, they update their own caches, and only after updating multiple entries, continue the propagation process.

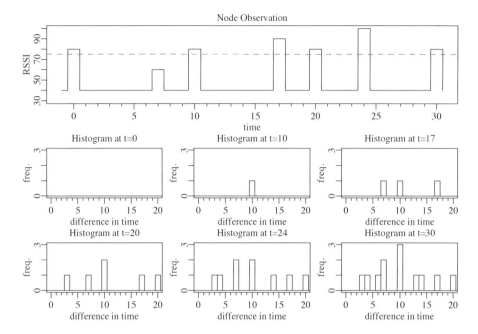

Fig. 2. Simplified illustration of our time-based classification approach. The top plot shows impulses in the RSSI that exceed the determined threshold (horizontal dashed line). Below, plots show the development of a histogram over time as the node detects impulses.

Whenever nodes receive new information, either by way of an announcement or an aggregate packet, they bin the cached classifications, sort the bins by count, and use the most common classifications to parameterize their MAC. When transmitting either an announcement or aggregate packet, a node attempts to synchronize with the interference (Sect. 4).

3.4   Comparison with Earlier Work

The described classification approach has significant advantages over our earlier frequency-domain technique [5, 6]. First, nodes can typically determine the frequency within six seconds versus approximately eighteen seconds. Second, the new approach allows nodes to classify the period with a precision of one millisecond; the earlier approach required approximately five times more memory to match this precision. Finally, the new approach allows communication to proceed during the classification process.

4   Pattern Synchronization

Four parameters allow a node to avoid periodic impulsive interference: (a) the period of each interferer, (b) the duration between each interferer's impulses,

(c) the duration of each interferer's impulses, and (d) the duration of the next packet transmission. After the classification and distribution phase, a node has (a) and (b), can measure (c), and can calculate (d). Note that naïvely sampling the RSSI and waiting for the first impulse is insufficient for synchronizing and predicting future impulses; a single impulse may be random or belong to any one of the identified interferers. A node must first temporally locate each interferer's impulses in order to predict the timing of future impulses.

The procedure described in this section only occurs when a node intends to transmit packets, possibly soon after a sleeping node wakes. It involves the collection of RSSI samples and requires energy, but that energy cost is much lower than with synchronization. Ideally, the node can amortize the cost across multiple sequential transmissions.

4.1   Temporally Locating the Impulses

Let the period of an interferer i be denoted as p_i. To associate an impulse at time t with i, a node must determine whether impulses occurred earlier at multiples of the period, i.e., at $t - j \cdot p_i$ where $j \in \mathbb{Z}^+$. To this end, a node creates one array per interferer for recording RSSI samples. Each element summarizes the RSSI values sampled within a time slice of duration s; this duration is selected with consideration to the desired synchronization accuracy, the operating system scheduler accuracy, and memory usage (the array for interferer i has $\lceil \frac{p_i}{s} \rceil$ elements).[1] Given arrays for n interferers, the synchronization process samples the RSSI every millisecond for a multiple m of the longest period, i.e., $m \cdot \max P$ where $P = \{p_i\}_{i=1}^n$. For example, if period $p_i = 39$ ms, time slice $s = 4$ ms, and multiple $m = 2$, then the array has $\lceil \frac{p_i}{s} \rceil = \lceil \frac{39 \text{ ms}}{4 \text{ ms}} \rceil = 10$ elements; if $p_i = \max P$, then the node will sample for $m \cdot p_i = 2 \cdot 39 = 78$ ms.

For each interferer array, the node maintains an iterator that wraps around upon reaching the end. During each time slice, the RSSI samples are averaged, and at the end of the time slice, the array element is incremented by the average less the background noise. In the case of no interference, the time slice will have a value close to zero. After sampling completes, the largest array element in each array is used as that interferer's temporal location.

For example, consider the illustration in Fig. 3 where a node detects one interferer with $p_1 = 10$. Using time slice $s = 2$, it creates a 5-element array for synchronization with the interferer. At the end of synchronization, element 2 has a sum of 60, and as the maximum, corresponds to the temporal location of the interferer.

This synchronization procedure will terminate after $m \cdot \max P$, even if one or more of the originally classified interferers is inactive. Given an inactive interferer, communication can still proceed, although transmissions will be unnecessarily delayed. To reduce the number of false positives, a node can test

---

[1] With a desire to reduce memory usage, we selected 4 ms; other values may improve or worsen the performance. The approach works when the period is indivisible by s at the cost of some error; it might improve performance to make the final bin smaller than the rest.

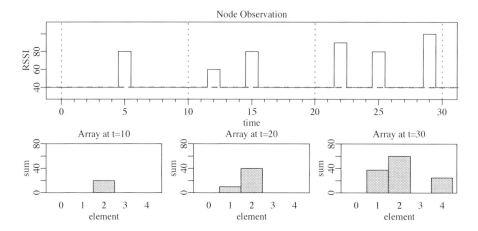

Fig. 3. Simplified illustration of our synchronization approach. The top plot shows impulses in the RSSI, where the horizontal dashed line is the background noise. The bottom plots show the development of an array over time as the node updates its elements.

the maximum array element against a minimum threshold. Similarly, it would be quite appropriate for nodes to return to the classification stage when they cannot synchronize with an interferer.

4.2   Tracking Located Interferers

After the localization phase, a node creates a separate thread for monitoring whether it currently has a large enough interference-free window for transmitting a packet. A hardware-based countdown timer (initialized prior to localization) helps track the expected start and end of each impulse; the previously localized time-stamps are relative to that timer. A function returns the number of milliseconds until the next expected interference impulse; the MAC uses this function to determine whether it has enough time to schedule a packet transmission.

When tracking the expected start and end of each impulse, the tracking process somewhat crudely attempts to account for slight errors in the classification and localization. To this end, it grows the prohibited window associated with each impulse by a number of milliseconds every period, and when the transmission window shrinks to the point where a node can no longer transmit, the node resynchronizes with the impulses if packets remain in the transmission queue. This approach of synchronization, growth, and resynchronization is quite suitable for the bursty traffic patterns inherent to many WSN applications.

4.3   Comparison with Earlier Work

The described synchronization approach has significant advantages over our earlier work [6]. The earlier interference tracking technique would assume, upon observing its first post-classification impulse, that the impulse belonged to the single classified pattern. It would then allow transmissions for slightly less than one period, and then it would prohibit further communication and start scanning for the next impulse again. It would carry on this approach indefinitely. In contrast, the current work allows a node to synchronize with multiple periodic interferers. Moreover, the actual synchronization is capable of detecting inactive interferers and not significantly impacted by random impulses.

5   Results

The proactive MAC was evaluated within the SIDE simulator [8], enhanced with VUE[2] [4] to support the PicOS [1] API. The simulations used its built-in shadowing channel model that we parameterized to match our hardware (EMSPCC11 wireless sensor nodes from Olsonet Communications). To evaluate the work, the simulator was extended to incorporate interference using the technique described in [6], which was enhanced to better randomize the simulations.

Each multi-hop configuration consisted of 39 source nodes and one destination node placed randomly within a 170 m × 170 m field. Nodes transmitted at 10 000 bps with a power of -10.0 dBm. Prior to running each simulation, the simulator ensured that the topology was connected, i.e., for every source node in the graph, it could find a path to the destination node where each link in the path could achieve a 95% packet reception rate (PRR). Given the transmit power and receive sensitivity, nodes that were separated by 33 m or less could achieve the desired PRR; farther nodes could still receive packets, just with lower reception rates. If the connectivity test encountered a disconnected node in the graph, the simulator would reject the entire topology and randomly generate a new one until it could find a connected topology.

In this first set of simulations, we randomly placed two interferers within the 170 m × 170 m environment. On start-up, each interferer randomly set its period to a value between 125 ms and 500 ms, corresponding approximately to frequencies between 8 and 2 Hz, respectively. After a random delay lasting for less than 125 ms, each interferer started producing periodic 5 ms impulses.

The simulations evaluated the impact of packet length on the PRR, the latency, and the average number of hops for (a) a traditional (reactive) LBT MAC, (b) the non-collaborative proactive MAC,[2] and (c) the collaborative proactive MAC. The proactive MAC used a time slice of 4 and localization multiple of 2. For these experiments, a per-node inter-arrival rate of 300 s reduced the impact of congestion, and nodes generated packets for 2000 s according to the exponential distribution. Packet lengths varied from 22 to 60 bytes, where 60 bytes is effectively the maximum for the hardware being simulated. For each packet length and MAC configuration, we ran 100 randomized simulations. Figure 4 summarizes the results.

---

[2] The non-collaborative proactive MAC uses the classification and synchronization approach described, but it skips the distribution phase.

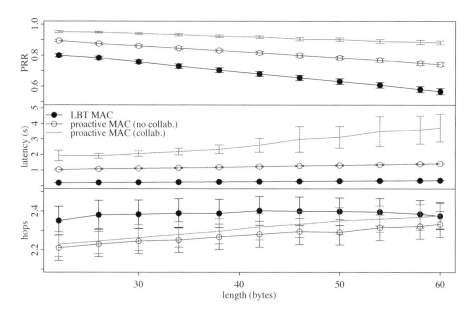

Fig. 4. A comparison of the traditional LBT MAC with the proposed MAC (both collaborative and not) given two interferers within the environment. Error bars indicate 95% confidence intervals.

Figure 4 (top) shows the impact on the PRR. The LBT MAC, oblivious to predictable interference, schedules transmissions upon detecting an idle channel. As the packet length increases, so does the probability of an impulse colliding with a transmission, and the PRR subsequently decreases. Interference impacts the non-collaborating proactive MAC to a lesser extent because it could identify and synchronize with one of the two interferers, and it could avoid some of the impulses. Finally, the collaborative proactive MAC experienced the highest reception rates, as nodes were able to avoid most of the impulses.

In terms of latency (Fig. 4, middle), the latency increases as the MAC is more aware of the interference in the environment. With the proactive MACs, synchronization occurs on demand, i.e., when a node needs to transmit one or more packets. A node's synchronization process lasts for approximately two times the longest period, and each node in a path must perform this synchronization. For the collaborative MAC, the transmitter must consider the impulses of two interferers, giving fewer opportunities to schedule the transmission. As the packet length increases, it becomes even more difficult for nodes to find appropriately-sized windows to schedule their transmissions.

Finally, the simulations showed an interesting increase in the average hop count as the packet lengths increased (Fig. 4, bottom). The probability of a bit error is related to the signal-to-interference-plus-noise ratio (SINR), which is related to the distance between the transmitter and receiver. As the packet length increases, the links between nodes separated by greater distances, where the probably of a bit error is highest, are the first to fail. Given a sufficiently dense network, a node closer to the transmitter may still successfully receive and rebroadcast these lost packet, increasing the average hop count. In the case of the LBT MAC, the rebroadcasts may be corrupted by interference, and a path that potentially had more hops may fail completely.

The simulations evaluated the impact of the per-node inter-arrival rate (μ) on the PRR, latency, and average number of hops for all three MAC approaches. For these experiments, a packet length of 60 bytes was selected to accentuate the result, and nodes generated packets for 2000 s according to the exponential distribution. μ varied from 27 s to 300 s. For each value of μ and MAC configuration, we ran 100 randomized simulations. Fig. 5 summarizes the results.

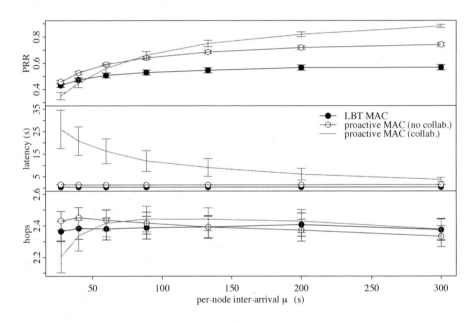

Fig. 5. A comparison of the traditional LBT MAC with the proposed MAC (both collaborative and not) given two interferers within the environment. Error bars indicate 95% confidence intervals.

When μ is 40 s or less, the collaborative MAC provides no benefit over even a traditional LBT MAC, while incurring significant latency. At this rate, the network is saturated with traffic, and memory constraints at some nodes cause them to drop packets without even attempting transmission. As μ increases to 60 s, the collaborative approach realizes some benefit in terms of the PRR, but these gains are at a high cost in terms of latency. When μ reaches 89 s, the collaborative approach finally realizes an improved PRR over the non-collaborative approach. When μ is less than 89 s (high levels of saturation), the collaborative approach, where nodes synchronized with two interferers, likely causes increased congestion during open windows, additional collisions, and higher packet losses. As μ approaches 300 s, all PRR and latency curves begin to stabilize. In terms of the average path length, as μ increases, fewer transmissions in the network likely cause fewer of the long links to fail, and fewer long links failing means fewer hops are necessary to reach the sink node.

A goal for the collaborative proactive MAC is to achieve the packet reception performance of a quiet environment while in an environment polluted by inter-ferers. To evaluate the extent to which it achieves this goal, Fig. 6 compares the performance of the collaborative MAC in two impulsive environments relative to a quiet environment. In the first case, a single interferer was placed randomly within the 170 m × 170 m environment, and it was configured with a random period of 0.125 s to 0.500 s and a 37 dBm transmit power. In the case of two in-terferers, they were configured as described earlier. Given the poor performance observed in saturated conditions, the simulations varied packet lengths.

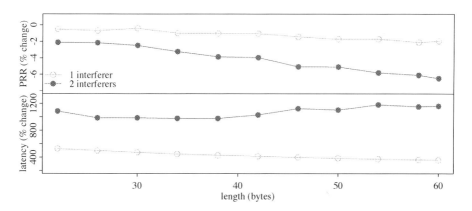

Fig. 6. The performance of the proposed MAC (with collaboration) in the presence of periodic impulsive interference compared against performance in the absence of the interference

In the presence of interference, the figure shows only a slight decrease in PRR from the same protocol operating in a quiet environment. As the packet length increases, the PRR suffers more in the presence of interference, which could be attributable to greater congestion during the quiet windows. This PRR perfor-mance in the presence of interference comes at a considerable cost, however. In terms of latency, the results show an extra 359 to 529% given a single interferer and an extra 976 to 1180% given two interferers. Note that our use of on-demand synchronization, where nodes do not synchronize with the impulsive interference until they have data to transmit, is a big contributor to this latency.

6   Conclusion

This paper presented a new technique that allows WSN nodes to better adapt their communication to their environment. It described a distributed time-domain approach to classifying periodic impulsive interference within WSNs. After classifying the strongest interferer individually, nodes can distribute their observations to their neighbours, and given knowledge gained from neighbours, nodes can better parameterize their MAC to proactively avoid periodic impulsive interference. Moreover, it described an approach for nodes to synchronize with multiple interferers within the constrained resources of a WSN node.

In an unsaturated network, the collaborative proactive MAC allows nodes to quite successfully navigate their transmissions around the impulsive interference. They perform this navigation at the cost of a significant increase to the latency, but for applications where this is acceptable, the gains to the PRR may outweigh the cost. In a saturated network, the avoidance of multiple impulsive interferers leaves nodes with short communication windows between impulses, which cause nodes to quickly fall behind in their transmissions. While saturated, latency increases more quickly than with a traditional LBT MAC protocol, and this protocol is not suitable for such environments.

In the future, we plan to evaluate (a) the existing proof-of-concept WSN node-based implementation of the work described here and (b) the energy consumption of the proposed approach. Additionally, we aim to survey a variety of urban environments around the city to better gauge the prevalence of such interference. Moreover, we plan to explore other opportunities that this type of interference affords, e.g., using powerful interference for synchronizing within a low-powered WSN.

Acknowledgements. The authors thank MacEwan University for financial support through the Research, Scholarly Activity and Creative Achievement Fund.

References

1. Akhmetshina, E., Gburzynski, P., Vizeacoumar, F.: PicOS: A tiny operating system for extremely small embedded platforms. In: Arabnia, H.R., Yang, L.T. (eds.) Embedded Systems and Applications, pp. 116–122. CSREA Press (2003)
2. Bertocco, M., Dalla Chiara, A., Gamba, G., Sona, A.: Experimental comparison of spectrum analyzer architectures in the diagnosis of RF interference phenomena. In: I2MTC 2009: Proceedings of the Instrumentation and Measurement Technology Conference, pp. 765–770 (2009)
3. Boers, N.M., Chodos, D., Huang, J., Stroulia, E., Gburzynski, P., Nikolaidis, I.: The Smart Condo: Visualizing independent living environments in a virtual world. In: PervasiveHealth 2009: Proceedings from the 3rd International Conference on Pervasive Computing Technologies for Healthcare, London, UK (April 2009)
4. Boers, N.M., Gburzynski, P., Nikolaidis, I., Olesinski, W.: Developing wireless sensor network applications in a virtual environment. Telecommunication Systems 45(2-3), 165–176 (2010)
5. Boers, N.M., Nikolaidis, I., Gburzynski, P.: Sampling and classifying interference patterns in a wireless sensor network. ACM Transactions on Sensor Networks 9(1), 2 (2012)
6. Boers, N., Nikolaidis, I., Gburzynski, P.: Impulsive interference avoidance in dense wireless sensor networks. In: Li, X.-Y., Papavassiliou, S., Ruehrup, S. (eds.) ADHOC-NOW 2012. LNCS, vol. 7363, pp. 167–180. Springer, Heidelberg (2012)

7. Chandra, A.: Measurements of radio impulsive noise from various sources in an indoor environment at 900 MHz and 1800 MHz. In: 13th IEEE International Symposium on Personal, Indoor and Mobile Radio Communications, vol. 2, pp. 639–643 (2002)

8. Gburzynski, P.: Protocol Design for Local and Metropolitan Area Networks. Prentice Hall PTR, Upper Saddle River (1995)

9. Lee, H., Cerpa, A., Levis, P.: Improving wireless simulation through noise modeling. In: IPSN 2007: Proceedings of the 6th International Conference on Information Processing in Sensor Networks, pp. 21–30. ACM, New York (2007)

10. Lomb, N.R.: Least-squares frequency analysis of unequally spaced data. Astrophysics and Space Science 39, 447–462 (1976)

11. Mitra, J., Lampe, L.: Sensing and suppression of impulsive interference. In: CCECE 2009: Proceedings of the Canadian Conference on Electrical and Computer Engineering, pp. 219–224 (2009)

12. Musaloiu-E, R., Terzis, A.: Minimising the effect of WiFi interference in 802.15.4 wireless sensor networks. Intl. Journal of Sensor Networks Journal of Sensor Networks 3(1), 43–54 (2008)

13. National Archives and Records Administration: Telecommunication: Definitions. Code of Federal Regulations (CFR), Title 47, Pt. 15.3 (October 1, 2010)

14. Press, W., Teukolsky, S., Vetterling, W., Flannery, B.: Numerical Recipes in C: The Art of Scientific Computing, 2nd edn. Cambridge University Press, Cambridge (1992)

15. Rusak, T., Levis, P.: Physically-based models of low-power wireless links using signal power simulation. Computer Networks 54(4), 658–673 (2010)

16. Srinivasan, K., Levis, P.: RSSI is under appreciated. In: EmNets 2006: Proceedings of the Third ACM Workshop on Embedded Networked Sensors (2006)

17. Srinivasan, K., Dutta, P., Tavakoli, A., Levis, P.: Understanding the causes of packet delivery success and failure in dense wireless sensor networks. In: SenSys 2006: Proceedings of the 4th International Conference on Embedded Networked Sensor Systems, pp. 419–420. ACM, New York (2006)

18. Srinivasan, K., Dutta, P., Tavakoli, A., Levis, P.: An empirical study of low-power wireless. ACM Transactions on Sensor Networks 6(2), 16:1–16:49 (2010)

19. Texas Instruments: Data sheet for CC1100: Low-power sub-1 GHz RF transceiver (October 2009)

20. Zhou, G., Stankovic, J.A., Son, S.H.: Crowded spectrum in wireless sensor networks. In: EmNets 2006: Proceedings of the Third IEEE Workshop on Embedded Networked Sensors. Harvard University, Cambridge (2006)

A Passive Solution for Interference Estimation in WiFi Networks

Claudio Rossi, Claudio Casetti, and Carla-Fabiana Chiasserini

Politecnico di Torino, Italy
{claudio.rossi,casetti,chiasserini}@polito.it

Abstract. Identifying the cause of an underperforming WLAN can be challenging due to the presence of a plethora of devices and networks operating in the ISM band. Such devices cause electromagnetic interference that can potentially undermine the throughput of residential and enterprise WiFi networks. In this paper, we propose an interference estimation and monitoring technique for 802.11 Access Points that can be implemented without specialized hardware and without any modification to wireless stations. We validate our technique with commercial hardware, evaluating its accuracy with different types of interferers.

Keywords: interference, estimation, 802.11, MAC, WiFi, driver, access point.

1 Introduction

In today's overcrowded, arbitrary deployment of home WLANs, interference is likely coming from a neighboring Access Point (AP), operating on either the same or a different frequency channel. In addition, there are many non-802.11 devices working on ISM bands and representing possible sources of interference. It follows that, in spite of the increasing availability of planning, deploying and managing tools, radio interference remains a key performance bottleneck for home and enterprise WLANs alike.

Very few tools are really helpful to understand how much interference affects the operation of a given wireless network, and how interference patterns evolve over time. To further compound this problem, whatever tools are available require expert usage and only operate as spectrum scanners, often providing little insight on the nature, causes and effects of interference. Among commercial solutions, Airmagnet Spectrum XT [1] and AirMaestro [2] are examples of custom hardware systems that integrate spectrum analyzer functionality to facilitate non-WiFi device detection. In the scientific literature, several examples of interference estimation tools based on available bandwidth testing [3–5] require traffic injection. The downside of these approaches is that they affect normal network operations and require certain traffic patterns to test interfering links, which may be incompatible with realistic traffic scenarios.

A different approach is based on trace collection and subsequent analysis. Proposed solutions aim at analyzing specific aspects of a 802.11 wireless network, ranging from physical and link-level behavior [6–8], wireless station (WS) location and coverage [9], to transport and network layer performance [10]. In [11], traces are captured using several sniffers in a WLAN and a state machine-based learning approach is proposed to

S. Guo et al. (Eds.): ADHOC-NOW 2014, LNCS 8487, pp. 156–168, 2014.

identify interference. Similarly, the authors of [12] exploit a large wireless monitoring infrastructure to monitor a production WLAN and perform a cross-layer analysis to diagnose performance problems. While some of these approaches appear to be effective, they only have offline applicability, as they require postprocessing of wireless traces to identify interfering signals. They fail to evaluate the accuracy and agility of interference estimation mechanisms, especially in presence of WS mobility and sophisticated bit rate adaption mechanisms. Also, they do not discuss the integration of their interference estimation mechanisms with applications like power control and channel assignment.

An example of online, passive interference estimation is given in [13], which presents a methodology to dynamically generate fine-grained interference estimates across an entire WLAN. However, the solution in [13] requires both a second wireless card on the APs and to compute the real-time graph of all interfering nodes. The latter implies the presence of a centralized controller for the entire WLAN, which may not be always available, especially in residential networks. Similarly, [19] uses a specific functionality provided by a recent WiFi chipset to perform online detection of multiple non-WiFi devices including fixed frequency devices (e.g., ZigBee), frequency hoppers (e.g., Bluetooth) and broadband interferers (e.g., microwave ovens).

In this work, we propose a MAC-layer approach to interference estimation by adopting passive measuring techniques. Our solution is implemented at the AP and accounts for all possible causes of interference, specifically: transmissions originated within neighboring Basic Service Sets (BSSs), either operating on the same or on a different frequency channel, and transmissions from non-802.11 devices. The key point of the proposed technique resides in the comparison that the AP performs, for each of the data frames it sends, between the expected time required to successfully transmit the frame and the actual time measured by the AP. In order to understand the impact of interference on the BSS throughput performance, we then extend the computation of saturation throughput [14,15] by accounting for the interference effects. We implement our solution in a testbed and validate it via experimental results. Unlike previous solutions, ours can be implemented on commercial APs with any WiFi chipset, without requiring either specialized hardware or modifications to the WS.

2   Inferring Interference

We consider an IEEE 802.11 BSS managed by an AP capable of monitoring local wireless resources. The AP and its associated WSs may transmit frames with different payload size and their data rate may vary according to the experienced channel propagation conditions. The AP collects statistics within its BSS and it carries out such measurements periodically over a time interval, hereinafter referred to as measurement period.

Inferring interference requires that the AP estimates whether the channel is sensed busy because of "legitimate" ongoing transmissions or because of interference. This procedure consists in computing, for each frame k sent by the AP, what is the *expected* time interval $T_e(k)$ that a transmission (either successful at the first attempt or subject to collisions/errors, hence repeated one or more times) should take, were it not affected by interference. This time computation should then be compared against actual *measurements* of transmission interval $T_m(k)$, taken by the AP driver, in order to infer whether unaccounted-for signals (interferers) are cluttering the channel. The latter unduly lengthen the transmission because the channel is sensed busy even if no legitimate BSS node is actively sending data. The estimation of the fraction of time taken by interferers, I, can be computed as follows:

$$I = \frac{\sum_{k=1}^{K} \left(\frac{T_m(k) - T_e(k)}{T_c(k)} \right)}{K}, \tag{1}$$

where K is the total number of frames transmitted by the AP within the current measurement period. $T_m(k) = t_a(k) - t_x(k)$, where $t_a(k)$ corresponds to the ACK notification time, while $t_x(k)$ is the time at which frame k is handed over to the driver for transmission. As for $T_e(k)$, for the sake of clarity, we describe how it is computed by referring to Fig. 1.

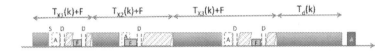

Fig. 1. Timeline of a repeated transmission by the AP. "S", "A", "D" and "F" stand for, respectively, SIFS, ACK, DIFS, and Freeze

The figure portrays the case where the AP repeats the transmission of a frame four times before success. Each attempt is renewed after the mandatory random backoff period. It is to be underlined that each retransmission attempt could occur at a lower bit rate than the previous one. Indeed, the MAC-layer rate control procedure implemented in most 802.11 drivers mandates for each MAC frame to be associated with a retry vector. This vector specifies the number of retries to be performed at decreasing bit rates, successively attempted in case a transmission fails (i.e., no ACK is received). These rates are known by the driver. Thus, the duration of each frame transmission can be easily computed by the AP itself. We denote by $T_d(k)$ the duration of the successful transmission, which depends on the data rate and the payload size of the frame and is computed as specified in the 802.11 standard [16]. As for the failed transmissions, we indicate the i-th transmission attempt of frame k as $T_{Xi}(k)$. Such a quantity is composed by: the transmission duration of the data frame, the retransmission timeout T_o (which is set equal to SIFS plus the ACK duration), the backoff time associated with the i-th attempt, and the DIFS time intervals needed to declare the channel idle. The contribution due to the backoff time is set to half the contention window used at the i-th attempt. However, it is important to recall that the backoff counter is frozen by an 802.11 interface whenever the channel is sensed busy. Carrier sensing may be triggered by: (i) transmissions from other WSs in the same BSS, (ii) transmissions from other BSSs operating on the same channel, (iii) transmissions from neighboring BSSs using a different channel or from non-802.11 devices. Note that the first case is a "legitimate" interruption and, as such, is not classified as interference. As for data frames transmitted within other BSSs operating on the same channel, they can be received by the AP of the tagged BSS through a virtual interface operating in monitoring mode. In our computation, we separately account for such contribution and denote it by $\delta(k)$.

Based on the description and definitions above, the expected transmission interval $T_e(k)$ can be obtained as:

$$
T_e(k) = \sum_{i=1}^{A} T_{Xi}(k) + T_d(k) + \text{SIFS} + \text{ACK} +
$$
$$
\sum_{j=1}^{N} (T_{RXj}(k) + T_{NRj}(k)) + \delta(k) + \epsilon, \tag{2}
$$

where:

- A is the number of failed transmissions for frame k;
- SIFS and ACK are, respectively, SIFS and ACK durations. The ACK duration is computed by considering its actual transmission rate;
- N is the number of WSs in the tagged BSS;
- $T_{RXj}(k)$ and $T_{NRj}(k)$ are the duration of, respectively, successful and failed transmission cycles by other WSs within the tagged BSS (during which the AP has to freeze its own backoff while trying to transmit frame k);
- $\delta(k)$ is the airtime taken by non-colliding transmissions from neighboring BSSs operating on the same channel, while the AP tries to transmit frame k;
- ϵ is the approximation error due to the granularity with which time intervals are detected by the AP's driver.

We remark that the AP has no knowledge of failed transmissions by WSs. However, assuming a symmetrical channel between AP and WSs, the same packet error rate (PER) may apply to any transmission, thus yielding a rough estimate of the percentage of WS transmissions that ultimately fail.

In conclusion, by computing $T_e(k)$ and measuring $T_m(k)$ for each data frame, the AP can obtain an estimate of I at each measurement interval. *This method can be effectively implemented at run time and it does not require any knowledge of the past.* Since per-frame processing and statistics are already included in any WiFi driver, our method adds only a negligible complexity. The technology we refer to can be any among a, b, and g; also, the proposed technique can be easily extended to DCF with handshake as well as to the 802.11e/n EDCA.

3 Computation of the Saturation Throughput

For a practical understanding of the impact of interference, we now introduce a simple methodology to compute the theoretical saturation throughput that would be achieved if the BSS operated in an unhindered scenario. We will then estimate the contribution of interferers and derive the *theoretical saturation throughput in presence of interference*. In Sec. 5, the latter quantity will be compared against actual live measurements.

Again, we consider an IEEE 802.11 BSS managed by an AP capable of monitoring local wireless resources at the MAC layer. As in [17, 18] during each measurement period, the AP computes: average size of the frame payload (P), maximum payload size (P_{max}), average data rate for data frames and for ACKs, average PER (p_e), and number of active WSs within the BSS (N). The AP considers a node in the BSS (either itself or a WS) to be active if the node has successfully transmitted at least one data frame within the last measurement period. The average PER could be computed based on the modulations used for the transmissions in the measurement period, their associated signal-to-noise (SNR) ratio, and assuming independent bit errors on the channel. Since this method gives poor results due to multipath effects and inaccurate SNR measurements by the hardware, we estimate p_e as the ratio of the number of erroneously received frames to the number of transmitted frames. To compute the numerator, at the receiver we count the CRC errors (at the PHY and MAC layer), while at the transmitter we count all unsuccessful transmission attempts at the physical layer. The latter results in a worst case PER estimation, as collisions are also included in the count[1]. Conversely, the computation at the receiver underestimates the actual number as the physical layer cannot always decode a corrupted frame and hand it to the MAC layer.

We now introduce the theoretical saturation throughput S_{th}, defined as the value of maximum achievable throughput in the BSS given the current traffic load. The saturation throughput is given in [15], which extends the original model in [14] to account for errors due to channel propagation conditions:

$$S_{th} = \frac{N\tau(1-\tau)^{N-1}P(1-p_e)}{E[T]}. \tag{3}$$

In (3), τ is the probability that a node (either a WS or the AP) accesses the medium at a generic time slot[2] and $E[T]$ is the average duration of a time interval in which an event occurs, namely, an empty slot, a successful transmission, a transmission failed due to channel errors, or a collision. $E[T]$ can be computed as:

$$\begin{aligned} E[T] = (1-\tau)^N \, \sigma + \\ [N\tau(1-\tau)^{N-1}(1-p_e)]T_s + \\ [1-(1-\tau)^N - N\tau(1-\tau)^{N-1}]T_c + \\ [N\tau(1-\tau)^{N-1}p_e]T_{err} \end{aligned} \tag{4}$$

where σ is the slot time duration as defined in the 802.11 standard. By assuming that the retransmission timeout is equal to SIFS plus ACK, the average duration of a successful transmission, T_s, and of an erroneous transmission, T_{err}, are equal and given by:

$$T_s = T_{err} = T_d + \text{ACK} + \text{SIFS} + \text{DIFS} \tag{5}$$

where T_d is the average frame duration computed as specified by the standard, according to the BSS type, and using the average payload size and average rate measured by the AP at the BSS level.

---

[1] Collisions cannot be discriminated from errors caused by harsh channel conditions without changing the WS software or the 802.11 protocol.

[2] Considering a slotted time is the main approximation of this model.

As far as the average collision duration is concerned, its exact computation would require the AP to be aware of the number of nodes that are hidden with respect to each other. The work in [14, 15] does not account for hidden WSs and the approaches proposed in the literature, e.g., [20], are not viable in our set up, as we do not require the AP to have knowledge of the users distribution within its coverage area. Thus, we approximate the average collision duration by making a worst-case assumption. Each collision involves a data frame of maximum payload size P_{max} among those observed by the AP during the measurement period. Clearly, the average collision time is overestimated in absence of hidden WSs, leading to underestimating the saturation throughput. It follows that T_c is computed as T_{err} but using P_{max} instead of P. We also observe that the AP can easily compute τ using the following equations [15]:

$$p = 1 - [(1-\tau)^{N-1}(1-p_e)]$$
$$\tau = \frac{2(1-2p)(1-p^{m+1})}{W(1-(2p)^{m+1})(1-p) + (1-2p)(1-p^{m+1})}$$

where p is the the conditional probability that a transmitted data frame encounters a collision or is received in error in saturation conditions; W is the minimum contention window; m is the retransmission limit. While deriving our results, we set $W = 31$ and $m = 5$. Note that p and τ have to be obtained through numerical methods, as described in [14, 15]. We can pre-compute all values of τ as a function of N and p_e, and perform at each measurement period a simple look-up. For instance, if we consider N varying from 1 to 30, τ from 0.05 to 1 with 0.05 resolution, and a half precision floating point representation (16-bit) for τ, we would need only 1.2 MB memory space to store all values. As τ ranges from 0 to 1, this requirement can be further reduced by using an ad hoc code.

We stress that, although S_{th} represents the saturation throughput considering the node average behavior, it accounts for the different air time that WSs take to transmit their frames. Indeed, the average payload size P and $E[T]$ in (3) depend on the payload, data rate and access rate of each single WS.

In order to reflect the effects of all types of interferers, we let the AP keep track of I, computed as in (1), and of the average δ, computed as $\left(\sum_{k=1}^{K} \delta(k)\right)/M$ where M is the measurement period duration. We then discount from S_{th} the portion of throughput that cannot be achieved due to the two components above and, finally, compute the saturation throughput in presence of interference, S_{in}:

$$S = (1-\delta)S_{th}$$
$$S_{in} = (1-I)S.$$

4   System Implementation: A MAC-Layer Approach

As mentioned, our solution has the following desirable properties: (i) it allows online interference detection, from both WiFi and non-WiFi devices; (ii) it does not need specialized hardware; (iii) it runs on an AP without additional software (or hardware) modification to WSs.

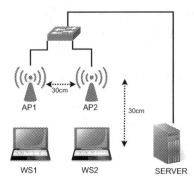

Fig. 2. Testbed deployment used for validation

We implement the estimation procedure described in Sec. 2 at the MAC layer, specifically within the mac80211 module of the Linux wireless driver *compact-wireless 2011-21-01* [21]. We select the MAC layer because it is the highest layer in the stack from which we can retrieve the data rate for every data transmission attempt as well as for ACKs.

The estimation procedure implies the implementation in the AP driver of: (i) the additional passive measurements described in Secs. 2 and 3, and (ii) the estimation of I. All required measurements, along with I, are made available to the application layer by using debugfs at each measurement period. Note that, since we modify only the mac80211 module and not the AP hardware or physical driver, such measurements can work on any wireless chipset. We then implement the computation of S and S_{in} with a simple user-space program that reads the measurements from file system and process them as described in Sec. 3. Given its low complexity, such computation can be implemented on any device.

5   Experimental Evaluation

We evaluate the validity of our approach through a testbed deployed in a university laboratory at Politecnico di Torino. In the laboratory, there are 18 detectable APs, which are part of 6 different SSID, whose signal is received at an average strength of −83 dBm. We use two stations, WS1 and WS2, each associated to a 802.11g AP. We name the access points, AP1 and AP2, and refer to their BSSs as BSS1 and BSS2, respectively. Each AP runs the modified driver for the computation of I and the application for the computation of S and S_{in}. We connect the APs to a switch, which, in its turn, is connected to a desktop PC that we use as a traffic sink. All equipment is placed on a desk, at approximately 1 m height. AP1, AP2, WS2 and WS1 are placed at the vertices of a square of 30 cm side, in clockwise order starting from the right-top vertex (see Fig. 2).

APs are implemented in embedded wireless nodes featuring an Alix PC Engines motherboard, equipped with an AMD Geode 500 MHz processor and a IEEE 802.11 b/g compliant Wistron DCMA-82 Atheros wireless card. Each Alix runs OpenWrt Backfire, a Linux distribution for embedded devices. WSs are represented by ASUS notebooks, model P52F, with Ubuntu 12.04. Both APs are powered through PoE (Power

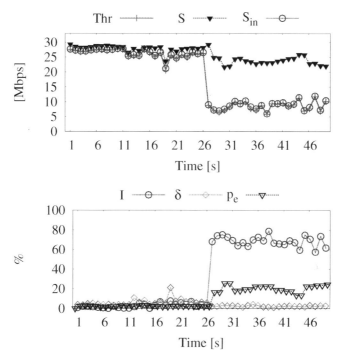

Fig. 3. 802.11 interferer enabled at second 26 on a non-overlapping channel (ch11). Top: comparison between BSS1 throughput (Thr), the estimated saturation throughput S and that accounting for interference (S_{in}). Bottom: average PER (p_e) and time fractions during which the channel is sensed as busy due to co-channel interference (δ) and to BSS2 operating on ch11 (I).

over Ethernet). BSS1 operates on channel 6 and is our tagged BSS, while BSS2 acts as an interferer. In particular, BSS2 operates either on the same channel as BSS1, or on another among the 802.11 standard ones (in the 2.4 GHz band). We set the transmit power of both APs and WSs to 20 dBm. Since we are interested in evaluating the throughput loss caused by interferers and the accuracy of our solution in estimating I and S_{in}, we saturate BSS1's capacity with a downlink UDP flow from the server to WS1 (through AP1), at a rate of 30 Mbps. The choice of UDP, rather than TCP, traffic allows us to precisely control the load without unpredictable effects due to congestion control mechanisms.

We start by setting BSS2 on channel 11 so as to assess the performance of our solution in presence of 802.11 interferers operating on a different frequency channel. Specifically, we consider a dynamic traffic scenario where no traffic is generated within BSS2 in the interval [0, 25 s], then the server starts transmitting a UDP downlink flow at 30 Mbps to WS2 (through AP2). The results are shown in the top plot of Fig. 3, which compares the throughput achieved by AP1 (Thr) to the saturation throughput estimations S and S_{in}. Recall that S_{in} differs from S as it accounts for the interference term I. The bottom plot of Fig. 3 instead depicts (i) the estimated time fraction I during which the channel within BSS1 is sensed as busy due to BSS2, (ii) the average PER (p_e),

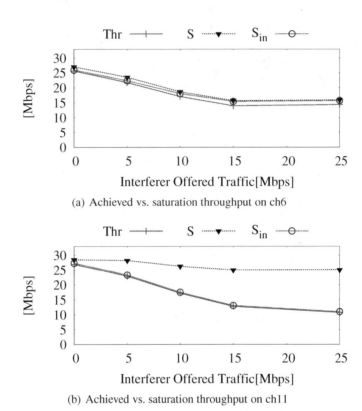

(a) Achieved vs. saturation throughput on ch6

(b) Achieved vs. saturation throughput on ch11

Fig. 4. Achieved vs. saturation throughput in BSS1 when BSS2 operates on the same channel (ch6) (a) and on a different channel (ch11) (b)

and (iii) the time fraction δ during which AP1 senses the medium as busy due to other BSSs operating on the same channel. Note that δ is determined by those BSSs, out of the 18 that are present, that operate on the same channel as BSS1 (channel 6). All these quantities are expressed as percentages.

We observe that S_{in} closely matches the throughput measured by AP1 (Thr) before and after the interfering flow is enabled within BSS2. This clearly indicates that I correctly reflects the negative effect of a flow activated within a BSS operating on a different frequency channel (with BSS2 achieving an average aggregate throughput of 15.7 Mbps). It is also important to remark that the quantitative impact of interferers operating on non-overlapping channels may be severe, especially in the case of devices in close proximity (as in our case). Furthermore, when the interferer is enabled, not only I but also p_e increases. This suggests that transmissions within BSS2 may cause collisions at AP1, beside making AP1 detect the channel as busy.

We now extend our evaluation by varying both the channel used by the interferer (BSS2) and its offered load. We set AP2 to operate on a different channel at each run, namely, 6, 7, 9 and 11. For each channel, we carry out several experiments (each lasting 50 s) so as to vary the generation rate of the interfering traffic (i.e., the rate of the UDP

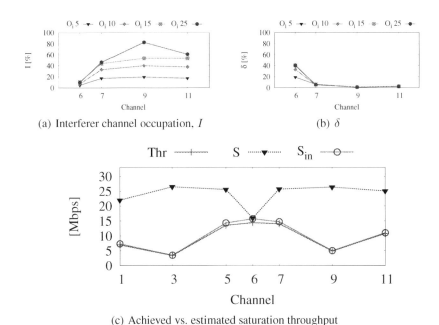

(a) Interferer channel occupation, I (b) δ

(c) Achieved vs. estimated saturation throughput

Fig. 5. (a) Interference estimation I and (b) time fraction during which AP1 detects the channel busy due to co-channel interference (δ). The results are plotted as the channel used by BSS2 and its traffic load vary. (c) BSS1 throughput (Thr) vs. estimated saturation throughput S and S_{in}, as the channel used by BSS2 varies.

downlink traffic flowing from the server to WS2). In all cases, we obtain an excellent match between the throughput achieved by AP1 and the estimated saturation throughput S_{in}. Due to lack of space, we only plot in Fig. 4 the comparison between the throughput measured by AP1 (Thr), S and S_{in} for channel 6 (top) and 11 (bottom). The results are presented as functions of the offered load in BSS2 (i.e., the interfering traffic genera-tion rate). Each point plotted in the figure is the average over time of the values obtained during a 50-s experiment. Clearly, Thr decreases as the interferer traffic load increases. Again, S accurately estimates the throughput of BSS1 only in case of co-channel in-terference, i.e., when BSS2 operates on channel 6. Indeed, in this case the majority of the frames transmitted by AP2 are received by AP1's monitoring interface, which can correctly account for it through the quantity δ. Conversely, when the interferer operates on channel 11, S cannot reveal its presence, while S_{in} perfectly reflects its effects.

Figs. 5 (a) and (b) show the value of the time fractions I and δ, as the channel on which BSS2 operates and BSS2's traffic load vary. In case of co-channel interference (i.e., BSS2 on ch6), I is low as this quantity does not account for it; conversely, δ well captures such interference. Furthermore, we observe that the traffic load within the interfering BSS has a significant impact and this is correctly represented by both I (for channels other than 6) and δ (for channel 6).

Fig. 5(c) presents the results obtained with BSS2 operating also on channels 1, 3, and 5 and setting the rate of its traffic flow to 25 Mbps. We note that the behavior of S_{in}

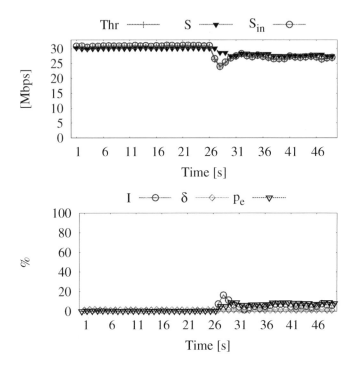

Fig. 6. Bluetooth file transfer enabled at second 26. Top: comparison between BSS1 throughput (Thr), estimated saturation throughput (S) and that accounting for the interference effect (S_{in}). Bottom: average PER (p_e), and time fractions during which AP1 senses the channel as busy due to co-channel interference (δ) and to the Bluetooth interferer (I).

closely matches that of the measured throughput Thr on all channels. The values of Thr however change significantly depending on the considered channel: this is due to the different multipath conditions affecting the channels. We repeated the experiments placing the devices at different locations and we obtained similar results, which we omit for brevity. The fact that S_{in} and Thr consistently match suggests that our methodology provides an accurate estimation no matter the working environment that is considered.

Next, we focus on non-802.11 interferers and employ a pair of Bluetooth nodes and an analog video sender. First, we start a file transfer between the Bluetooth nodes, with the sender and the receiver placed, respectively, at the left and right side of WS1 and equally spaced from WS1 by 30 cm. As before, the file transfer starts at second 26 and the experiment lasts 50 s. As shown in Fig. 6, initially the interfering traffic flow causes a noticeable throughput degradation at AP1, but after a couple of seconds the performance improves again. This effect is due to the Bluetooth adaptive frequency hopping (AFH) scheme, which tends to avoid channels characterized by high PER. Again, S_{in} closely follows the behavior of Thr in all phases of the experiments, confirming the validity of our technique.

We then let the video sender act as interferer. It is placed on the left of WS1, at 30 cm distance, turned on at second 26. The interference generated is so strong that the

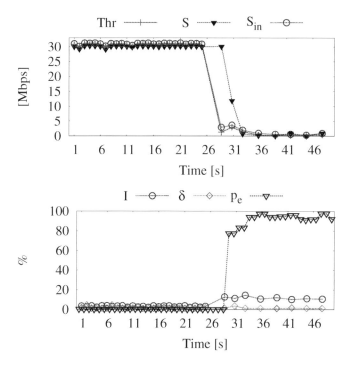

Fig. 7. Analogue video sender enabled at second 26. Top: comparison between BSS1 throughput (Thr), estimated saturation throughput (S), and that accounting for the interference effect (S_{in}). Bottom: average PER (p_e), and time fractions during which AP1 senses the channel as busy due to co-channel interference (δ) and to the interference caused by the video sender (I).

throughput drops almost to zero after a couple of seconds. Again, S_{in} is able to quickly reflect the behavior of the measured throughput, as highlighted by the results in Fig. 7.

6   Conclusions

We have designed and implemented a technique for interference estimation in 802.11 WLANs, which accounts for all possible sources of interference. It can be implemented at the access point and does not require any specialized hardware, changes in the 802.11 standard or in the wireless stations. We have validated our technique in a 802.11g network with different types of interferers. Experimental results show that our solution can estimate the impact of interference with excellent accuracy, under different scenarios.

Acknowledgements. This paper was made possible by NPRP grant $\sharp/5-782-2-322$ from the Qatar National Research Fund (a member of Qatar Foundation). The statements made herein are solely the responsibility of the authors.

References

1. AirMagnet Spectrum XT, www.airmagnet.net/products
2. Bandspeed AirMaestro spectrum analysis solution, http://www.bandspeed.com/
3. Niculescu, D.: Interference map for 802.11 networks. In: IMC (2007)
4. Padhye, J., et al.: Estimation of link interference in static multi- hop wireless networks. In: IMC (2005)
5. Ahmed, N., et al.: Online estimation of RF interference. In: ACM CoNext (2008)
6. Sheth, A., et al.: MOJO: a distributed physical layer anomaly detection system for 802.11 WLANs. In: MobiSys (2006)
7. Aguayo, D., et al.: Link-level measurements from an 802.11b mesh network. In: SIGCOMM (2004)
8. Vutukuru, M., Jamieson, K., Balakrishnan, H.: Harnessing exposed terminals in wireless networks. In: NSDI (2008)
9. Chandra, R., Padhye, J., Wolman, A., Zill, B.: A location- based management system for enterprise wireless LANs. In: NSDI (2007)
10. Cheng, Y.-C., et al.: Automating cross-layer diagnosis of enterprise wireless networks. In: ACM SIGCOMM (2007)
11. Mahajan, R., et al.: Analyzing the MAC-level behavior of wireless networks in the wild. In: ACM SIGCOMM (2006)
12. Cheng, Y.-C., et al.: Jigsaw: solving the puzzle of enterprise 802.11 analysis. In: ACM SIGCOMM (2006)
13. Shrivastava, V., Rayanchu, S., Banerjee, S., Papagiannaki, D.: PIE in the sky: online passive interference estimation for enterprise WLANs. In: USENIX (2011)
14. Bianchi, G.: Performance analysis of the IEEE 802.11 Distributed Coordination Function. In: IEEE JSAC, vol. 18(3), pp. 535–547 (2000)
15. Chatzimisios, P., Boucouvalas, A.C., Vistas, V.: Performance analysis of the IEEE 802.11 DCF in presence of transmission errors. In: IEEE ICC (2004)
16. 802.11 standard, http://standards.ieee.org/about/get/802/802.11.html
17. Rossi, C., Casetti, C., Chiasserini, C.-F.: Bandwidth Monitoring in Multi-rate 802.11 WLANs with Elastic Traffic Awareness. In: GLOBECOM (2011)
18. Rossi, C., Casetti, C., Chiasserini, C.-F.: Energy-efficient Wireless Resource Sharing for Federated Residential Networks. In: WoWMoM (2012)
19. Rayanchu, S., Patro, A., Banerjee, S.: Airshark: Detecting non-WiFi RF devices using commodity WiFi hardware. In: IMC (2011)
20. Mahani, A.K., Naderi, M., Casetti, C., Chiasserini, C.-F.: MAC layer channel utilization enhancements for wireless mesh networks. IET Communications 3(5), 794–807 (2009)
21. Wireless driver, http://linuxwireless.org/

Adaptive Duty-Cycled MAC for Low-Latency Mission-Critical Surveillance Applications

Ehsan Muhammad and Congduc Pham

LIUPPA Laboratory,
University of Pau, France
{muhammad.ehsan,congduc.pham}@univ-pau.fr

Abstract. Mission-critical surveillance applications such as intrusion detection or disaster response have strong requirements in communication delays. We consider a Wireless Image Sensor Network (WISN) with a scheduling of image sensor node's activity based on the application criticality level. Sentry nodes capable of detecting intrusions with a higher probability than others will alert neighbor nodes as well as activating cover sets member for image disambiguation or situation-awareness purposes. At the access level, we consider duty-cycled Medium Access Control to periodically set nodes in sleep mode for energy preservation. However, in doing so, care must be taken to also preserve the quality of event detection and sentry nodes must still be able to propagate quickly alert messages. We propose an original approach to dynamically determine the duty-cycle values of image sensor nodes to increase the probability of matching active period between nodes, thus reducing the alert latency while globally reduce the energy consumption.

1   Introduction

In addition to traditional sensing network infrastructures, a wide range of emerging wireless sensor network applications can be strengthened by introducing a visioning capability. The vision capability is a more effective means to capture important quantity of richer information and vision constitutes a dominating channel by which people perceive the world. Nowadays, such applications are possible since low-power sensors equipped with a visioning component already exist. Wireless Image Sensor Networks (WISN) where sensor nodes are equipped with miniaturized visual cameras to provide visual information is a promising technology for intrusion detection or search&rescue applications.

Our research considers surveillance applications where image sensor nodes are thrown in mass, randomly, to start the surveillance process, e.g. intrusion/anomaly detection, situation awareness,... Figure 1 shows the scenario of a random deployment of image sensor nodes which is typical of the kind of applications we want to address in this paper. When an intrusion is detected by a node it will (*i*) send a number of images to the sink and (*ii*) alert a number of neighbor nodes. Alerts could be propagated at k hops but this is not illustrated in the figure. Figure 2 shows a close-up view of the dashed square area in

S. Guo et al. (Eds.): ADHOC-NOW 2014, LNCS 8487, pp. 169–182, 2014.

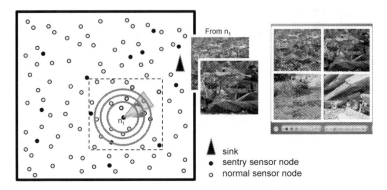

Fig. 1. Mission-critical intrusion detection system

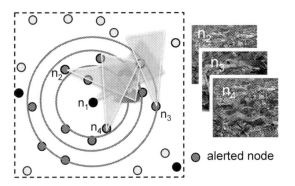

Fig. 2. Alert propagation

Figure 1 to illustrate the alert process in which neighbor nodes are put in alert mode (red nodes).

Our work in this article focuses on the Medium Access Control layer for providing low energy consumption and low latency for alert propagation. In figure 2, it is desirable that neighbor nodes can receive the alert indication as soon as possible in order to propagate the alert towards the sink. However, as event detection in WSN can be quite sporadic and nodes be idle for a long period of time MAC layers usually adopt a duty-cycled behavior in order to save the energy consumption of maintaining the radio module awake listening for incoming packets: an active or listening period alternates with an inactive or sleep period. A simple approach for duty-cycling such as the one proposed by the 802.15.4 standard can be improved with synchronization features to have common active periods (SMAC [1], TMAC [2] to name a few) or with low-power listening (LPL) capabilities and preamble transmissions (B-MAC [3], X-MAC [4], TP-MAC [5] to name a few). Reader can refer to [6] to have a survey of MAC protocols for WSN. While synchronous approaches are not scalable for large networks, LPL and preamble-based approaches still suffer from high latencies when node's sleeping period is large. We propose to adapt the active period of sensor nodes to provide low-latency alert communication. In figure 2, neighbor nodes for node n_1 will set their listening period according to the criticality level of node n_1: the higher the criticality level, the longer the listening period. Additionally, the node's listening period will also depend on the node's redundancy level in order to determine a listening period which will not compromise the node's lifetime. The contribution in this article is based on a criticality model we developed in [7] for image sensors: (i) each sensor node n has a frame capture rate which depends on the criticality level and node n's redundancy level, then (ii) each neighbor node n_i will set its listening period according to node n's frame capture rate and its own redundancy level. Although our approach has been designed from the beginning for image sensor the proposition can also work with traditional scalar sensors with disk coverage. Our contribution can also be used with LPL and preamble approaches to determine the receiver periodic channel sampling interval, thus reducing the cost of preambles.

The remainder of the paper is structured as follows. Section 2 reviews our criticality-based scheduling. Our proposed MAC approach is explained in Section 3. Simulations and results are shown in Section 4 and we conclude in Section 5.

2   Criticality-Based Node Scheduling

Early surveillance applications involving WSN have been applied to critical infrastructures such as production systems or oil/water pipeline systems [8, 9]. There have also been some propositions for intrusion detection applications [10–16] but most of these studies focused on coverage and energy optimizations without explicitly having the application's criticality in the control loop which is the main concern in our work. For instance, with image sensors, the higher the capture rate is, the better relevant events could be detected and identified. However, even in the case of very mission-critical applications, it is not realistic to consider that video nodes should always capture at their maximum rate when in active mode. In randomly deployed sensor networks, provided that the node density is sufficiently high, sensor nodes can be redundant (nodes that monitor the same region) leading to overlaps among the monitored areas. Therefore, a common approach is to define a subset of the deployed nodes to be active while the other nodes can sleep. One obvious way of saving energy is to put in sleep mode nodes whose sensing area are covered by others. However, in mission-critical applications where it is desirable to increase responsiveness, nodes that possess a high redundancy level (their sensing area are covered many times by other nodes so that they have many cover sets) could rather be more active than other nodes with less redundancy level. As illustrated in Figure 1 sensor nodes can self-organize themselves to designate a number of nodes to act as sentry nodes (nodes in black) to better detect intrusions and to trigger alerts. This hierarchical organization is not mandatory but it helps increasing network lifetime, by putting some nodes in low consumption mode, which is one important issue when autonomous sensor nodes are considered, even with mission-critical applications.

In [7] the idea we developed is that when a node has several covers, it can increase its frame capture rate because if it runs out of energy it can be replaced by one of its cover sets. Then, depending on the application's criticality, the frame capture rate of those nodes with large number of cover sets can vary: a low criticality level indicates that the application does not require a high image frame capture rate while a high criticality level does. We proposed to link the criticality level to the number of cover sets by concave and convex curves as illustrated in figure 3 with the following interesting properties:

- a concave curve has most projections of x values on the y-axis close to 0 (figure 3 box A). Such curve could represent "low criticality" applications that do not need high frame capture rate;
- a convex curve where most projections of x values on the y-axis are close to the maximum frame capture rate (figure 3 box B). Such curve could represent "high criticality" applications that need high frame capture rate;

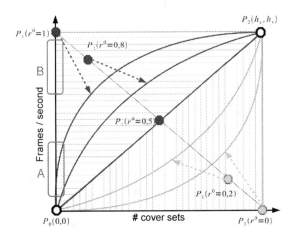

Fig. 3. The Behavior curve functions

We proposed in [7] to use a Bezier curve to model the 2 application classes. 3 points can define a convex (high criticality) or concave (low criticality) curve: $P_0(0,0)$ is the origin point, $P_1(b_x, b_y)$ is the behavior point and $P_2(h_x, h_y)$ is the threshold point where h_x is the highest cover cardinality and h_y is the maximum frame capture rate determined by the sensor node hardware capabilities. As illustrated in figure 3, by moving the behavior point P_1 inside the rectangle defined by P_0 and P_2, we are able to adjust the curvature of the Bezier curve, therefore adjusting the criticality level: according to the position of point P_1 the Bezier curve can move from a convex to a concave form. P_1 therefore defines a criticality level r^0 which is between 0 and 1, 1 being the highest criticality level which requires fast frame capture rate.

Assuming $P_0(0,0)$, $P_1(b_x, b_y)$ and $P_2(h_x, h_y)$ we can define the Bezier curve (BV) as follows:

$$BV : [0, h_x] \longrightarrow [0, h_y]$$
$$X \longrightarrow Y$$

$$BV_{P_1, P_2}(X) =$$
$$\begin{cases} \frac{(h_y - 2b_y)}{4b_x^2} X^2 + \frac{b_y}{b_x} X & if \ (h_x - 2b_x = 0) \\ (h_y - 2b_y)(\propto (X))^2 + 2b_y \propto (X), \ if \ (h_x - 2b_x \neq 0) \end{cases}$$

$$Where \ \propto (X) = \frac{-b_x + \sqrt{b_x^2 - 2b_x * X + h_x * X}}{h_x - 2b_x} \land \begin{cases} 0 \leq b_x \leq h_x \\ 0 \leq X \leq h_x \\ h_x > 0 \end{cases}$$

We then define the Rk function such that varying r^0, the dynamic risk level, between 0 and R^0 gives updated positions for P_1 thus obtaining corresponding values for b_x and b_y:

$$Rk : [0, R^0] \longrightarrow [0, h_x] * [0, h_y]$$
$$r^0 \longrightarrow (b_x, b_y)$$
$$Rk(r^0) = \begin{cases} b_x = -h_x \times r^0 + h_x \\ b_y = h_y \times r^0 \end{cases}$$

If we set the maximum cover set cardinality to 12 and the maximum frame capture rate to 3fps then we have $P_2(h_x, h_y) = (12, 3)$. b_x and b_y can then be determined with the equation above, for a given value of r^0. Table 1 shows the corresponding capture rate for some relevant values of r^0 when the number of cover-sets is varied.

Table 1. Capture rate in fps when P_2 is at (12,3)

r^0	1	2	3	4	5	6	7	8	9	10	11	12
0	.01	.02	.05	0.1	.17	.16	.18	.54	.75	1.1	1.5	3
.1	.07	.15	.15	.17	.51	.67	.86	1.1	1.4	1.7	2.1	3
.4	.17	.15	.55	.75	.97	1.1	1.4	1.7	2.0	2.1	2.6	3
.6	.16	.69	1.0	1.1	1.5	1.8	2.0	2.1	2.4	2.6	2.8	3
.8	.75	1.1	1.6	1.9	2.1	2.1	2.5	2.6	2.7	2.8	2.9	3
1	1.5	1.9	2.1	2.4	2.6	2.7	2.8	2.9	2.9	2.9	2.9	3

With this criticality-based scheduling approach at the application layer, nodes with high number of cover-set will implicitly become sentry nodes by having a higher frame capture rate. The motivation of the work we present in this paper is to allow fast propagation of alert messages from the sentry nodes to the sink. For that purpose, we propose an adaptive listening period for the MAC layer, making each neighbor node of a sentry node ready to receive the alert message and to forward it to the sink.

3   Adaptive MAC Protocol Design

In our application scenario a node detecting an event will first send an alert to the sink. Note that any node can detect an event but according to the previously described criticality-based scheduling approach, some nodes will have higher capture rates than others and therefore will act as sentry nodes because they have a higher probability of detecting an intrusion or any changes in the environment (under the assumption that events occur uniformly in the covered area). As indicated previously, all nodes have a radio duty-cycled behavior where the radio module is put to sleep for some time, and then waked up to listen for other nodes wanting to communicate with it, e.g. transmission of an alert message for instance. In our scenario, it is desirable that neighbor nodes of a sentry can receive the alert message as soon as possible to (i) increase their criticality level and, most importantly, (ii) propagate the alert towards the sink.

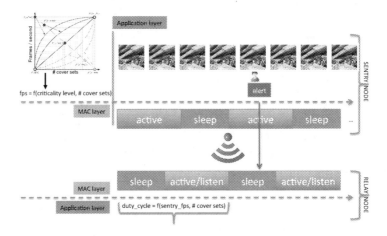

Fig. 4. Active and Sleep periods of the MAC layer

Figure 4 shows how at the application layer, the frame capture rate of a node can be determined based on an application criticality level and the node's number of cover-sets using the criticality scheduling approach reviewed in Section 2. The node's activity at the application level can actually be independent from the radio activity which is the focus of this work. However, while the radio duty-cyle value of the sentry node can be quite simple to determine according to its frame capture rate, it is more difficult to set the listening period of neighbor relay nodes, especially when there are many relay nodes as illustrated previously in Figure 2. In order to avoid difficult and costly synchronization mechanisms, we propose a probabilistic approach where a neighbor relay node will set its listening period according to the sentry's frame capture rate and its own redundancy level (i.e. number of cover-sets). Our contribution works in 2 phases. The first phase is to determine for each image sensor node its associated sentry node, i.e. the image node in its neighborhood with the highest frame capture rate. A node with an associated sentry node will be called a follower node. Then in a second phase, we adapt the follower node's listening period to increase its responsiveness in case of alerts: be ready to receive and quickly relay data to the sink.

3.1   Sentry Selection Phase

In this first phase, after having determined its cover-set and frame capture rate [7], every node broadcasts these information. Once all the nodes have finished broadcasting, each node can identify the node with the highest capture rate in its neighborhood. That node is termed as sentry node or master node in its neighborhood. Remind that the capture rate of any node is calculated using the Bezier curve model described previously with examples shown in Table 1. Figure 5 depicts the end of phase 1 where a sentry node (the black node with the highest frame capture rate) has been identified and associated to follower nodes in a given neighborhood.

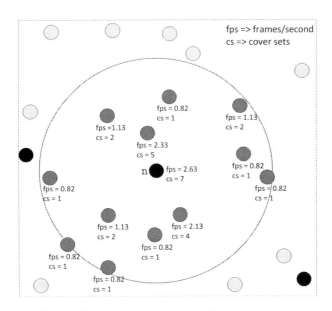

Fig. 5. Sentry node selection at the end of phase 1

Once the sentry node has been identified and its frame capture rate known, the second phase is to set the follower node's radio duty-cycling pattern. We propose that the listening period of the follower nodes be calculated in relation to the frame capture rate of their sentry node. However another important factor to consider is the follower node's redundancy level. Similar to ideas we developed in [7], when a node has several cover sets, if it runs out of energy it can be easily replaced by one of its cover sets. Therefore this follower node can afford to have a duty cycling pattern with longer listening time. This is the purpose of phase 2 described below in more details.

3.2   Determining Duty-Cycling Pattern

If a follower node has a small number of cover sets then it is preferable that it preserves its energy because it can hardly be replaced. This means that each follower node of a given sentry node may have different listening time depending on the size of its cover-sets. We propose that the duty-cycling pattern of a follower node follows the convex/concave model previously described in figure 3 in order to maintain the properties of criticality-based scheduling. However, the y axis will now give the corresponding duty-cycle value (between 0 and 1, corresponding to the listening period ratio) based on the cardinality of the cover sets of the follower node itself expressed on the x axis and the sentry node's frame capture rate. Actually, the sentry node's capture rate value is normalized against the maximum frame capture rate and is used as a new criticality level for the node, whose duty-cycle value is being calculated. In this duty-cycle model, we therefore now have $P_2(h_x, h_y) = (12, 1)$: maximum considered number of cover sets is 12 and duty-cycle ratio is between 0 and 1. The concave curves will represent the smallest capture rates (normalized) where most duty-cycle values will give have smaller listening periods, i.e. values are near to zero, unless if a node has high number of cover sets in which case it will have a larger listening period. Similarly, the convex curves represent the highest capture rates (normalized). In this case the duty-cycle values calculated for the follower nodes will be longer. Follower nodes with larger number of cover sets will have duty-cycle values longer than those with smaller number of cover sets.

Figure 6 illustrates the entire duty-cycling computation process at follower nodes. In this example, with an application criticality level of 0.8 (which gives a convex curve), a node having 9 cover sets will capture at a frame rate of 2.75 fps. Assuming that this node is selected as a sentry node, then its neighbors will use its capture rate to compute their own duty cycle value. Therefore we see in figure 6 how the capture rate is normalized (against the maximum capture rate defined by hardware constraints, 3fps in the example) and used as a new criticality level for computing the duty cycle value at a follower node, taking its number of cover sets into account. Here, this new criticality level gives a curve which is more convex, i.e. most values on the y axis will be in the upper half of the curve even with smaller number of cover sets, which means longer duty-cycle values for follower nodes.

As mentioned previously, our approach can also be a very effective method for the preamble length calculation in preamble-based MAC protocols, like B-MAC [3], X-MAC [4]. We will describe in the next section the performance evaluation of our method and comparisons with a static duty-cycle approach.

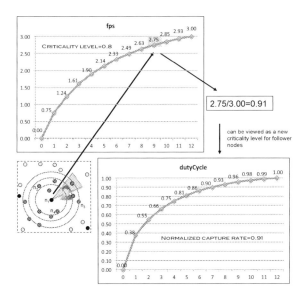

Fig. 6. Criticality curve example

4   Preliminary Simulation Results

To evaluate our approach we conducted a series of simulations using the OM-NET++/Castalia simulator. For these set of experiments, we randomly deployed 110 sensor nodes in a 400mx400m area.

Each sensor node captures with a given number of frames per second (between 0 fps and 3 fps). We set the maximum number of cover sets for a node to be 12: nodes with higher number of cover sets will only consider 12 cover sets. Minimum duty cycle is fixed at 0.1 and the criticality level is set at 0.8. Random intrusions are introduced in the simulation model and nodes can detect an intrusion if the intruder is covered by their FoV at the time of the image capture. Upon intrusion detection, a node will broadcast an alert message.

We compared our approach with a traditional static duty-cycled MAC protocol with varied duty cycle values: 0.5, 0.6, 0.7 and 0.8. For instance, a cycle duration of 1s with a duty-cycle value of 0.8 will give 0.8s of radio activity (e.g. can receive) followed by a 0.2s period of inactivity (e.g. can not receive). In figure 8, the duty cycle values of all nodes in our simulation scenario are shown after calculation from the criticality model in descending order. It should be kept in mind that the follower node's duty cycle values varies depending on its number of cover sets and capture rate of their sentry node. A sentry node does not need to keep its radio active for a period of time, hence its duty cycle is kept at minimum, i.e. 0.1 (shown in red in figure 8).

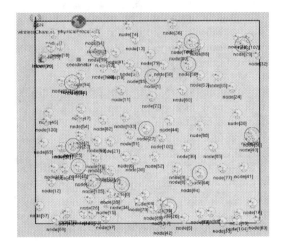

Fig. 7. Snapshot of the Omnet++ Simulator

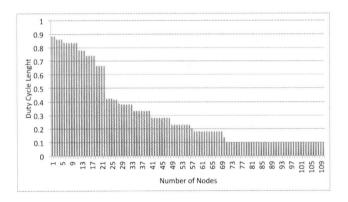

Fig. 8. Duty cycle lengths of all the nodes

For verification of our approach, we designed all the nodes to respond with an acknowledgment message on reception of an alert, confirming the reception of the alert. The responses received from the followers confirm that the alert was successfully propagated. On the other hand, no response from the followers goes on to show that the follower nodes were on the sleep mode and they did not received the alert message. Now this means that the alert was sent but none of the follower nodes was available to hear that communication, alert was not propagated and it was not relayed to the sink, which can have severe consequences for applications of critical nature.

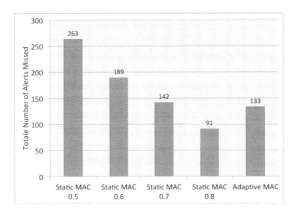

Fig. 9. Number of alerts missed

The total number of alerts sent by all the sentry nodes is approximately 1070. Figure 9 shows the total number of alerts sent by sentry nodes to which no responses were received, i.e. the number of alerts which were not propagated through the network.

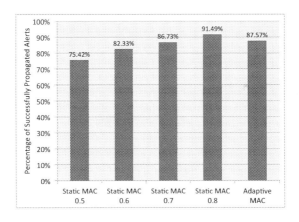

Fig. 10. Received and successfully propagated alert messages

Figure 10 shows the percentage of received and successfully propagated alert messages. We see in figure 9 and 10 that the results of the MAC protocol proposed in this paper, are second only to a static MAC with 80% duty cycle. Criticality adaptive MAC proposed in this paper shows better results in comparison to duty cycled static MAC with duty cycle of 0.7. As figure 8 showed that the number of nodes working at 0.7 duty-cycle or above is small, the results shown in figure 9 clearly illustrate the benefit of our criticality adaptive MAC approach: fewer nodes working on high duty cycle values but better

responsiveness. Non-sentry nodes also sent 353 alerts and in response received
358 acknowledgements.

Figure 11 shows the comparison of total energy consumption of all the nodes in
the network in Joules. In the figure we can see that our criticality adaptive MAC
protocol consumed 48% less energy in comparison to a static MAC with duty
cycle of 0.8, for a static with duty cycle of 0.7, the energy saved was around 44%,
and the corresponding values were 38% and 32% respectively for static MAC 0.6
and static MAC 0.5.

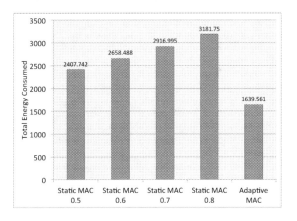

Fig. 11. Comparison of energy consumed

Taking the global energy consumption of the network, Figure 12 shows the
energy consumed per successfully propagated alert message for the various MAC
protocols.

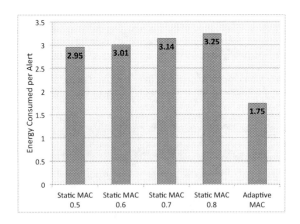

Fig. 12. Comparison of energy consumed per alert message

We see in figure 12 that the adaptive MAC approach gives significantly better results in comparison to Static MAC with different duty cycle durations. The energy is efficiently utilised to increase the network lifetime.

5 Conclusions

In this paper, we proposed a duty-cycled MAC protocol for low latency alert propagation and low energy consumption. In a network all the nodes work for the same purpose so if a node has high redundancy, it can be used more extensively for detection purposes. The key point in our approach is to link the duty cycle of nodes with image capture rate and number of cover sets.

Simulations have shown the efficacy of our approach. The results have shown that our approach was responsive to high number of alerts in comparison to various duty cycle lengths for a static duty-cycled MAC. At the same time the energy consumed for the whole network was minimum for our approach, which are very encouraging results. In future we want to extend our sentry node selection to two-hops and we want the duty cycle of nodes on the route to sink to be calculated with this approach, to receive images at the sink with minimum latency.

Acknowledgment. This work is partially supported by the Aquitaine-Aragon OMNIDATA project.

References

1. Ye, W., Heidemann, J., Estrin, D.: Medium access control with coordinated adaptive sleeping for wireless sensor networks. IEEE/ACM Trans. Netw. 12(3), 493–506 (2004)
2. van Dam, T., Langendoen, K.: An adaptive energy-efficient mac protocol for wireless sensor networks. In: Proceedings of the 1st International Conference on Embedded Networked Sensor Systems, SenSys 2003, pp. 171–180 (2003)
3. Polastre, J., Hill, J., Culler, D.: Versatile low power media access for wireless sensor networks. In: Proceedings of the 2nd International Conference on Embedded Networked Sensor Systems, SenSys 2004, pp. 95–107 (2004)
4. Buettner, M., Yee, G.V., Anderson, E., Han, R.: X-mac: a short preamble mac protocol for duty-cycled wireless sensor networks. In: Proceedings of the 4th International Conference on Embedded Networked Sensor Systems, SenSys 2006, pp. 307–320 (2006)
5. Grilo, A., Macedo, M., Nunes, M.: An energy-efficient low-latency multi-sink mac protocol for alarm-driven wireless sensor networks. In: García-Vidal, J., Cerdà-Alabern, L. (eds.) Euro-NGI 2007. LNCS, vol. 4396, pp. 87–101. Springer, Heidelberg (2007), http://dl.acm.org/citation.cfm?id=1961386.1961396
6. Bachir, A., Dohler, M., Watteyne, T., Leung, K.K.: Mac essentials for wireless sensor networks. IEEE Communications Surveys and Tutorials 12(2), 222–248 (2010)
7. Makhoul, A., Saadi, R., Pham, C.: Risk management in intrusion detection applications with wireless video sensor networks. In: IEEE WCNC (2010)

8. Stoianov, L.N.I., Madden, S.: Pipenet: A wireless sensor network for pipeline monitoring. In: ACM IPSN (2007)
9. Albano, S.C.M., Pietro, R.D.: A model with applications for data survivability in critical infrastructures. Journal of Information Assurance and Security 4 (2009)
10. Dousse, O., Tavoularis, C., Thiran, P.: Delay of intrusion detection in wireless sensor networks. In: ACM MobiHoc (2006)
11. Zhu, Y., Ni, L.M.: Probabilistic approach to provisioning guaranteed qos for distributed event detection. In: IEEE INFOCOM (2008)
12. Czarlinska, A., Kundur, D.: Wireless image sensor networks: event acquisition in attack-prone and uncertain environments. Multidimensional Syst. Signal Process. 20, 135–164 (2009)
13. Freitas, E., Heimfarth, T., Pereira, C., Ferreira, A., Wagner, F., Larsson, T.: Evaluation of coordination strategies for heterogeneous sensor networks aiming at surveillance applications. In: IEEE Sensors (2009)
14. Keally, M., Zhou, G., Xing, G.: Watchdog: Confident event detection in heterogeneous sensor networks. In: IEEE Real-Time and Embedded Technology and Applications Symposium (2010)
15. Alaei, M., Barcelo-Ordinas, J.M.J.M.: Priority-based node selection and scheduling for wireless multimedia sensor networks. In: IEEE WiMob (2010)
16. Paniga, S., Borsani, L., Redondi, A., Tagliasacchi, M., Cesana, M.: Experimental evaluation of a video streaming system for wireless multimedia sensor networks. In: Proceedings of the 10th IEEE/IFIP Med-Hoc-Net (2011)

How to Improve CSMA-Based MAC Protocol for Dense RFID Reader-to-Reader Networks?

Ibrahim Amadou, Abdoul Aziz Mbacké, and Nathalie Mitton

Inria Lille, Nord Europe, France
`firstname.lastname@inria.fr`

Abstract. Due to the dedicated short range communication feature of passive radio frequency identification (RFID) and the closest proximity operation of both tags and readers in a large-scale dynamic RFID system, when nearby readers simultaneously try to communicate with tags located within their interrogation range, serious interference problems may occur. Such interferences may cause signal collisions that lead to the reading throughput barrier and degrade the system performance. Although many efforts have been done to maximize the throughput by proposing protocols such as NFRA or more recently GDRA, which is compliant with the EPCglobal and ETSI EN 302 208 standards. However, the above protocols are based on unrealistic assumptions or require additional components with more control packet and perform worse in terms of collisions and latency, etc. In this paper, we explore the use of some well-known *Carrier Sense Multiple Access (CSMA)* backoff algorithms to improve the existing CSMA-based reader-to-reader anticollision protocol in dense RFID networks. Moreover, the proposals are compliant with the existing standards. We conduct extensive simulations and compare their performance with the well-known state-of-the-art protocols to show their performance under various criteria. We find that the proposals improvement are highly suitable for maximizing the throughput, efficiency and for minimizing both the collisions and coverage latency in dense RFID Systems.

1 Introduction

Most radio frequency identification applications, such as supply markets, localisation and objects tracking, activity monitoring and access control and security, etc., use passive RFID tags, which communicate with the RFID reader by modulating its reflection coefficient (backward link) to incoming modulated RF signal from the reader (forward link). However, unlike the traditional radio communication systems, in such systems, the RF signal does not provide reciprocity between forward and backward links because the reflected RF signal from a tag is inversely proportional to the fourth power of the distance between reader and tag [18]. For example, in the European Regulation [20], the reader output power of 2 Watt Effective Radiated Power (ERP) limits the *reader-to-tags* read range to a maximum distance of 10 meters while *reader-to-reader* interference range may

S. Guo et al. (Eds.): ADHOC-NOW 2014, LNCS 8487, pp. 183–196, 2014.
© Springer International Publishing Switzerland 2014

reach 1000 meters [19]. This link unbalance requires in above large-scale applications of RFID systems to deploy a large number of RFID readers allowing the coverage of the interested environment. A direct consequence of this feature deployment is the operation within the closest proximity of several tens or hundreds of readers in order to overcome the shortcoming of the backward communication distance. However, due to readers close proximity, when nearby readers simultaneously try to communicate with tags located within their *interrogation range*, serious interference problems may occur. This is mainly due to the overlapping of readers' field. Such interferences may cause signal collisions that lead to the reading throughput barrier and degrade the system performance. Although collision problems can also appear during the tag communication [12], [4] most of these problems are considered to be solved and are part of patents developed by EPCglobal standard [4]. While the *reader-to-reader* collision (RRC) problem obtained few attention because previous applications of RFID systems considered only a reader with several tags, the design of an efficient reader-to-reader anti-collision protocol has emerged as the most interesting research issues in recent years. The key to make the RFID system efficient is to schedule readers activities so that neither interferences nor collisions may occur.

The state-of-the-art protocols can be broadly classified as CSMA-based [2,11, 20] and activity scheduling based [1,3,6,7,15] through time division, frequency or by putting together both approaches. The former approach is considered as an efficient and more adaptive approach in large-scale RFID reader networks because it is *full-distributed* algorithm and it does need neither synchronization nor additional resource (e.g. server) like in the latter approach. However, the existing protocols still suffer from traditional *backoff scheme* in dense RFID networks as it is recently observed in NFRA [3] and GDRA [1]. Therefore, the backoff algorithm for CSMA-based may need to be investigated in depth in order to maximize the network throughput but also to minimize the collisions which still exist as they observed in [1].

In this work, we investigate the use of an adaptation of some well-known adaptive backoff algorithms proposed in mobile ad hoc networks (MANETs) in order to improve the performance of CSMA-based anti-collision protocol for dense RFID networks. Our objective is to use protocols such as *Idle Sense* [9], *Sift* [14] and *Reverse Backoff* [13] to perform channel access through broadcasting message with almost zero-collision or in collision-free fashion. Once the channel is reserved by a reader, it can communicate with tags located in its interrogation area in collision-free fashion. Furthermore, our aim is, based on observed characteristics of above protocols, to propose a novel medium access control protocol which can significantly improve the performance metrics such as throughput, fairness behavior during channel access and with the minimum latency in large-scale RFID system. Unlike the majority of previous works, in this work, we extend our performance evaluation in terms of qualitative and quantitative criteria.

The rest of this paper is organized as follows. Section 2 introduces the system model. Section 3 reviews the state-of-the-art anti-collision protocols for RFID networks. In Section 4, we provide a general overview of the basic adaptive backoff algorithms of MANETs investigated in this work in order to improve the performance of CSMA-based reader-to-reader anti-collision protocols. In Section 5, we investigate their performance via extensive simulation under various RFID environment scenarios and according to several criteria. Finally, Section 6 concludes by discussing future research direction.

2 System Model

We consider a large-scale RFID system with multiple readers and homogeneous local density of RFID tags within the interrogation area. The readers are randomly deployed with uniform distribution on 2D plane. Readers are assumed to have homogeneous properties. Therefore, their communication range is assumed to be the same. Similar to ETSI EN 302 208-1 regulation [20], in this paper, we assume the use of multichannel network scheme. We assume the existing of an overlapping area in their interrogation areas. Note that the knowledge of these overlapping areas is not a necessity in this work. We assume that readers are able to accurately estimate their neighborhood reader size in the case of adaptive backoff schemes that are based on the number of active readers. Note also that this assumption can be easily removed because, as in [1], we also make use of two bistatic antennas.

But in the future, our investigation will focus on designing a novel full distributed medium access control protocol which can not involve any additional requirement such as bistatic antennas. Therefore, in the next step of this study, according to our target objective in terms of performance metrics and the selected approach, we will either investigate the design of an efficient neighbor estimation mechanism or the adaptive approach with similar performance. Furthermore, it should be able to deal with mobility of both readers and tags in RFID environment.

3 Related Work

In this section, we present the main characteristics of the state-of-the-art proposed reader-to-reader anti-collision protocols in RFID environment. According to the taxonomy previously introduced, we first overview the CSMA-based approaches. Thereafter, we present the activity scheduling approaches proposed in the literature.

3.1 CSMA-Based Approach

LISTEN BEFORE TALK (LBT) is proposed by the European Telecommunication Standard Institute under ETSI EN 302 208-1 [20] for *ultra high frequency (UHF)* dense RFID environment. With listen before talk concept, each reader must listen for a specified minimum time equal to $5ms$ before confirming that the selected channel is not busy. At the end of this fixed period of listening, if the selected channel is unoccupied, reader begins the tags interrogation process, which is called reader-to-tag communication phase, over this channel. Otherwise, it must randomly select a new channel and repeat the same process. However, as the number of readers increases in the system, LBT performs poorly in terms of throughput [1].

PULSE [2] is CSMA-based protocol which is proposed to resolve *hidden terminal* issue. PULSE proposes the use of two non-interfering channels: *control channel* and *data channel*. It consists of periodically sending a beacon on control channel while the reader is communicating with the tags on the data channel. In such a way, the beaconing transmission prevents any another reader to access the channel. This leads to an increase of latency because the reader may need to wait thrice the beaconing transmission before any other channel access attempt. Moreover, its effectiveness in terms of collision avoidance comes at cost of high overhead of periodic beacon, which drastically impacts the energy-efficiency behavior when energy is a constraint. Furthermore, the channel access backoff parameter is set without any investigation. This does not mitigate collisions due to the simultaneous channel access which leads to the omission of some tags in the overlapping zone of collide readers. Even if it is presented as a solution to resolve the hidden terminal issue, PULSE does not completely resolve this problem.

Similar to PULSE protocol, DICA [11] is a distributed CSMA-based protocol with two channels, which aims to provide an energy-efficient collision avoidance protocol. During the channel access process, the reader may not need neither to periodically send beacon message nor to wait additional delay. To completely resolve hidden terminal and exposed terminal problems, DICA adjusts the control channel communication range by doubling the radius from the data channel communication range. As in PULSE protocol, the backoff algorithm is designed without any investigation inducing misreading tags during the reader-to-tags communication.

3.2 Activity Scheduling Approach

DISTRIBUTED COLOR SELECTION (DCS) [15] is one of the first distributed TDMA-based reader-to-reader anti-collision protocols. It is also known as *Colorwave* in the literature. In DCS, communication is organized in groups of color timeslots, that are called frames or rounds. Each timeslot is composed of two phases: *kick* and *transmission*. The kick phase is used by readers that collided in the previous round in order to prevent a new collision in the current frame, while the latter phase is used by readers to read tags located in their interrogation range. Basically, it performs as follows : each reader randomly selects a color timeslot where it can send the request to read tags during its transmission phase. If a collision is observed by colliding readers, they randomly select a new color and reserve them for the next frame. In the next frame, during the kick phase, colliding readers broadcast their color to all their neighborhood in order to prevent a new collision. Based upon information received from collide readers, neighborhood readers adjust their color so that a new collision could be avoided. Unfortunately DCS experiences a performance degradation due to the use of a fixed number of colors. Because in a dense network, when the number of colors is too low, DCS experiences many collisions, while, in a low dense network, when the number of used colors is too large, the throughput is degraded because of the a large number of colors. To overcome this inflexibility, authors proposed *Variable-Maximum DCS* which dynamically adjusts the maximum number of colors regarding the percentage of successful transmissions monitored by the reader in order to improve the performance. A more efficient but centralized version has been proposed in [8].

PROBABILISTIC DSC [7] is an improvement version of DCS which introduces a new parameter p, representing the probability to change color after a collision. It is proposed to reduce the number of consecutive collisions due to too many changes in neighborhood of colliding readers. This approach improves DCS. Moreover, PDCS is a multichannel protocol. Distributed Color Noncooperative selection (DCNS) [6] is also TDMA-based protocol, which aims to improve the system throughput even in high density networks.

NEIGHBOR FRIENDLY READER ANTICOLLISION (NFRA) [3] is TDMA-based protocol, which uses a polling server to synchronize and coordinate readers activity. It improves the efficiency of RFID readers networks in term of throughput by allowing readers to operate in collision-free manner during tags reading process. Similar to *Colorwave* [15], in NFRA, a reader-to-tag communication is scheduled in round by polling server. The round is formed by a fixed number of contention slots, where a reader can contend for reading tags around its communication range, plus a period of time necessary to read tags (e.g. 0.46s). NFRA performs as follows: at the beginning of each round, the server first sends an *Arrangement Command (AC)*, which holds the number of slots (e.g. MN) in the round and advertises also the beginning of round. Upon reception of *AC*, each reader chooses k, an integer distributed uniformly in [1, MN] and waits for receiving the *OC* from the server, which advertises the slot k, before attempting to gain the channel throughout the transmission of beaconing message. If the beaconing is sent without collision, it continues to send an *Overriding Frame (OF)* message to suppress its active interfering neighbors that have a high value of k. The aims of this suppression mechanism is to ensure that the reader-to-tag communication will operate in *collision-free* fashion. However, if a collision is observed during the beaconing transmission, the collide readers cancel the transmission scheduling and wait for the next round to attempt. Unfortunately, NFRA implicitly assumes that when a collision occurs, collide readers can detect the collision. While in practice, a collision of broadcasting message can not be detected. NFRA++ [5] aims to improve the fairness of NFRA which performs poorly when the reader neighborhood is too large.

In order to remove unpaired assumption introduced in NFRA, GEOMETRIC DISTRIBUTION READER ANTICOLLISION (GDRA) [19] is proposed with the aims to be compliant with the EPCglobal standard and ETSI EN 302 208-1 regulation and to minimize the reader collision problem by using Sift [14] geometric probability distribution function to choose the slot value k. To be compliant with ETSI EN 302 208-1 regulation, GDRA is based on multichannel and also proposes to use a $5ms$ as slot size which is the minimum time required to listen the channel before transmission attempting. Instead of assuming the collision detection during the beaconing step, GDRA assumes the use of two bistatic antennas by reader. This avoids the OF transmission. Moreover, it also suppresses the use of OC by assuming that readers are able to synchronize themselves in a round, which requires additional resources and do not fit dynamic environments.

Unlike CSMA-based protocols, these protocols necessitate a high level of synchronization among the backbone network formed by readers or an additional resource such as centralized server which can address the synchronization problem through wired or wireless connection. In addition to the high level of synchronization requirement, in TDMA-based protocol, the use of dedicated server for explicit coordination restricts the use of these solutions into a limited environment. Moreover, if energy is part of performance requirements, the performance can gradually decrease.

4   Overview of Investigated Adaptive Protocols

In this section, we detail the adaptive backoff algorithms taken from wireless MANETs in order to improve existing CSMA-based reader-to-reader anticollision protocols, which still suffer from the *backoff scheme*. Because the vast majority of observed collisions in RFID system is due to the use of inappropriate *backoff scheme*. Moreover, how to set the backoff algorithm during channel access in order to make them efficient, however, is far less investigated. [5] has recently proved that by increasing the backoff value of these protocols, they improved the fairness behavior.

So far, many research efforts have been made to improve the performances or the reliability of IEEE 802.11 DCF technique for MANETs by using an adaptive backoff scheme. Here, we look at these well-known solutions that proved their efficiency in high congestion wireless networks. Although there are many full distributed MAC protocols that have been proposed to improve the efficiency of CSMA-based techniques in MANETs, in the following, we mainly focus on particular interest protocols that involve less additional requirements with respect to the RFID reader constraints.

4.1   Maximum Backoff or Reverse-Backoff

Generally speaking, the probability of a collision decreases when the contention window size increases. Based on this, [13] introduces *reverse backoff* (RB). RB starts with a backoff with maximum value decreased every time a beacon expires until some maximum backoff counter, which does not induce the beaconing expiration. It copes with collision problem during the channel access because the backoff counter is large. However, the achieved value depends on the beacon lifetime. As our goal is to minimize the channel access collisions and that message expiration is not part of this work, in the following, we set the backoff counter to be equal to the maximum value of IEEE 802.11 protocol ($CW = 1023$).

4.2   Idle Sense

Idle Sense [10] aims at maximizing the throughput and providing a short-time fairness for IEEE 802.11 DCF. Instead of performing the *binary exponential backoff* algorithm (BEB) of IEEE 802.11 after a collision or failed transmission due to channel behavior, Idle Sense dynamically adjusts the contention window size so that the contention window size of all nodes converges in a full distributed way to similar values. It performs as follows: each host estimates n_i, the number of consecutive idle slots between two transmission attempts. According to the observed values of n_i, every *maxtrans* transmissions, it computes $\hat{n}_i$, which represents the average. Then, it uses the average value to adjust its contention window size to target value n^{target} computed numerically based on the IEEE 802.11 PHY and MAC layers parameters such as the slot size and average collision size. Idle Sense is based on *Additive Increase Multiplicative Decrease* (AIMD) mechanism to adjust the window. The heuristic is performed as follows :

- if $\hat{n}_i \geq n^{target}$ $CW \longleftarrow \alpha CW$
- if $\hat{n}_i < n^{target}$ $CW \longleftarrow CW + \epsilon$

where α and ϵ are parameters of the AIMD algorithm. While in the first version of Idle Sense [10], *maxtrans* was chosen by simulation and fixed, in [9] they proposed adaptive algorithm to update the value of *maxtrans* in order to speed up the convergence of the average $\hat{n}_i$ to the target value. Idle Sense is designed for unicast data transmission, because nodes may need to detect a collision in order to adjust their contention window, while in our work, the communication is performed through broadcast transmission. As we assume that each reader can track its right neighborhood size and according to the value of slot size selected in this work and formulas (4)-(5) and (10)-(12) of [10], we provide in Table 1 for each neighborhood size its optimal contention window for the target value of 2.49.

Table 1. Values of CW for $n_i^{target} = 2.49$

Neighbor Size	CW	Neighbor Size	CW	Neighbor Size	CW
2	10.86	12	70.16	22	129.46
4	22.72	14	82.02	24	141.32
6	34.58	16	93.88	26	153.18
8	46.44	18	105.74	28	165.04
10	58.30	20	117.60	30	176.90

4.3 Sift

Sift [14] is motivated by the limitation of classical CSMA-based approach in wireless sensor networks (WSNs) when several nodes with spatially-correlated data simultaneously attempt to report data to the sink node. It aims at reducing the latency and the collisions with a competitive throughput by using a suppression mechanism at MAC Layer. Instead all nodes report data, only few nodes called *the first R nodes* report their data in a collision-free fashion, while the remaining other nodes inhibit their report. Instead of using BEB algorithm to access the channel, it uses a fixed-size contention window with a non-uniform geometrically-increasing probability distribution for picking a transmission slot in the window. This avoids the shortcoming due to the adaptive increasing of window size when collision occurs. This distribution function is given as:

$$p_r = \frac{(1 - \beta)\beta^{CW}}{1 - \beta^{CW}} \cdot \beta^{-r}, \text{ for } r = 1 \dots CW \tag{1}$$

where $0 < \beta < 1$ is a distribution parameter. This distribution function p_r increases exponentially with r. As its objective is to reduce the number of data reports by allowing only few nodes to report their data, when the transmission or collision occurred, all contending nodes select new random contention slots and repeat the same process. Although Sift is proposed with the aim of reducing the number of collisions through the suppression mechanism, collisions do still exist. But, they are drastically reduced. Our objective is to use Sift in order to allow collision-free channel access for all contending nodes.

As our goal is to improve the existing CSMA-based and in order to be compliant with ETSI EN 302 208-1 [20], we adapt them to follow its requirements, which means, that they are used in a multichannel system with the unit of backoff equal to $0.5ms$. Note also that these approaches can be applied to any other CSMA-based reader-to-reader anti-collision in order to improve its performance.

5   Performance Evaluation

In order to highlight the benefit brought by these mechanisms, we implemented *Idle Sense, Maximum Backoff, Sift, NFRA* and *GDRA* approaches using WSNet [16], an event-driven simulator for large scale WSNs and fairly evaluate their performance under various network scenarios. We consider a dense RFID system where 100 to 500 readers are randomly deployed with uniform distribution on a network of dimension 4000m × 4000m. According to the reader maximum transmission power defined in [20] as 3.2 Watts EIRP, we set the reader-to-tag communication range to 10m while their interference range is 1000m. For each scenario 100 simulations are run. Each simulation lasts 500s. The results are presented with 95% confidence intervals. Table 2 sums up all parameters.

5.1   Performance Metric

Before presenting our simulation results, in this section, we briefly introduce our performance evaluation metrics:

Table 2. Simulation Parameters

Protocol Parameters	
Mechanism	Parameters
NFRA	AC Packet = 2.83ms, OC Packet = 1ms
	OF Packet = 0.3ms, Beacon = 0.3ms
	Slot size = 1.3ms
GDRA	AC Packet = 2.83ms, Beacon = 0.3ms
	Contention Window Size = 32, Slot Size = 5ms
Idle Sense	Slot Size = 0.5ms, Contention Window = N/A
Max. Backoff	Slot Size = 0.5ms, Contention Window = 1024
SIFT	Slot Size = 0.5ms, Contention Window = 32
Common Parameters	
Parameter	Value
Reader-to-tag communication length	0.46 s
Number of tags	800

- *Collision*: We can distinguish two types of collisions: *channel access collision* and *reading collision*. When the former occurs, we consider that the tags reading process is unsuccessful while the latter problem will just impact the throughput of tags reading. According to the transmission power used in this work, the latter will generally happen only if there is carrier-sense errors. Note that due to some of our assumptions in this work, reading collisions can not happen.

- *Throughput*: We define the throughput as the ratio of the average number of successful query sections per reader over the simulation duration. The higher system throughput, the more efficient protocol.

- *Efficiency*: It defines the ratio of the number of successful query sections over the total number of attempted query sections (i.e. successful and lost query sections).

- *Jain's Fairness Index*: In the literature, the fairness property is generally evaluated with the Jain's fairness index, which describes how similar and fair the resources allocated to each reader are. It is computed as follows:

$$J = \frac{\mid \sum_{i=1}^{N} X_i \mid^2}{N \sum_{i=1}^{N} X_i^2} \qquad (2)$$

where N and X_i are respectively the number of readers in the network and the throughput of the i-th reader.

- *Latency*: It describes the average time to read all tags.

- *Coverage*: It defines the average number of tags read by readers during 100s of simulation.

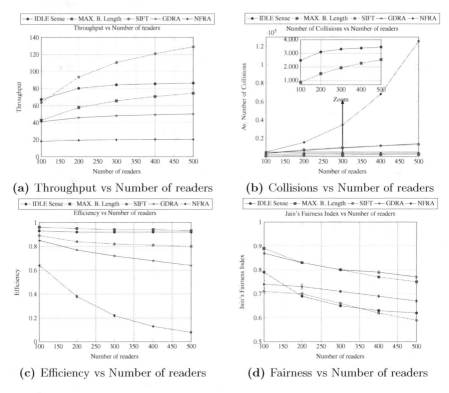

(a) Throughput vs Number of readers   **(b)** Collisions vs Number of readers

(c) Efficiency vs Number of readers   **(d)** Fairness vs Number of readers

Fig. 1. Performance of GDRA, Idle Sense, Maximum Backoff, NFRA and Sift schemes. (1a) Throughput, (1b) Number of collisions, (1c) Protocol efficiency and (1d) Jain's Fairness Index vs The number of readers.

5.2   Results

Fig. (1a) illustrates the throughput according to the number of readers in the system. The improvement of the proposed backoff schemes over the activity scheduling approaches is obvious. It is mainly explained by the polling process happening prior to any transmision in these protocols. However GDRA still display a higher throughput thanks to the use of the geometric distribution upon the choice of an adequate timeslot and the multichannel setting. The gap shown between the proposed backoff schemes is explained by their contention delay, the shorter it is, better the throughput is, which explains Sift the highest performances. Fig. (1b) shows the number of collisions based on the number of readers. The high number of collisions recorded by NFRA is due to the uniform distribution used for the selection of a timeslot. The use of the geometric distribution in GDRA and Sift allows them to show similar performances, not totally eliminating collisions but reducing them. Thanks to the use of a large contention window in Maximum Backoff, the number of collisions is highly reduced offering here the best performances. Fig. (1c) displays the efficiency according to the number of readers, it is the number of successful transmissions on the total number of transmissions attempted. The high number of collisions recorded with NFRA explains its poor performances here, while the low number of collisions observed with Maximum Backoff or Idle Sense, though they are not offering the best throughput, allow them to be the most efficient protocols. The mid-range efficiency obtained by Sift and GDRA is explained by the way they both deal with the geometric distribution to reduce the collisions. Fig. (1d) exhibits the Jain's fairness Index based on the number of readers in the system. This index allows to show how fair is the access to the transmission medium among readers. GDRA and Maximum Backoff display almost similar performances but for different reasons. The large contention window used in Maximum Backoff allows each reader to access to the channel with a fair number of collisions. Though Sift uses the same geometric distribution as GDRA does, the difference observed here is explained by the use of the polling server in GDRA which ensures in a way that readers get a fair access to the channels. The high number of collisions noticed in NFRA is also the reason why such a high drop of performance is observed as the number of readers in the system goes up.

5.3 Coverage and Latency

Fig. (2a) illustrates the average number of covered tags according to the number of readers in the system. The high performances observed with GDRA and Maximum Backoff is due to their high fainess behavior, since access to the channels is made in a fair way then the number of covered tags is maximized. Other protocols display growing performances except NFRA, which presents a poor performance. This is mainly due to the use of the uniform distribution in order to select the transmission slot, which records a high number of collisions in addition to the suppression mechanism introduced by the OF message. The average

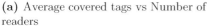

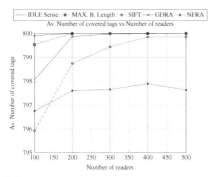
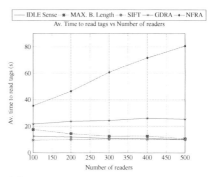

(a) Average covered tags vs Number of readers

(b) Average latency vs Number of readers

Fig. 2. Performance of GDRA, Idle Sense Maximum Backoff, NFRA and Sift when 800 tags are deployed. (2a) Number of covered tags and (2b) Time necessary to cover tags vs The number of readers.

being obtained with 300 readers in the system. Fig. (2b) shows the average time to read all tags in the coverage area based on the number of readers. This graph results can be explained by looking at the Fig. (1a) where Sift and Idle Sense got the best results in terms of throughput thus resulting in them obtaining the shortest time needed to read all tags. NFRA due to the collisions registered and the low throughput takes a longer time to access all tags informations. Noticing also the fact that as the number of readers increases the delay needed for Maximum Backoff decreases accordingly which is due to the state of convergence being reached faster with the high number of contending readers.

6   Discussion and Conclusion

Before the introduction of [19], NFRA, though it is based on unrealistic assumptions, was considered as one of the best protocols to overcome the issue of reader-to-reader collisions problem in RFID systems. The study presented in this paper shows that it is possible to improve the performance of existing CSMA-based approaches by introducing a novel backoff scheme. Indeed, we can witness the significant improvement in terms of throughput, efficiency and fairness while being able to reduce the number of collisions and shorten the delay needed for reading the tags. Though it seemed like the best performing protocol, Sift is not a viable solution in real deployments, indeed it has been presented in a interference free environment, its performances considerably dropped when an ambient noise plus interference were introduced in the simulation settings as we have shown in our previous study. The use of Idle Sense implies the use of bistatic antennas on readers, same as in GDRA, which naturally induces a hardware modification of readers. While the approach can be appropriate in case of hardware dependent study, in our future study, our aim is to go toward a novel reader-to-reader anti-collision protocol design, which can be able to perform efficiently regardless of the target reader characteristics with no additional complexity or hardware. The second approach consists of combining Idle Sense scheme with an efficient *neighbor estimation algorithm*. However, when the neighbor estimation error is higher than 5%, as we have already observed in [21], Idle Sense performs poorly. While Maximum Backoff offers the best performance regarding the number of collisions and efficiency, it is also the one displaying one of the lowest throughput due to the delay needed for readers to reach a convergence state with the right contention window size. In light of the study made in this paper, we can state that the exploration of a distributed approach, making a right estimation on the size of the contention window, would offer the best solution for reducing the number of collisions while maximizing the throughput and efficiency of the readers. Based on the results shown in this paper, we assume that an approach combining both the maximized contention window size of Renverse Backoff and the research made by Idle Sense to reach the theoretical optimal contention window size should offer the best tradeoff. In many large-scale RFID applications, the use of mobility devices is a strong requirement, because the unbalanced link behavior can induce uncover tags or tags are attached to mobile things, that should be taken into account in the design of an efficient protocol. To deal with this challenge issue, our future research should be also oriented to mobility aspect of both readers and tags.

Acknowledgments. This work is partially supported by a grant from CPER Nord-Pas-de-Calais/FEDER Campus Intelligence Ambiante.

References

1. Bueno-Delgado, M.V., Ferrero, R., Gandino, F., Pavon-Marino, P., Rebaudengo, M.: A Geometric Distribution Reader Anti-Collision Protocol for RFID Dense Reader Environments. IEEE Transactions on Automation Science and Engineering 10(2), 296–306 (2013)
2. Birari, S.M., Iyer, S.: PULSE: a MAC protocol for RFID networks. In: Enokido, T., Yan, L., Xiao, B., Kim, D.Y., Dai, Y.-S., Yang, L.T. (eds.) EUC Workshops 2005. LNCS, vol. 3823, pp. 1036–1046. Springer, Heidelberg (2005)
3. Eom, J.-B., Yim, S.-B., Lee, T.-J.: An Efficient Reader Anticollision Algorithm in Dense RFID Networks With Mobile RFID Readers. Trans Ind. Electron 56(7), 2326–2336 (2009)
4. EPCglobal Standard specification. EPC TM radio-frequency identity protocols class-1 generation-2 UHF RFID protocol for communications at 860 Mhz - 960 Mhz version 1.2.0 (2007)
5. Ferrero, R., Gandino, F., Montrucchio, B., Rebaudengo, M.: A Fair and High Throughput Reader-to-Reader Anticollision Protocol in Dense RFID Networks. IEEE Trans. Industrial Informatics 8(3), 697–706 (2012)
6. Gandino, F., Ferrero, R., Montrucchio, B., Rebaudengo, M.: DCNS: An Adaptable High Throughput RFID Reader-to-Reader Anti-collision Protocol. IEEE Trans. on Parallel and Distributed Systems 24(5), 893–905 (2013)
7. Gandino, F., Ferrero, R., Montrucchio, B., Rebaudengo, M.: Probabilistic DCS: An RFID reader-to-reader anti-collision protocol. J. Network and Computer Applications 34(3), 821–832 (2011)
8. Hamouda, E., Mitton, N., Simplot-Ryl, D.: Reader Anti-Collision in Dense RFID Networks With Mobile Tags. In: IEEE Intern. RFID-TA (2011)
9. Grunenberger, Y., Heusse, M., Rousseau, F., Duda, A.: Experience with an implementation of the *Idle Sense* wireless access method. In: Proc. ACM CoNEXT, pp. 24:1–24:12 (2007)
10. Heusse, M., Rousseau, F., Guillier, R., Duda, A.: Idle sense: an optimal access method for high throughput and fairness in rate diverse wireless LANs. In: Proc. Conf. on Applications, Technologies, Architectures, and Protocols for Computer Communications, pp. 121–132 (2005)
11. Hwang, K.-i., Kim, K.-t., Eom, D.-S.: DiCa: Distributed Tag Access with Collision-Avoidance Among Mobile RFID Readers. In: Zhou, X., Sokolsky, O., Yan, L., Jung, E.-S., Shao, Z., Mu, Y., Lee, D.C., Kim, D.Y., Jeong, Y.-S., Xu, C.-Z. (eds.) EUC Workshops 2006. LNCS, vol. 4097, pp. 413–422. Springer, Heidelberg (2006)
12. Simplot-Ryl, D., Stojmenovic, I., Micic, A., Nayak, A.: A hybrid randomized protocol for RFID tag identification. Sensor Review 26(2), 147–154 (2006)
13. Stanica, R., Chaput, E., Beylot, A.-L.: Enhancements of IEEE 802.11p Protocol for Access Control on a VANET Control Channel. In: Proc. of ICC (2011)

14. Jamieson, K., Balakrishnan, H., Tay, Y.C.: Sift: a MAC protocol for event-driven wireless sensor networks. In: Römer, K., Karl, H., Mattern, F. (eds.) EWSN 2006. LNCS, vol. 3868, pp. 260–275. Springer, Heidelberg (2006)
15. Waldrop, J., Engles, D.W., Sarma, S.E.: Colorwave: An anticollision algorithm for the reader collision problems. In: Proc. of ICC (2003)
16. WSNet. Wsnet, http://wsnet.gforge.inria.fr/
17. Lazaro, A., Girbau, D., Villarino, R.: Effects of interferences in UHF RFID systems. Progress in Electromagnetics Research 98, 425–443 (2009)
18. Kim, D.-Y., Yoon, H.-G., Jang, B.-J., Yook, J.-G.: Interference analysis of UHF RFID systems. Progress in Electromagnetics Research B 4, 115–126 (2008)
19. Leong, K.S., Leng Ng, M., Cole, P.H.: The reader collision problem in RFID systems. In: IEEE International Symposium on Microwave, Antenna, Propagation and EMC Technologies for Wireless Communications (MAPE 2005), pp. 658–661 (2005)
20. ETSI EN 302 208-1 Version 1.4.1 (2011), http://www.etsi.org/
21. Amadou, I., Mitton, N.: Revisiting Backoff algorithms in CSMA/CA based MAC for channel Reservation in RFID reader Networks through broadcasting. In: IEEE 9th International Conference on Wireless and Mobile Computing, Networking and Communications (WiMob), pp. 452–457 (2013)

Revisiting the Performance of the Modular Clock Algorithm for Distributed Blind Rendezvous in Cognitive Radio Networks

Michel Barbeau[1], Gimer Cervera[2], Joaquin Garcia-Alfaro[3],
and Evangelos Kranakis[1]

[1] School of Computer Science, Carleton University, K1S 5B6, Ottawa,
Ontario, Canada
{barbeau,kranakis}@scs.carleton.ca

[2] Universidad Tecnológica Metropolitana, 97279, Merida, Yuc., Mexico
gimer.cervera@utmetropolitana.edu.mx

[3] Telecom SudParis, 91000, Evry, France
joaquin.garcia-alfaro@acm.org

Abstract. We reexamine the *modular clock algorithm* for distributed blind rendezvous in cognitive radio networks. It proceeds in rounds. Each round consists of scanning twice a block of generated channels. The modular clock algorithm inspired the creation of the *jump-stay rendezvous* algorithm. It augments the modular clock with a *stay-on-one-channel* pattern. This enhancement guarantees rendezvous in one round. We make the observation that as the number of channels increases, the significance of the stay-on-one-channel pattern decreases. We revisit the performance analysis of the two-user symmetric case of the modular clock algorithm. We compare its performance with a random and the jump-stay rendezvous algorithms. Let m be the number of channels. Let p be the smallest prime number greater than m. The expected time-to-rendezvous of the random and jump-stay algorithms are m and p, respectively. Theis et al.'s analysis of the modular clock algorithm concludes a maximum expected time-to-rendezvous slightly larger than $2p$ time slots. Our analysis shows that the expected time-to-rendezvous of the modular clock algorithm is no more than $3p/4$ time slots.

1 Introduction

The cognitive radio network approach aims at a more intense use of the radio spectrum. Indeed, segments of radio spectrum allocated to communications services are often underused. Dynamic spectrum access has been proposed to address this issue. For instance two classes of users, primary and secondary, may be defined and have simultaneous access to a shared segment of radio spectrum. Priority is granted to the primary users. They may access and use their allocated radio spectrum segment anytime. Secondary users may be active and use the residual air time left when primary users are not active.

We assume that the radio spectrum segment is channelized. Secondary users can communicate over idle channels of the radio spectrum segment as long as

S. Guo et al. (Eds.): ADHOC-NOW 2014, LNCS 8487, pp. 197–208, 2014.

they do not create interference to the primary users. Primary users may jump in anytime. Secondary users know what the channels are, but they do not know which ones among them are available. For a group of secondary users, dynamically finding idle channels and making rendezvous on a common channel, available to all, are challenging issues.

Assuming a number of possible channels, how can a group of secondary users make rendezvous on a communication channel? The problem can be addressed using a central controller, a distributed approach with a dedicated common control channel or a distributed blind rendezvous approach. We focus on the latter. Secondary users hop over a set of channels attempting to make rendezvous. Secondary users may have a common channel set (the symmetric model) or different, but non disjoint, channel sets (the asymmetric model).

Time is divided into equal length intervals called time slots. There are two conditions for a successful rendezvous: being on the same channel during a time slot and a successful protocol handshake. These two conditions can be modeled individually and independently. The probability of a successful rendezvous is the product of the probability of being on the same channel during a time slot and a successful protocol handshake. In this paper, the focus is on achieving the condition *being on the same channel during a time slot*.

We are interested in minimizing the time required by two secondary users to make rendezvous. To achieve the condition *being on the same channel during a time slot*, we consider the *modular clock rendezvous algorithm* [17]. It inspired the authors of the jump-stay rendezvous algorithm [10,14,12], augmenting modular clock with a *stay-on-one-channel* pattern. This addition guarantees rendezvous in one round, in the symmetric case. We make the following observation. In these algorithms, channel hopping is done according to a randomized step increment. As the number of channels increases, the probability that two different users generate different step increments grows, a requirement to make rendezvous happen during hopping. The significance of the *stay-on-one-channel* pattern in the jump-stay rendezvous algorithm drops.

Let m denote the number of channels (a positive integer). Let p be the smallest prime number greater than m. The modular clock rendezvous algorithm proceeds in rounds consisting of two hopping phases of p time slots each. It generates blocks of p channels in accordance with the jump-stay rendezvous algorithm (*stay-on-one-channel* pattern omitted). After each round, a new block of p channels is generated. We revisit the performance analysis of the modular clock algorithm. The expected time-to-rendezvous (TTR) of the random and jump-stay algorithms are m and p time slots, respectively. Theis et al.'s analysis of the modular clock algorithm concludes a maximum expected TTR slightly larger than $2p$ time slots [17]. Our analysis shows that the expected TTR of the modular clock algorithm is no more than $3p/4$ time slots.

In Section 2, we review related work. The *modular clock rendezvous* algorithm used for our analysis is described in Section 3. The estimation of the expected TTR is done in Section 4. Simulation results are presented in Section 5. We conclude with Section 6.

2   Related Work

The performance of the channel hopping algorithms is evaluated using the TTR metric. In the two users case, from the moment both users are running, it is the number of time slots required to achieve rendezvous. An algorithm with a finite maximum TTR is said to be *guaranteed rendezvous*.

Related works include the random channel and orthogonal-sequence-based algorithms of Theis et al. [17,7]. The random channel algorithm visits all channels in a random order. For each time slot, a channel is selected among the m channels with uniform probability. The user is tuned to that channel for the whole time slot. Under the symmetric model, the expected TTR is m time slots. Under the asymmetric model, the expected TTR is m^2/g time slots. In both models, rendezvous is not guaranteed.

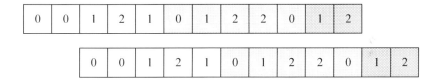

Fig. 1. Orthogonal-sequence-based channel hopping

With the orthogonal-sequence-based algorithm, channels are visited according to the same pattern by all nodes. By construction, two hopping users are eventually on the same channel. Rendezvous is guaranteed. Let $s_0, s_1, \ldots, s_{m-1}$ be a permutation of the m channels, the hopping pattern is

$$s_0, s_0, s_1, \ldots, s_{m-1}, s_1, s_0, s_1, \ldots, s_{m-1} \cdots s_{m-1}, s_0, s_1, \ldots, s_{m-1}.$$

Two hopping users are illustrated in Figure 1. In that example, m is three. The nodes make rendezvous in the third time slot, from the start of the second user. Rendezvous is guaranteed within $m(m+1)$ time slots.

Shin et al. have proposed the channel rendezvous sequence algorithm [16]. Rendezvous is guaranteed to take place. The asynchronous user ring-walk algorithm has been proposed by Lin et al. [11,13]. Preference is given to channels with low interference to primary users. Rendezvous is not guaranteed to take place.

Bahl et al. proposed an approach for WiFi/802.11 networks [1]. Rendezvous is guaranteed to take place under the symmetric model. Krishnamurthy et al. proposed a two-phase algorithm [9]. The first phase is for neighbor discovery. It is conducted on common local channels. In the second phase, a global common channel is determined among the participating users. Bian et al. use a quorum principle [6,4,5]. Rendezvous is guaranteed. They have a solution for a two-channel case. Yang et al. have proposed an algorithm based on the k-shift-invariant concept that guarantees rendezvous [19].

Lin et al. authored the *enhanced jump-stay rendezvous* algorithm [10,14,12], hereafter called the *jump-stay rendezvous* algorithm. It is designed for multiple synchronous users with guaranteed rendezvous. We illustrate the principle with two users. Each secondary user implements a cyclic behavior consisting of four equal length phases. The first three are identical. The secondary user hops from channel-to-channel. All channels are visited. Each hop lasts for the duration of one time slot. During the last phase, the secondary user stays on the same channel for the whole duration.

Channel hopping is performed according to a pattern determined by the following procedure. Let p be the smallest prime number greater than m (the number of channels). For instance, if there are four channels, then p is five. Hopping is performed in steps of r units, with $r \in \{1, \ldots, m\}$, and starting index $i \in \{0, \ldots, p-1\}$. Each phase consists of p time slots. In the first three phases, hopping is performed for p time slots. During the fourth phase, the secondary user stays on channel r for p time slots. The total length of a cycle, called a *round*, is therefore $4p$ time slots. Let us index the time slots with variable $t = 0, 1, 2, \ldots, 4p-1$. As a function of p, r, i and t; a *channel number pattern* is generated according to the formulae

$$j = (i + tr) \mod p \text{ for } t = 0, 1, 2, \ldots, 3p-1, \tag{1}$$

$$j = r \text{ for } t = 3p, 3p+1, \ldots, 4p-1. \tag{2}$$

In the hopping phases, defined by Equation 1, the sequence of generated channel numbers is such that any window of length p time slots is a permutation of the numbers $0, \ldots, p-1$. Channel indices range from zero to $m-1$. The indices of the corresponding channels are obtained as $c = j \mod m$. Every channel is visited at least once during any interval of p time slots.

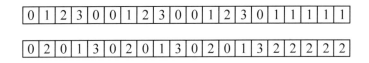

Fig. 2. Two jump-stay rendezvous sequences, round synchronized users

Two example jump-stay rendezvous patterns are shown in Figure 2. Let us say, the upper sequence is performed by User 1 and the lower one by User 2. In both examples, m is equal to four and p is equal to five. Each line represents a cyclic behavior. Each number corresponds to a channel visited during a time slot. Each phase consists of five time slots. The channels of the three hopping phases are listed first. The constant channel of the stay phase follows. In the first example, r is equal to one. It is equal to two in the second example. The start index (i) is zero in both examples. The users make rendezvous when they are on a common

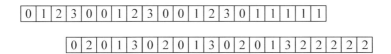

| 0 | 1 | 2 | 3 | 0 | 0 | 1 | 2 | 3 | 0 | 0 | 1 | 2 | 3 | 0 | 1 | 1 | 1 | 1 | 1 |

| 0 | 2 | 0 | 1 | 3 | 0 | 2 | 0 | 1 | 3 | 0 | 2 | 0 | 1 | 3 | 2 | 2 | 2 | 2 | 2 |

Fig. 3. Two jump-stay rendezvous sequences, non round synchronized users

channel number during the same time slot, which occurs in the first time slot in the example of Figure 2. The TTR is one. Note that in this example, users are round synchronized. Figure 3 shows another example where the sequences are the same as in Figure 2, but users are not round synchronized. With respect to User 1, User 2 starts in the fourth time slot. They make rendezvous in the third time slot time from the start of User 2. The TTR is three.

The initial values of the step increment r and start index i are selected at random. The index is incremented to the successor value, modulo p, after each round. Given a sequence generated with $r = r_1$ and another sequence generated with $r = r_2$, with $r_1 \neq r_2$, then any *jump pattern* window of p time slots of the first sequence has a common channel time slot with an overlapping *jump pattern* window of p time slots of the second sequence [12].

Lin et al. address the followings cases: two symmetric users, two asymmetric users and multiple-users. In companion papers, we have improved the analysis of the jump-stay rendezvous algorithm under the symmetric model [2] and developed a new analysis for the asymmetric model in [3]. Under the two-user symmetric model, the expected TTR of the jump-stay rendezvous algorithm is equal to p time slots [2]. Under the two-user asymmetric model, assuming that g is the number of common channels (less than or equal to m), the expected TTR of the jump-stay rendezvous algorithm is [3]

$$\frac{p+1}{1+g} \text{ time slots.} \tag{3}$$

The modular clock algorithm has been originally proposed by Theis et al. [17]. It is based on ideas initially introduced by DaSilva and Guerreiro [7]. It is analogous to the jump-stay rendezvous algorithm, but the stay pattern is not performed. Two-node rendezvous is guaranteed when they hop using different step increments, i.e., different values for r. Because of the absence of the stay pattern, rendezvous does not occur when they start hopping on different channels with identical step increments. When a node fails to rendezvous for $2p$ time slots, it switches to a different step increment. In the modular clock algorithm, described by Theis et al. [17], the step increment r is in $\{0, \ldots, p-1\}$. In the jump-stay rendezvous algorithm it is in $\{1, \ldots, m\}$. In both cases, the generated sequences of p channels share the same aforementioned mathematical properties. Practical evaluations of the modular clock and random algorithms have been conducted by Robertson et al. using the GNU radio framework [15].

3   The Modular Clock Algorithm

The modular clock rendezvous algorithm proceeds in rounds. Each round consists of two phases, of p time slots each. In the sequel, we use the channel number pattern formula of the jump-stay rendezvous algorithm, i.e., Equation 1. It is mathematically equivalent to the formula used for the original presentation of the modular clock algorithm. In other words, every user generates blocks of p channels, following the jump-stay algorithm, but the stay pattern is omitted. Each round consists of two times p jumps (a block of p channels). After each round, each user randomly generates a new starting index i and step length r.

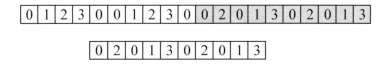

Fig. 4. Two modular clock rendezvous sequences, non round synchronized users

An example is shown in Figure 4. The upper band represents the sequence of channels visited by User 1, the lower band the ones scanned by User 2. The sequences are as in Figures 2 and 3. The users are not *round synchronized*. In this example, User 2 starts after User 1 has started. Two rounds for User 1 are shown and one round for User 2. The first round of User 2 overlaps the first and second rounds of User 1. User 1 uses different step increments (r) for each round. The TTR is one.

The following can be observed. In the jump-stay rendezvous algorithm, for each round the probability that two users generate different step increments (their r) is proportional to the number of channels, i.e., m. The larger m is, the more likely that two users pick different step increments. As a consequence, the usage of the stay pattern becomes less significant in the performance of the algorithm. This is confirmed by the upcoming analysis and simulation results.

4   Estimation of the Expected TTR

We assume that there are two users: User 1 and User 2. They respectively use step increments r_1 and r_2. Without loss of generality, we assume that User 2 starts when or after User 1 has started. A round is made of two modular clock sequences performed by User 2. We start counting the TTR from the time slot when User 2 starts. We first state the main result of this work.

Theorem 1. *The expected TTR of the modular clock rendezvous algorithm is at most $\frac{3p}{4}$ time slots.*

Proof. In the upcoming Lemma 2, it is shown that the expected number of rounds is equal to one. This is because the probability of success P of a round is the parameter of a geometric random variable with mean $1/P$. $1/P$ is equal to one asymptotically in m. Furthermore, asymptotically in m there are only two cases with non-null probability (Cases 1.1 and 2.2). Their expected number of time slots required to make rendezvous are $\frac{p+1}{2}$ and $\frac{2p+1}{2}$, respectively. Using their respective probability these translate to

$$\frac{p+1}{2p} \cdot \frac{m-1}{m} \cdot \frac{p+1}{2} + \frac{p-1}{2p} \cdot \frac{m-1}{m} \cdot \frac{2p+1}{2} \text{ time slots.}$$

Asymptotically in m, this is equal to $3p/4$ time slots. We make this statement mathematically precise in the following two Lemmas.

We define the following function that is used in several mathematical expressions in the sequel:

$$S_m(k) := \sum_{l=1}^{k} \left[1 - \left(\frac{m-1}{m} \right)^l \right] \tag{4}$$

Lemma 1. *For any m, let p be the smallest prime number bigger than m. Then*

$$\frac{S_m(2p)}{2p} \approx \frac{1}{2} - \frac{1}{2e^2}, \tag{5}$$

asymptotically in m.

Proof. Elementary calculations on the function $S_m(k)$ yield the following identity

$$S_m(k) = k - (m-1) + (m-1) \left(\frac{m-1}{m} \right)^k. \tag{6}$$

We are interested in deriving the asymptotic of $S_m(k)$ when $k = 2p$. Recall that p was chosen to be the smallest prime number greater than m. Using well-known results in number theory concerning the difference between consecutive primes, it is easily seen that p is lower than $m + m^{6/11}$ (see [8], Section A9 for additional bounds and discussion). Therefore, since $\left(\frac{m-1}{m} \right)^m \to \frac{1}{e}$, as $m \to \infty$, we have

$$S_m(2p) \approx p - \frac{m-1}{e^2}, \tag{7}$$

asymptotically in m, where e denotes Euler's constant. Since $\frac{m}{p} \to 1$ as $m \to \infty$, Lemma 1 follows.

Lemma 2. *The probability of success of a round is:*

$$P \geq \frac{p+1}{2p} \cdot \frac{m-1}{m} + \frac{p-1}{2p} \cdot \frac{m-1}{m}$$

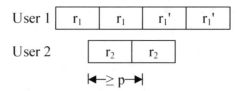

Fig. 5. Overlap is greater than or equal to p

Proof. The analysis is structured into two main cases, with respect to the overlap, in time slots, between the current rounds of two users. In the first case, it is assumed that the overlap is greater than or equal to p. In the second case, it is assumed that the overlap is less than p.

Case 1 (overlap is greater than or equal to p): The overlap is greater than or equal to p time slots, but lower than or equal to $2p$ time slots. This case occurs with probability

$$\frac{p+1}{2p} \tag{8}$$

because User 2 starts from the first to the $p+1$-th time slot from the beginning of User 1. This case is illustrated in Figure 7. The round of User 2 partially overlaps over the first and second rounds of User 1. In the first round of User 1, the step increment is r_1. In the second round, it is r_1'. The step increment of User 2 is r_2. There are four subcases.

Case 1.1 ($r_1 \neq r_2$): In their current round, both users select different step increments, i.e., ($r_1 \neq r_2$). The users make rendezvous in a maximum of p time slots. On average, they make rendezvous in

$$\frac{1}{p} \sum_i^p i = \frac{p+1}{2} \text{ time slots.}$$

The probability of this subcase is $\frac{p+1}{2p} \cdot \frac{m-1}{m}$, because two users pick different step increments with that probability.

Case 1.2 ($r_1 = r_2$) and ($r_1' = r_2$): Both users select the same step increment. Rendezvous is not guaranteed to happen. However, in User 1's round each hop with index i in $1, \ldots, 2p$ can be seen as a Bernoulli trial with probability of success, i.e., rendezvous, $1/m$ (the two users pick the same channel) and probability of failure $\frac{m-1}{m}$ (the two users pick different channels). The two users meet with probability $\frac{S_m(2p)}{2p}$. This subcase occurs with probability $\frac{p+1}{2p} \cdot \frac{1}{m^2}$.

Case 1.3 ($r_1 = r_2$) and ($r_1' \neq r_2$) and overlap is equal to p: This subcase is illustrated in Figure 6. Rendezvous is guaranteed to occur during the second half of User 1's round, i.e., in a maximum of $2p$ time slots. We may assume that they are equally probable. Rendezvous is made with an average of $\frac{2p+1}{2}$ time slots. This subcase occurs with probability $\frac{1}{2p} \cdot \frac{1}{m} \cdot \frac{m-1}{m} = \frac{1}{2p} \cdot \frac{m-1}{m^2}$.

Case 1.4 ($r_1 = r_2$) and ($r_1' \neq r_2$) and overlap is greater than p: If ($r_1 = r_2$) and ($r_1' \neq r_2$), then rendezvous is not guaranteed to happen. In User 1's round each

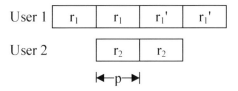

Fig. 6. Overlap is equal to p, r_1 and r_2 are equal, and r'_1 and r_2 are different

hop with index i in $1, \ldots, 2p$ can be seen as a Bernoulli trial with probability of success $1/m$ and probability of failure $\frac{m-1}{m}$. The two users meet with probability $\frac{S_m(2p)}{2p}$. This subcase occurs with probability $\frac{p}{2p} \cdot \frac{p}{m} \cdot \frac{m-1}{m} = \frac{p}{2p} \cdot \frac{m-1}{m^2}$.

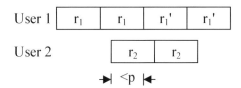

Fig. 7. Overlap is lower than p

Case 2 (overlap is lower than p): The overlap is lower than p time slots. This case occurs with probability

$$\frac{p-1}{2p} \tag{9}$$

because User 2 starts from the first to the $p + 2$-th time slot from the start of User 1. There are two subcases.

Case 2.1 ($r'_1 \neq r_2$): Rendezvous is guaranteed to occur during the second half of User 1's round, i.e., in a maximum of $2p$ time slots. We may assume that they are equally probable. Rendezvous is made with an average of $\frac{2p+1}{2p}$ time slots. This subcase occurs with probability $\frac{p-1}{2p} \cdot \frac{m-1}{m}$.

Case 2.2 ($r'_1 = r'_2$): In User 1's round, each hop with index i in $1, \ldots, 2p$ can be seen as a Bernoulli trial with probability of success $1/m$ and probability of failure $\frac{m-1}{m}$. The two users meet with probability $\frac{S_m(2p)}{2p}$.

The probability of success of a round is:

$$
\begin{aligned}
P = {} & \frac{p+1}{2p} \left(\frac{m-1}{m} + \frac{1}{m^2} \cdot \frac{S_m(2p)}{2p} \right) \\
& + \frac{m-1}{m^2} \left(\frac{1}{2p} + \frac{p}{2p} \cdot \frac{S_m(2p)}{2p} \right) \\
& + \frac{p-1}{2p} \left(\frac{m-1}{m} + \frac{1}{m} \cdot \frac{S_m(2p)}{2p} \right)
\end{aligned}
$$

Asymptotically in m, only Cases 1.1 and 2.2 are significant. We can therefore derive the following lower bound on P:

$$P \geq \frac{p+1}{2p} \cdot \frac{m-1}{m} + \frac{p-1}{2p} \cdot \frac{m-1}{m}$$

5   Evaluation

Figure 8 plots the TTRs obtained with an OMNeT++ [18] simulation of the jump-stay, random and modular clock algorithms, for two-user scenarios. 95% confidence intervals are shown as small horizontal bars. On the x-axis, the number of channels m ranges from 10 to 100 channels. On the y-axis, the mean TTR is plotted as a function of the number of channels for two users. Numbers obtained through simulations are labelled *Jump-stay (simulations)*, *Random (simulations)* and *Modular clock (simulations)*. The expected TTR (ETTR), calculated using the analytical models, is also plotted for the jump-stay, random and modular clock algorithms. The analytical expected TTR for the jump-stay algorithm is labelled *Jump-stay (ETTR)*. It is calculated using expression $\frac{p+1}{1+g}$ time slots, i.e., Equation 3. Simulations results are slightly better. The analytic expected TTR for the random algorithm, i.e., m time slots (Section 2), is labelled *Random (ETTR)*. The analytical expected TTR for the modular clock

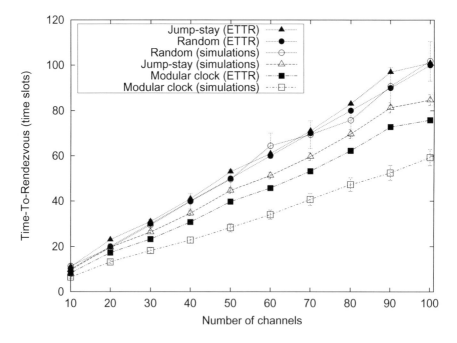

Fig. 8. Mean and expected TTR for the jump-stay, random and modular clock algorithms (with two users, 10 to 100 channels)

algorithm is labelled *Modular clock (ETTR)*. It is calculated using equation $\frac{3p}{4}$ time slots, i.e., Theorem 1. The simulations results are slightly better than the analytic model. For the jump-stay and modular clock algorithms, simulations yield better results than the analytic models. It means that there are slightly more rendezvous opportunities than what the analytical models can capture. Analytical models provide upper bounds. Our simulation confirms that the expected TTR of the modular clock algorithm is no more than $3p/4$ time slots. Simulations performance from worst to best are with random, jump-stay and modular clock algorithms.

6   Conclusion

We have revisited the performance of the *modular clock rendezvous* algorithm. We compared with the performance of the jump-stay rendezvous algorithm. In contrast, the modular clock algorithm does only two hopping phases, of p time slots each. Each round consists of two phases. Rendezvous is not guaranteed. However, our analysis and simulation confirm that as the number of channels increases, the relevance of the stay pattern in the jump-stay rendezvous algorithm drops. Better performance can be expected with the modular clock algorithm. Theis et al.'s analysis of the modular clock algorithm concludes a maximum expected TTR slightly larger than $2p$. Our analysis concludes that the expected TTR of the modular clock algorithm is no more than $3p/4$. This has been confirmed through simulation.

Acknowledgments. We acknowledge financial support from Natural Sciences and Engineering Research Council of Canada, Spanish Ministry of Science (projects CONSOLIDER INGENIO 2010 CSD2007-0004 ARES and TIN2011-27076-C03-02 CO-PRIVACY) and Innovation and Ministry of Education of Mexico (PROMEP).

References

1. Bahl, P., Chandra, R., Dunagan, J.: SSCH: slotted seeded channel hopping for capacity improvement in IEEE 802.11 ad-hoc wireless networks. In: Proceedings of the 10th Annual International Conference on Mobile Computing and Networking, MobiCom, New York, NY, USA, pp. 216–230. ACM (2004)
2. Barbeau, M., Cervera, G., Garcia-Alfaro, J., Kranakis, E.: A new analytic model for the cognitive radio jump-stay algorithm. In: IFIP Wireless Days, pp. 1–3 (November 2013)
3. Barbeau, M., Cervera, G., Garcia-Alfaro, J., Kranakis, E.: A new analysis of the cognitive radio jump-stay algorithm under the asymmetric model. In: IEEE ICC 2014 - Cognitive Radio and Networks Symposium (2014)
4. Bian, K., Park, J.-M.: Asynchronous channel hopping for establishing rendezvous in cognitive radio networks. In: Proceedings of IEEE INFOCOM, pp. 236–240 (2011)

5. Bian, K., Park, J.-M.: Maximizing rendezvous diversity in rendezvous protocols for decentralized cognitive radio networks. IEEE Transactions on Mobile Computing 12(7), 1294–1307 (2013)
6. Bian, K., Park, J.-M., Chen, R.: A quorum-based framework for establishing control channels in dynamic spectrum access networks. In: Proceedings of the 15th Annual International Conference on Mobile Computing and Networking, MobiCom, New York, NY, USA, pp. 25–36. ACM (2009)
7. DaSilva, L.A., Guerreiro, I.: Sequence-based rendezvous for dynamic spectrum access. In: 3rd IEEE Symposium on New Frontiers in Dynamic Spectrum Access Networks (DySPAN), pp. 1–7 (2008)
8. Guy, R.: Unsolved problems in number theory, 3rd edn. Springer (2004)
9. Krishnamurthy, S., Thoppian, M., Kuppa, S., Chandrasekaran, R., Mittal, N., Venkatesan, S., Prakash, R.: Time-efficient distributed layer-2 auto-configuration for cognitive radio networks. Computer Networks 52(4), 831–849 (2008)
10. Lin, Z., Liu, H., Chu, X., Leung, Y.-W.: Jump-stay based channel-hopping algorithm with guaranteed rendezvous for cognitive radio networks. In: Proceedings of IEEE INFOCOM, pp. 2444–2452 (April 2011)
11. Lin, Z., Liu, H., Chu, X., Leung, Y.W.: Ring-walk rendezvous algorithms for cognitive radio networks. Ad-Hoc and Sensor Wireless Networks 16(4), 243–271 (2012)
12. Lin, Z., Liu, H., Chu, X., Leung, Y.-W.: Enhanced jump-stay rendezvous algorithm for cognitive radio networks. IEEE Communications Letters, 1–4 (2013)
13. Liu, H., Lin, Z., Chu, X., Leung, Y.-W.: Ring-walk based channel-hopping algorithms with guaranteed rendezvous for cognitive radio networks. In: IEEE/ACM Int'l Conference on Cyber, Physical and Social Computing (CPSCom), Green Computing and Communications (GreenCom), pp. 755–760 (2010)
14. Liu, H., Lin, Z., Chu, X., Leung, Y.-W.: Jump-stay rendezvous algorithm for cognitive radio networks. IEEE Transactions on Parallel and Distributed Systems 23(10), 1867–1881 (2012)
15. Robertson, A., Tran, L., Molnar, J., Fu, E.-H.F.: Experimental comparison of blind rendezvous algorithms for tactical networks. In: IEEE International Symposium on a World of Wireless, Mobile and Multimedia Networks (WoWMoM), pp. 1–6 (2012)
16. Shin, J., Yang, D., Kim, C.: A channel rendezvous scheme for cognitive radio networks. IEEE Communications Letters 14(10), 954–956 (2010)
17. Theis, N.C., Thomas, R.W., DaSilva, L.A.: Rendezvous for cognitive radios. IEEE Transactions on Mobile Computing 10(2), 216–227 (2011)
18. Varga, A., Hornig, R.: An overview of the OMNeT++ simulation environment. In: 1st International Conference on Simulation Tools and Techniques for Communications, Networks and Systems & Workshops, Simutools (2008)
19. Yang, D., Shin, J., Kim, C.: Deterministic rendezvous scheme in multichannel access networks. Electronics Letters 46(20), 1402–1404 (2010)

A Preventive Energy-Aware Maintenance Strategy for Wireless Sensor Networks

Skander Azzaz and Leila Azouz Saidane

RAMSIS Team, CRISTAL Lab, ENSI, Manouba University, Tunis, Tunisia
{skander.azzaz,leila.saidane}@ensi.rnu.tn

Abstract. In this paper, we propose proactive maintenance strategies useful to preserve the coverage of a Wireless Sensor Network (WSN) and protect its connectivity from the eventual sensor failures. Failures are handled before they happen. To achieve this goal, we introduce an analytical model to represent the energy dissipation of a node sensor. The proposed model is useful to estimate the life time of each sensor in the network. Once identified, the anticipated faulted sensors are replaced with new ones by a set of mobile robots. The maintainer robots are scheduled using two different approaches. The Heuristic Centralized Proactive Maintenance Strategy (HCPMS) uses a centralized approach to schedule a set of maintainer robots when dealing with the expected sensor failures. However, the Fixed Distributed Proactive Maintenance Strategy (FDPMS) and the Market based Distributed Proactive Maintenance Strategy (MDPMS) use distributed algorithms to select and designate the maintainer robot of each anticipated failure. Simulation results show that HCPMS is adapted only for the small-scale WSNs. However, MDPMS can provide a good compromise between the communication cost and the provided network dysfunction time in large-scale ones.

1 Introduction

Due to the limited energy resources of sensors, Wireless Sensors Networks (WSNs) are susceptible to many failures. For this reason, WSNs must deploy a set of mechanisms to restore the coverage and the connectivity of the network upon an occurred failure. Many solutions have been introduced in literature such as: Exploiting redundancy nodes to fill holes induced by a sensor failure or the use of mobile sensors [1]. To reduce the deployment cost of the WSN maintenance solution, Mei et al. in [2] propose using a small number of mobile robots to replace the detected failed sensors in a static WSN. Three robot coordination algorithms have been proposed: The Centralized Manager Algorithm (CMA), the Fixed Distributed Manager Algorithm (FDMA) and The Dynamic Distributed Manager Algorithm (DDMA). With CMA, a guardian guardee relationship is established between the network sensors to detect an occurred failure. Indeed, each sensor (guardee node) broadcasts periodically a signaling message. If a one-hop neighbor (a guardian node) has not received any messages from a guardee node for a certain amount of time, it deduces that the guardee has failed and

S. Guo et al. (Eds.): ADHOC-NOW 2014, LNCS 8487, pp. 209–222, 2014.

notifies a manager robot. In CMA, one robot is selected to operate as a manager. Upon the reception of the failure report, the manager robot schedules the closest maintainer robot (to the occurred failure) to handle the failed sensor. However, FDMA splits statically the network map in a set of subareas with equal surface. Each subarea is assigned to a single robot to deal with its detected sensor failures. FDMA can be viewed as CMA per subarea where the assigned robot is both a manager and a maintainer. With DDMA, the manager robot is selected dynamically by the guardian node as the closest robot to the failed guardee. CMA, FDMA and DDMA are considered as reactive maintenance strategies; since, with these strategies, each occurred failure must be detected and repaired in a following step. During the reaction time elapsed to repair a given failure, the network coverage (and connectivity) can degrade enormously, especially if the fault occurs in an overloaded relay point nearest the sink node. To remedy this problem and provide a continuous service without interruption, we propose to replace the expected failures in the network with functional ones before they happen. To achieve this goal, we use: (i) an analytical model to estimate the life time of each sensor and (ii) a robot coordination algorithm to schedule the maintainer robot activity. Based on the proposed analytical model representing the energy dissipated by each sensor to deduce its life time, we introduce three proactive (preventive) maintenance strategies. Indeed, like CMA, the Heuristic Centralized Proactive maintenance strategy (HCPMS) uses a centralized architecture to synchronize the robot motion. However, the Fixed Distributed Proactive Maintenance Strategy (FDPMS) the proactive version of FDMA and the Market based Distributed Proactive Maintenance Strategy (MDPMS) use a distributed approach to schedule the available maintainer robots.

The present paper is organized as follows: in section 2, we detail our proposed analytical energy model. In section 3, we present the centralized proactive maintenance strategy HCPMS. However, in section 4, we detail our proposed distributed maintenance strategies: FDPMS and MDPMS. Proposed strategies are evaluated in section 5 in terms of the provided network dysfunction time, the robot traveling distance and the induced signaling cost. And finally, we conclude the paper in section 6.

2   An Analytical Energy Dissipation Model

Generally, the energy dissipated by a sensor node can be divided in several parts [3]: the energy spent for radio communications by the sensor radio interface denoted in this paper by ε_{Tot}^{Rad}, the sensing energy (the energy it costs to take measurements) denoted by ε_{Tot}^{Sens}, and the energy consumed by the sensor processing unit.

Assumption: The energy consumed by the sensor processing unit is negligible.

Based on the study presented by [4] and [5], we claim that the status of the MAC layer of the radio interface of a sensor node can be modeled by a Markov Chain. The states of this Markov Chain depend on the MAC layer used by the sensor node. For example, if we use the IEEE 802.11 MAC layer [6], we have three operational modes (states): Transmit, Receive and Idle. Each state corresponds to a different power consumption level:

- **State 1 (Idle)**: even when no messages are being transmitted over the medium, the node stays idle and listening the medium with idle power ε_{Id}.

- **State 2 (Transmit)**: node is transmitting a frame with the transmission power ε_{Tx}.

- **State 3 (Receive)**: node is receiving a frame with the reception power ε_{Rx}.

If the sensor node uses IEEE 802.15.4 or SMAC (Sensor MAC) [7] as a MAC protocol, we have a fourth state:

- **State 4 (Sleep)**: The radio is turned off, and the node is not capable to detect signals. We suppose that the node uses a power ε_{Sl} in this state.

[4] shows that some transitions (the transitions: Idle - Transmit, Idle - Receive and Sleep - Idle) arent instantaneous. These particular transitions are achieved with an additional amount of energy in a requested amount of time. To take account of the transition delay, we introduce three virtual states representing the state of the radio module during the transitions.

- **State 5 (Pre-Idle)**: a virtual state that inevitably precedes the Idle state in the transition sleep - Idle. ε_{Id}^{T} represents its dissipation energy rate:

$$\varepsilon_{Id}^{T} = E_{TI}/d_{TI} \tag{1}$$

E_{TI} is the transition energy of the Sleep - Idle transition and d_{TI} is the transition time.

- **State 6 (Pre-Transmit)**: This virtual functional mode models the state of the radio module during its transition to the state Transmit. The energy dissipation rate in this state, denoted by ε_{Tx}^{T}, is given by:

$$\varepsilon_{Tx}^{T} = E_{TT}/d_{TT} \tag{2}$$

E_{TR} is the transition energy of the Idle - Receive transition and d_{TT} is the time requested by the transition.

- **State 7 (Pre-Receive)**: An intermediate state acquired by the radio module when transiting to the functional mode Receive. ε_{Rx}^{T} representing the energy dissipation rate during this virtual state is equal to:

$$\varepsilon_{Rx}^{T} = E_{TR}/d_{TR} \tag{3}$$

With E_{TR} is the amount of energy dissipated by the Idle - Receive transition and d_{TR} is its delay.

We note S_i the random variable related to the state of the MAC layer at time i. We choose a time step (unit) such that the duration of any action (e.g; transmission/reception of a frame) is a multiple of this time step and we suppose that all state transitions occur at the beginning of the time step.

The notation $(S_n = i)$ means that the MAC layer is in state i at time step n. Let P_{ij} be the probability that a node in state i will enter in state j at the next transition.

$$P_{ij} = P\{S_{n+1} = j/S_n = i\}, 1 < i, j < 7 \tag{4}$$

[4] has demonstrated that many transitions such as: Receive-Transmit or Sleep-Receive are forbidden. These transitions arent direct, while the module radio must transit first to the Idle state before the target state. We will obtain, consequently, a set of null transition probabilities (such as: P_{32} or P_{43}).

Any sensor can determine the transition probability P_{ij} based on his activity history. In fact, the sensor must divide firstly a fixed interval Ti of his past time in a set of unitary time slots TS. The size of the TS is taken so small to avoid that a sensor MAC layer transits from more than two states in a single TS, and we suppose also that all MAC states transition occurs in the beginning of the TS. For each time slot of the interval time Ti, the sensor determines the corresponding status MAC layer. And, P_{ij} can be computed as the number of time slots with state i followed by a time slots with the state j ($\left|TS^T_{i\to j}\right|$ in equation 5) divided by the total number of time slots with state i ($\left|TS^T_i\right|$ in equation 5)

$$P_{ij} = \frac{\left|TS^T_{i\to j}\right|}{\left|TS^T_i\right|} \tag{5}$$

P_{ij} is recomputed periodically each a fixed time interval (T_r).

In conclusion, we model any MAC layer of a sensor node as a discrete time Markov Chain with the transition matrix P (with P_{ij} represents the (i, j)th element of P) between the seven states: Idle, Transmit, Receive, Sleep, Pre-Idle, Pre-Transmit and Pre-Receive.

Let $\Pi^{(n)} = [\Pi^{(n)}_1, \Pi^{(n)}_2, \Pi^{(n)}_3, \Pi^{(n)}_4, \Pi^{(n)}_5, \Pi^{(n)}_6, \Pi^{(n)}_7]$ be the probability vector with $\Pi^{(n)}_i$, ($1 \leq i \leq 7$) represents the probability that the MAC layer is in state i at time step n ($P\{S_n = i\}$). Then:

$$\Pi^{(n)}_i = \sum_{k=1}^{7} \Pi^{(n-1)}_k P_{ki} \tag{6}$$

Knowing that at time step 0 the node is generally in the state 1 (Idle), we have:

$$\Pi^{(0)} = [1, 0, 0, 0, 0, 0, 0] \tag{7}$$

Let $cons^{(n)}_j$ the energy consumed by radio communication at the nth transition given that the node is in state j at time step n:

$$cons^{(n)}_j = \sum_{i=1}^{7} \Pi^{(n-1)}_i P_{ij} E_j \tag{8}$$

Where: E_j is the energy consumed by radio communication during a step given that the node is in state j. E_j is the element of column j of the vector E.

$$E = \Delta t.[\varepsilon_{Id}, \varepsilon_{Tx}, \varepsilon_{Rx}, \varepsilon_{Sl}, \varepsilon^T_{Id}, \varepsilon^T_{Tx}, \varepsilon^T_{Rx}] \tag{9}$$

Where Δt is the length in seconds of a time step.

We note $cons^{(n)}$ the energy consumed for radio communication by a node during the step n:

$$cons^{(n)} = \sum_{j=1}^{7} \Pi_j^{(n)} cons_j^{(n)} \tag{10}$$

Then, $\varepsilon_{Tot}^{Rad^{(n)}}$ that represents the total energy consumed by the radio interface up to the end of the step n is:

$$\varepsilon_{Tot}^{Rad^{(n)}} = \sum_{i\in[1..n]} cons^{(i)} \tag{11}$$

In addition to radio communication energy, the sensor consumes an amount of energy for event sensing. We note by $cost_{sens}$ the average consumption energy in sensing mode (to take measurements).

$$cost_{sens} = \frac{\lambda}{\mu}\, \varepsilon_{sensing} \tag{12}$$

Where λ is the average rate of events detected by sensor node, $\frac{1}{\mu}$ is the sensing mean time of an event and $\varepsilon_{sensing}$ is the energy consumed by node in sensing mode (the energy it costs to take one measurement).

We note $\varepsilon_{Tot}^{Sens^{(n)}}$ the average sensing energy consumed by the sensor until the time-step n:

$$\varepsilon_{Tot}^{Sens^{(n)}} = n\, \Delta t\, cost_{sens} \tag{13}$$

Finally, the total energy consumed by a sensor node until the time step n is designated by $\varepsilon_{Tot}(n)$:

$$\varepsilon_{Tot}(n) = \varepsilon_{Tot}^{Rad^{(n)}} + \varepsilon_{Tot}^{Sens^{(n)}} \tag{14}$$

The estimated sensor lifetime, denoted by (ttl), corresponds to:

$$\begin{cases} \varepsilon_{Tot}(ttl) < \varepsilon_0 \\ \varepsilon_{Tot}(ttl+1) > \varepsilon_0 \end{cases} \tag{15}$$

Where ε_0 is the sensor residual energy (at $t = 0$).

3   The Centralized Proactive Maintenance Strategy

We focus in this section on the centralized maintenance approach [8]. We present first the role of each WSN component to detect, anticipate and handle the sensor failures. In a following step, we detail the algorithm adopted to schedule the available maintainer robots.

3.1   Coordination Architecture

In HCPMS, we distinguished three WSN components types: a selected manager robot, a set of maintainers robots and evidently the sensors nodes. One robot is manually configured to function as a manager. The robot manager broadcasts periodically its identity in a Robot Manager Advertisement Message (RMAM). And upon the reception of the RMAM:

- The sensor node saves the robot manager identity to be notified in case of an observed failure on a guardee node detected due to the established guardian guardee relationship between network's sensors. Indeed, many types of failures (e.g. hardware failures) are unpredictables. For this reason, like the classical maintenace strategies, we recommend to build a guardian gardee relation useful to detect this kind of failures.

- The maintainer robot replies with a unicast message (Robot Maintainer Reply Message (RMRM)) containing its identity and its current position. The RMRM messages allows to the manager to identify the list of his maintainers robots and their initial positions. After the initialization stage, each network sensor must measure the different parameters of the proposed analytical model (P_{ij}, λ and $\frac{1}{\mu}$) to be communicated with its position cordinates and its residual enegy ($E0$) to the robot manager using a Sensor Information Message (SIM). With the SIMs received from the different network sensors, the robot manager can estimate the deadlines of the N expected failures with their correspondent positions. And, knowing the list of the available robots, the manager robot must schedule the robot reparation activity ensuring the WSN coverage and connectivity preserving. In others words, the manager must resolve two sub-problems: (i) a repairing task allocation problem which is the maintainer robot of each estimated failure? and (ii) a planning problem when the reparation order is sent to the selected maintainer robot?. The quality of the robot scheduling solution is evaluated based on the induced network dysfunction time. We designate by the network dysfunction time the sum of the off-service time of failed sensors (the time difference between the sensor failure time and its reparation time). The scheduling solution must also optimize of the traveled distance achieved by the maintainer robots.

3.2   The HCPMS Robot Scheduling Algorithm

Optimizing the robot scheduling solution is an NP Hard problem. Many approaches can be used to solve such a scheduling problem. HCPMS uses a scheduling algorithm based on an heuristic that optimizes locally the dysfunction time (the off-service time) of each expected failure. In HCPMS, failures are processed according to their estimated chronological order (estimated apparation order on the network). To determine the maintainer robot for a failure f^i ($i \in [1, N]$), the robot manager identifies firstly the robots list able to handle f^i at his required replacement time $f^i.t$ (with a null off-service time). Indeed, a robot R_m ($m = [1, NR]$, NR is the total number of maintainer robots) is considered as a candidate maintainer robot for the failure f^i, if and only if, the time required to travel from its current position (corresponding to the position of the last failure handled by R_m denoted by $f^{l^i_m}$) to the failure f^i position with a speed S_{R_m} is less than $(f^i.t - f^{l^i_m}.t)$. if f^i is the first failure handled by R_m, the initial position of R_m is considered instead of that of $f^{l^i_m}$.

From the determined candidate scheduling solutions list, the central manager robot selects the nearest robot as the f^i maintainer robot. Thereafter, the procedure is re-executed for the failure $f^{(i+1)}$ until building the robot scheduling solution for the N expected failures. In a case of scheduling blocking, when any robot satisfying the replacement time constraints of the failure f^i can't be found, HCPMS selects the maintainer robot able to replace the expected failure as soon as possible. HCPMS selects usually the robot that minimizes locally the node dysfunction time insofar as the considered heuristic optimizes the global network dysfunction time (sum of the node dysfunction time).

After determining the robot scheduling solution used to handle the N expected failures in the network, the manager robot must plans in a next step the failure reparation operations. We denote by $(d^i, i \leq 1 \leq N)$ the scheduling time of the robot M^i designated to repair the estimated failure f^i. d^i is computed as the difference between: (i) the estimated time of the ith failure: f^i and (ii) the requested time by the robot M^i to travel to the f^i position. At d^i a reparation order message is sent to the robot M^i including the failure position. Upon the reception of the reparation request, the maintainer robot moves to the correspondent position, replaces the estimated faulted sensor and notifies the manager robot the operation success.

4   The Distributed Proactive Maintenance Strategy

Dividing the total network map in subarea set, as FDMA proceeds, may be a candidate solution to distribute the maintenance operation over the available robots [9]. We propose, in a first step, a proactive version for FDMA denoted by FDPMS. In a following step, we introduce a novel Market based Distributed Proactive Maintenance Strategy denoted by MDPMS, the distributed version of HCPMS.

4.1   The Fixed Distributed Proactive Maintenance Strategy

Similarly to FDMA, FDPMS divides statically the WSN sensing area in subareas. Each robot will deal with the detected and anticipated failures in a single subarea. After measuring the model parameters, each network sensor communicates the measured values and its position coordinates to a known maintainer robot. The network map is split statically in subareas with equal surface and the sensors belonging to the same subarea share the same robot maintainer identity. With the received SIM messages, the maintainer robot can determine the expected failures in its assigned subarea and plan its reparation activity to repair the anticipated failure before they occur.

FDPMS consists to split the total network map in a fixed number of sub-areas. And, CPMS is applied in each subare where a single robot operates as both a manager and a maintainer.

4.2   The Market Based Distributed Proactive Maintenance Strategy

Generally, in a distributed maintenance approach, robots use local information and inter-robot communication to selects the maintainer robots of each estimated failures without the need of a central element. It's true that this type of maintenance strategy is more complex, since robots have to choose the convenient maintainer robot without having access to all the information (all estimated failures positions), but such a maintenance strategy can offer a good compromise between the communication cost and the quality of scheduling solution in terms of dysfunction time and robots traveled distance. By comparing different task allocation methods, Dias et al. show in [10] that distributed market-based approaches are the best option to provide good performances with reduced signaling requirements. Market-based approaches are inspired from the economy system. In this type of approaches, many auctions are established between participants. A set of items are exposed on the market by a particular auctioneer. And, the participants can make offers for these items by submitting bids. Once all bids are received during specified limit deadline, the auctioneer decides the winner (the participant offering the lowest cost). In our context, the participants are the maintainer robots and the items are the sensor repairing requests. Indeed, in MDPMS a sensor reports its model parameters with a Sensor Information Message to a selected robot (random robot among a set of manually configured robot list). Upon the reception of a SIM message, the maintainer robot (the auctioneer robot) estimates the correspondent estimated failure deadline. Then, it submits the determined expected with its deadline and position in an auction using an Announced Failure Message (AFM). The goal of auction is to select the convenient maintainer robot of the announced failure. The AFM is sent to the closest maintainer robot. Upon the AFM reception, the maintainer robot saves its cost for the announced failure in the received message and forwards it to the following robot (usually the closest robot). For each robot, the cost of an announced estimated failure is the pair (the induced dysfunction time, the induced traveled distance). The induced dysfunction time (respectively the induced traveled distance) is the difference between: (i) the dysfunction time (respectively the traveled distance) induced by the pre-assigned failures and (ii) that induced by adding the announced failure to the list of the pre-assigned failures. In other words, each robot supposes that itself is the maintainer robot of the announced failure and computes in consequence the novel estimated dysfunction time and the new estimated traveled distance induced by its assigned failures. The pair (the estimated dysfunction time difference, the estimated traveled distance difference) is considered as the announced cost for the announced failure. When the AFM message returns to the auctioneer robot, it designates with a Maintainer robot Designed Message (MRDM) the robot offering the lowest failure repairing cost (the induced dysfunction time is considered first) as the maintainer robot for the announced failure.

5   Experiments

In this section, we evaluate our proposed proactive maintenance strategies with simulation. The performances of the proactive maintenace strategies are compared to the classical strategies (CMS and FDMA).

5.1   Experimental Setup

We have implemented HCPMS, FDPMS and MDPMS in the $NS-2$ Simulator. And, we have selected the following simulation parameters: IEEE 802.15.4 is used as a sensor MAC Layer in the WSN; the routing protocol AODV (Ad hoc On Demand Distance Vector)[11] integrated by default in the Zigbee alliance; the network area is $(1600x1600)$ m^2 covered by (32×32) sensor nodes with a communication radius equal to 75 meters, a coverage radius equal to 60 meters and an intial energy equal to 1000 Joules. To guarantee the total coverage of the network map, we use a grid deployment scheme for the availible sensors. We deploy one sensor each 50 meters and we denode by $N(i,j)$ the node with the position coordinates: $(50.i, 50.j)$, $0\leq$ i <32 and $0\leq$ j <32); all sensors nodes generate the same traffic load destined to a sink node (node $N(0,0)$); the robot's speed is $1m/s$, based on the specification of Pioneer 3DX robots [2].

Table 1. IEEE 802.15.4: Transition energy and delay

The transition	Consumed Energy	Delay
Idle − Transmit	$9.93\mu j$	$194\mu s$
Idle − Receive	$6.63\mu j$	$194\mu s$
Sleep − Idle	$691 pj$	$970\mu s$

We use the energy vector E : $[0.1404, 0.1404, 0.0018, 0.000018]$ representing the power consumption in each sensor MAC layer state (*sleep*, *transmit*, *receive* and *idle*) taken from a ZigBee node implementing IEEE 802.15.4 medium access. The energy dissipated by all possible transitions with their correspondent delays (the instantaneous transitions are omitted) are presented in the table 1 reported from [4].

5.2   Model Validation

We begin first with the validation of the introduced analytical model. We compare, in this section, the residual energy of the network sensors obtained with the simulation versus the ones computed with the proposed model.

We have considered two node traffic models: a Poisson and a CBR (Constant Bit Rate) Pattern with constant packet sizes (128 bytes). The Poisson pattern

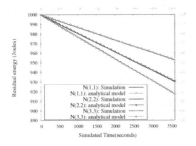

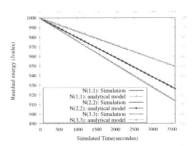

Fig. 1. The variation of the node residual energy versus the simulated time in a 802.15.4 WSN with a Poisson traffic model

Fig. 2. The variation of the node residual energy versus the simulated time in a 802.15.4 WSN with a CBR traffic model

traffic is usually used to model the generation of data packets of a sensor detecting a set of independent events (both spatially and temporally) occurring in a given area. However, the CBR one can represent a periodic traffic generated by a sensor (i.e. measures of the temperature, pressure,...) sent to a base station. We have represented in figure 1 and 2 the residual energy of the nodes N(1,1), N(2,2) and N(3,3) versus the simulation time. In our simulation scenarios, we have supposed that all network nodes generate the same load of traffic (equal to 5Kb/s) following respectively a Poisson and CBR model. Presented values are compared to the estimated residual energy computed with the analytical model. We have chosen a time step equal to $194\mu s$ representing the lowest transition time (especially the Idle-Receive transition). The delay of any action done by the sensor radio module (i. e. packet reception, packet transmission, any type of transitions) can be them considered as a multiple of this time step.

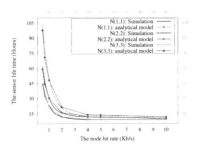

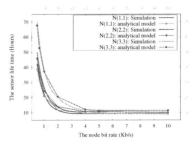

Fig. 3. The sensor life time versus the node bit rate generated according to a Poisson process in a 802.15.4 WSN

Fig. 4. The sensor life time versus the node bit rate generated according to a CBR process in a 802.15.4 WSN

Obtained results show that the proposed model can effectively estimate the residual energy of the considered sensor nodes and gives similar values as the simulation for the two traffic models. Presented figures proof also that nodes located in the sink vicinity (case of the node N(1,1)) consume more energy than the edge ones (case of the node N(3,3)). These particulars nodes are considered as relay points for routing the traffic to the sink node. For this reason, any eventual failure on these nodes can damage severely the service provided by the network. Using the introduced analytical model, we can deduce an anticipated life time for each node sensor. This deadline corresponds to the estimated time when the evaluated residual energy reaches a given threshold (zero by default). We have represented in figure 3 the life time of the nodes: N(1,1), N(2,2) and N(3,3) (obtained respectively with the analytical model and the simulation) versus the node bit rate. We have supposed, in this step, that packets are generated according to a Poisson Process. Moreover, we have validated the estimated life time (presented in figure) provided by the analytical model with a CBR traffic model. With the two node traffic patterns, we can remark the similarity between the real life time (obtained with simulation) and the anticipated one computed with our model. Based on this generic model representing the sensor energy dissipation and validated by simulation, we built our proposed proactive maintenance strategies evaluated in the following section.

5.3   Performance Evaluation

Three parameters are considered in the evaluation of the proposed proactive maintenance strategies: the Network Dysfunction Time, the Robot Traveling Distance and the Signaling Cost. Any WSN maintenance strategy must first reduce, as possible, the Network Dysfunction Time (service interruption time) induced by the occurred failures using a minimal number of robots. Moreover, the robot traveled distance and the signaling cost can be considered relevant criteria to judge the maintenance strategy. In the following simulation scenarios, we consider only a CBR traffic model with a node bit rate equal to 5Kb/s.

A- Network Dysfunction Time: The Network Dysfunction Time reflects the quality of service provided by a WSN maintenance strategy. The figure 8 shows the variation of the network dysfunction time percentage versus the number of maintainer robots with a traffic node bit rate equal to 10 Kb/s and a variable number of maintainer robots for the studied maintenance strategies.

With a higher failure rate exceeding the robots capacity, the maintainer robots become unable to guarantee the persistence of the WSN. Figure 8 shows that the proactive maintenance strategies request a smaller number of maintainer robots than the reactive ones. Indeed, HCPMS (respectively FDPMS) uses a minimum of 3 robots instead of 4 (respectively 5) robots in case of CMA (respectively FDMA) to guarantee a proper functioning of the WSN. In addition, contrary to the reactive maintenance strategies (CMA and FDMA), and with a sufficient number of robots (6 robots), obtained results show that the proactive maintenance strategies can provide a null network dysfunction time. Indeed, in case of a reactive maintenance strategy, the network dysfunction time of given failure remains usually greater than the detection time plus the robot traveled time, even if we increase the robot number. Among the proposed WSN maintenance strategies, figure 8 shows also that HCPMS and MDPMS provide the smallest network dysfunction.

With the FDMPS (or FDMA) partitioning technique, we risk an unbalanced load sharing among the available maintainer robots. We report in figure 6 the repairing load percentage of each robot designated by (ρ) and corresponding to the number of repaired sensors per robot with a node bit rate equal to 1 and 6 Kb/s. We designate by robot I (respectively II, III) the maintainer robot of the sink (respectively the middle, last) subarea.

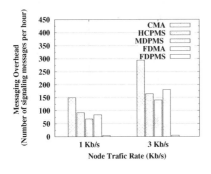

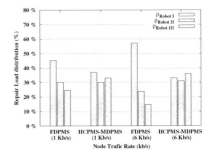

Fig. 5. The Messaging overhead versus the node bit rate

Fig. 6. The Robot Load distribution versus the node bit rate

Figure 6 shows that FDPMS overloads the robot I compared to the robot II and III. And, we can remark that the repairing load of a given robot depends on the supervised subarea position relative to that of the sink node: the repairing load increases for the subarea nearest the sink node. Indeed, in our simulation scenarios, like the general case in WSNs, a many-to-one traffic model is adopted. For this reason, we obtain a non-uniform distribution of failures over the network map and consequently a high failure rate neighboring the sink node. This behavior explains the obtained result: $(\rho_{RobotI} > \rho_{RobotII} > \rho_{RobotIII})$ for FDPMS. The unbalanced load sharing justifies the greater dysfunction time obtained for FDPMS or FDMA compared to HCPMS or CMA by considering the same number of robots. Since it selects the robot that minimizes the off-service time of the faulted sensor, figure 6 shows that MDPMS have solved the unbalanced sharing load problem introduced by FDPMS and provides approximately a fair load sharing among the robots. MDPMS can be considered as the distributed version of HCPMS since it gives a similar network dysfunction time.

B - The Robot Traveling Distance: In figure 7, we report the average distance traveled by robots to repair a single anticipated failure for the three proactive WSN maintenance strategies: HCPMS, FDPMS and MDPMS using 4 maintainer robots (the minimal number of robots requested by FDPMS).

Compared to HCPMS, figure 7 shows that FDPMS increases the induced robot traveling distance. However, MDPMS has produced the same traveled distance as HCPMS, since they use the same robot scheduling solution to repair a set of expected failures. On the other hand, with the FDPMS partitioning technique that tries only to equalize the subarea surface ignoring the failures rate distribution on the network map, we obtain, in addition to the poor network dysfunction time (figure 8), a greater robot motion load (figure 7).

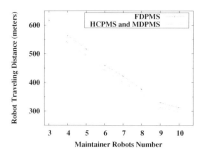

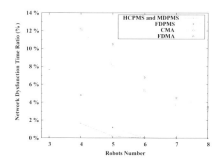

Fig. 7. The robot traveling distance versus the maintainer robots number

Fig. 8. The network dysfunction time ratio versus the robots number

C - The Signaling Cost: The CMA reactive maintenance strategy introduces two types of signaling messages:(i) the robots-robots signaling messages representing the messages exchanged between robots: RMAM, SPMM, Repair Order Message (ROM): messages sent by the manager to schedule a selected maintainer robot to repair a sensor failure and Repair Finish Message (RFM) sent by the maintainer robot to the manager announcing the success of the repair operation and (ii) the sensors-robots signaling messages generally sent by sensors to report an occurred failure: the Reporting Failure Messages (RFM). HCPMS adds a new sensors-robots signaling message type: the SIM messages sent by each sensor to its manager robot to report its energy model parameters. For the centralized proactive strategy (HCPMS), ROM and RFM messages are absents since sensor failures are anticipated. For this reason, we obtained a reduced signaling cost for HCPMS compared to CMA, as the figure 5 shows. However, FDMA and FDPMS don't need any robots-robots signaling messages, since each robot has to handle autonomously the occurred or anticipated failures in its assigned subarea. In case of FDMA, only the ROM messages transmitted by the guardian nodes to report an occurred failure are present. But with FDPMS, the failure repairing anticipation, achieved using a reduced number of messages (a single SIM message per one sensor), can avoid the transmission of this message type (ROM messages) and decreases consequently the signaling cost, as the figure 5 shows, compared to FDMA. With MDPMS, the robots-robots signaling messages is limited to a reduced number of message to establish the auction and to design the maintainer robot for an estimated failure. It's true that the signaling cost induced by MDPMS is greater than FDPMS but it remains significantly lower than HCPMS or GCPMS, as figure 5 shows.

6   Conclusion

In this paper, we have proposed proactive strategies to maintain the sensor failures of a WSN using a limited number of maintainer robots. Two approaches are adopted. The centralized maintenance strategy (HCPMS) opts for a selected manager robot to schedule the available robots in order to optimize, as possible, the robot scheduling solution in terms of network dysfunction time and robot traveled distance. However, the distributed maintenance strategies introduce distributed mechanisms to select the maintainer robot of an expected sensor failures. HCPMS selects the robot that minimizes the off-service time of an estimated failure with the minimal travel distance as its maintainer robot. But, despite that HCPMS has reduced the signaling cost compared the classical centralized reactive maintenance strategy (CMA), it stills inadequate for large-scale WSNs. Indeed, similar to any centralized approach, this type of WSN maintenance introduces a heavier signaling cost to synchronize the maintainer robots with their manager. With a fixed partitioning of the WSN map, FDPMS can provide a negligible signaling overhead since any maintainer robot can handle autonomously the expected failures in its subarea. However, simulation results have demonstrated that FDMPS can degrade the provided network dysfunction time compared to CPMS, since it overloads some particular robots: robots assigned to a subarea with high failure density. MDPMS has solved the problem using a market based approach to select the convenient maintainer robot with a reduced signaling cost. Simulation results have shown that MDPMS optimizes the exploitation of maintainer robots and it minimized consequently the network dysfunction time.

References

1. Azzaz, S., Saidane, L.: Maintenance strategies for wireless sensor networks: from a reactive to a proactive approach. Transactions on Emerging Telecommunications Technologies (2013)
2. Mei, Y., Xian, C., Charlie, Y.: Repairing Sensor Network Using Mobile Robots. In: Workshop on Wireless Ad hoc and Sensor Networks (2006)
3. Azzaz, S., Saidane, L., Minet, P.: Repairing sensors strategies in fault-tolerant wireless sensor networks. In: PE-WASUN (2011)
4. Bougard, B., Daly, D.C., Chandrakasan, A., Dehaene, W.: Energy Efficiency of the IEEE 802.15.4 Standard in Dense Wireless Microsensor Networks: Modeling and Improvement Perspectives. In: DATE (2005)
5. Azzaz, S., Azouz Saidane, L.: Fault Repair Schemes for Static Wireless Sensor Networks Driven by an Analytical Energy Dissipation Model. In: Cichoń, J., Gębala, M., Klonowski, M. (eds.) ADHOC-NOW 2013. LNCS, vol. 7960, pp. 50–62. Springer, Heidelberg (2013)
6. Puthal, D.K., Sahoo, B.: A Finite State Model for IEEE 802.11 Wireless LAN MAC DCF. In: ICETET (2008)
7. Yang, O., Heinzelman, W.: Modeling and throughput analysis for SMAC with a finite queue capacity. In: ISSNIP (2009)
8. Azzaz, S., Saidane, L.: A centralized preventive maintenance strategy for wireless sensor networks. In: HP-MOSys (2012)
9. Azzaz, S., Saidane, L.: A distributed preventive maintenance strategy for wireless sensor networks. In: PE-WASUN (2012)
10. Dias, N., Zlot, R., Kalra, N., Stentz, A.: Market-Based Multirobot Coordination: A Survey and Analysis. Proceedings of the IEEE (2006)
11. Perkins, C., Belding-Royer, E., Das, S.: Ad hoc on demand Distance Vector (AODV) Routing. IETF RFC 3561 (2003)

Extending Network Tree Lifetime with Mobile and Rechargeable Nodes

Dimitrios Zorbas and Tahiry Razafindralambo

Inria Lille, Nord Europe, France
dimitrios.zormpas@inria.fr

Abstract. In this paper, we assume network trees consisting of mobile, energy constrained and rechargeable nodes as well as a static sink which collects the monitoring data and it is the root of the tree. Almost exhausted nodes can autonomously move towards a charging point to recharge their battery. However, this action leads to network disconnections and reduced lifetime since one or more predecessor nodes cannot forward their data to the sink. To alleviate this problem and extend network lifetime we examine the feasibility of replacing almost exhausted nodes using nodes with higher remaining energy. Based on this idea we propose a localized algorithm to autonomously replace nodes with high communication burden by the leaves of the tree. Both theoretical and simulation results show a big improvement in terms of network lifetime extension compared to the case where no replacement is performed and to the case where rerouting is considered.

1 Introduction

Wireless networks – such as sensor networks – are usually organized in trees or the applied routing/clustering protocols follow a tree-based scheme [1–3]. The leaves of the tree correspond to the nodes which monitor the environment and forward the data to the sink. A number of intermediate nodes act as relays if there is no direct communication between the monitoring nodes and the sink. These kind of networks are usually energy constrained since the nodes use batteries as power source.

An unavoidable problem is the loss of data when a relay node runs out of energy and, thus, the communication between its neighbors is lost. The nodes that are closer to the sink shoulder the most of the communication burden since they forward data of multiple leaves. A possible loss of a node which is close to the sink leads to the disconnection of the entire branch and, thus, to a huge loss of information.

The idea of using mobile nodes to control connectivity is not new, however, none of the previous works related to robot networks mentions the problem of lifetime extension [4–8]. To mitigate the aforementioned problem we assume that nodes with high energy consumption can be replaced by other tree members with high remaining energy. Since at each instance of time some nodes consume less energy than others, they will likely have much more after a certain period

S. Guo et al. (Eds.): ADHOC-NOW 2014, LNCS 8487, pp. 223–236, 2014.

of time. The idea is to use part of this energy for traveling in order to replace nearly depleted nodes and keep this particular branch of the tree alive for the rest of its members.

Since all the nodes are considered mobile and energy constrained, we assume that they can move towards a recharging point to recharge their battery if there is such a need. This action may shorten the lifetime of a node in the tree, but ensures that the same node will come back with more energy to replace other exhausted nodes keeping the network alive for longer time.

Several applications can benefit from the use of the proposed scheme. Ad-hoc network applications such as the coverage of points of interest using robots [9, 10] and the area surveillance using sensors or UAVs [11, 12] consist of nodes which some of them are deployed to provide coverage and some to provide relaying. Due to the multimedia nature of the transmitted data, the relay nodes shoulder a heavy communication load which leads to a fast depletion of their battery. Moreover, wireless sensor networks are often organized in clusters, where each cluster consists of many sensing nodes. Sensing nodes can be used to replace exhausted cluster-heads in case the re-election process is either power consuming or some nodes have been isolated after a cluster-head's failure [13]. Above all, recharging and replacement can lead to autonomous networks without the need of human supervision or manual replacement. In the case where multiple nodes exist in different sites of the network, the proposed solution can be used to redeploy the network and balance the energy resources between the sites [14].

This paper contributes in the following aspects. First, we analyze the conditions under which a node replacement is feasible taking into account the energy consumption of the nodes, their distance from the recharge point, the distance between the nodes, their initial energy, and the movement cost. Second, we present "CoverMe", a localized algorithm that extends the network lifetime taking advantage of node replacement and recharge. Third, we present theoretical and simulation results and we compare against the case where no replacement is done and against the case where rerouting is chosen to connect isolated sensing nodes with the sink.

The rest of the paper is organized as follows. In Section 2, some important works on node replacement and network lifetime extension are mentioned along with their advantages and disadvantages. In Section 3, we formulate the problem and we give several conditions for a feasible node replacement, while in Section 4 we present "CoverMe". In Section 5, we discuss the theoretical and simulation results, and Section 6 concludes the paper.

2   Related Work

Many researchers have focused their works on wireless sensor networks lifetime extension problem. This section will not review the whole literature on energy management and lifetime extension of this kind of networks. Instead, this section will focus on strategies proposed in the literature to offset the negative impact of energy depletion.

The literature proposes many definitions of the "network lifetime". However, in general, the network lifetime is upper bounded by the energy of the nodes composing the network. Some nodes of the network are more prone to energy exhaustion. These critical nodes are the nodes that support a huge amount of network traffic. In the case of a network tree, where the root of the tree is the data sink, these critical nodes are located close to the sink. Whatever the definition of "network lifetime" used, these critical nodes are the bottlenecks for network lifetime extension [15, 16].

The death of a critical node in the network can lead to different levels of malfunction ranging from increased data delay to network partition and data loss [17]. To alleviate that problem, a number of solutions has been proposed in the literature. These solutions can be categorized in two common approaches: (a) the replacement of nearly exhausted nodes with new ones [18–21], and (b) the data rerouting through other nodes with higher remaining energy [22–24]. Both approaches have some advantages and disadvantages. We will comment these two approaches in the following paragraphs.

The first approach can infinitely extend the network lifetime if the replacement process is well scheduled [21]. Different replacement strategies can be used depending on the application's level of criticality [18]. However, this approach assumes that an external entity or the nodes themselves can replace a dead node. This implies the mobility of the nodes. Moreover, sometimes this approach requires extra nodes which is translated to extra cost as well as to extra techniques related to network discovery and replacement.

In the second approach, the rerouting process sends the data to the sink by using the remaining nodes without any replacement. This approach is simple. It targets to balance the uneven traffic load between the relay nodes [23] and tries to avoid the appearance of holes and bottlenecks. This method does not require the use of any external entity to replace nodes nor the mobility of the nodes and works well in dense networks or in uniform networks where it is easy to find an alternative way towards the sink [24]. However, in any case, this second approach cannot extend the network lifetime indefinitely and will eventually lead to the death of the network.

We think that the first approach is more suitable for the applications described above due to network traffic intensity. Therefore, our solution is mainly based on the first approach borrowing some features from the second approach. We compare our strategy to a modified version (to have a fair comparison) of the rerouting approach proposed in [22].

3   Network Tree Lifetime

We assume that a node spends E_s energy units per time unit for sensing, E_t for transmitting, and E_r for receiving. It also spends E_m energy units per traveling meter and E_{id} energy units for the rest of the functions. All the nodes initially have the same amount of energy E_0. We also use $d(i, i')$ to refer to the Euclidean distance between nodes i and i'.

Depending on the functionality and the position of a node in the tree, it has one of the following four roles. It may be sensing node, relay node, moving node, or both sensing and relay. The energy consumption per time unit for each of these roles, when no data aggregation is used, is:

$$
E_i^{cons} = \begin{cases} E_s + E_t + E_{id} & \text{if } i \text{ is sensing node,} \\ (E_r + E_t)\nu + E_{id} & \text{if } i \text{ is relay node,} \\ E_s + (E_r + E_t)\nu + E_t + E_{id} & \text{if } i \text{ is both relay \& sensing node,} \\ E_m & \text{if } i \text{ is moving,} \end{cases} \tag{1}
$$

where ν is the number of predecessor sensing nodes of i. In the next paragraphs we present conditions explaining the feasibility of replacing a relay node using a sensing node. The overall extension time of the network branch is computed.

First of all, the energy consumption of the relay must be higher than the energy consumption of the sensing node. It follows that:

$$
(E_r + E_t)\nu - (E_t + E_s) > 0. \tag{2}
$$

Each node i leaves its current position at time t_i^r to recharge itself. The moment at which it must leave depends on its distance from the recharging point, so it will not run out of energy before reaching its target:

$$
t_i^r = \frac{E_0 - E_{i,rp}^{mov}}{E_i^{cons}} \tag{3}
$$

$E_{i,rp}^{mov}$ corresponds to the energy consumption for traveling from i to the recharging point rp and it is equal to $E_m d(i, rp)$.

Moreover, in order to replace a relay node p, a sensing node q must start moving t_q^{alar} time units before t_p^r, where

$$
t_q^{alar} = \frac{d(q, p)}{U} \tag{4}
$$

and U is the speed of the node in meters per time unit.

Using Formulas (2), (3), and (4), it follows that the sensing node will have ΔE amount of energy when it reaches the relay node. ΔE includes sensing node's energy consumption for $t_p^r - t_q^{alar}$ amount of time as well as the energy it needs to travel to the relay node:

$$
\Delta E = E_0 - (t_p^r - t_q^{alar})E_q^{cons} - E_{q,p}^{mov} \tag{5}
$$

The higher the ΔE, the longer the tree branch survives and the later the next replacement (if there is any) takes place. Combining all the previous equations, the extension time Xt^0 of the branch by a single replacement is given by:

$$
Xt^0 = \frac{\Delta E - E_{p,rp}^{mov}}{(E_r + E_t)\nu' + E_{id}} =
$$

$$
\frac{E_0 - (\frac{E_0 - E_m d(p, rp)}{(E_r + E_t)\nu + E_{id}} - \frac{d(q, p)}{U})(E_s + E_t + E_{id}) - E_m d(q, p) - E_m d(p, rp)}{(E_r + E_t)\nu' + E_{id}},
$$

$$
\tag{6}
$$

where ν' is the number of predecessor sensing nodes of the new relay node after the replacement. If a node from another branch has been used for the replacement, then ν' equals ν, otherwise ν' equals $\nu - 1$. In Formula (6) is considered that i is a relay node only. In case i is a relay and a sensing node at the same time, the formula must be updated accordingly.

Since the relay node must go to the recharging point and return back to its initial position after being fully recharged, Xt^0 must be higher than the time it needs to go to the recharging point plus the time it needs to get fully recharged plus the time it needs to travel back to its position:

$$Xt^0 \geq \frac{E_0}{E_{rech}} + \frac{2d(rp, p)}{U}, \tag{7}$$

where E_{rech} is the recharge energy per time unit.

If Condition (7) holds true, the branch will stay connected, while achieving the minimum possible loss of information until another relay or sensing node dies. Apparently, as more relay nodes go to recharge, the loss of information increases, since more sensing nodes are needed to replace the relay nodes.

On the other hand, if Condition (7) does not hold true, the recharging node may leave before it gets fully recharged in order to replace the node it was previously replaced by. In this case, the node must have enough energy to support the network at least till the recharging node comes back to its previous position and returns back to the base. This condition is given by the following formula:

$$\left(Xt^0 - \frac{2d(p, rp)}{U}\right)E_{rech} > 2E_m d(p, rp). \tag{8}$$

The new extension time after a partial recharge is given by:

$$Xt^1 = \frac{\left(Xt^0 - \frac{2d(p, rp)}{U}\right)E_{rech} - 2E_m d(p, rp)}{(E_r + E_t)\nu' + E_{id}}. \tag{9}$$

Generalizing, the overall accumulated extension time after k replacements is:

$$Xt^k = \sum_{j=1}^{k} \frac{\left(Xt^{j-1} - \frac{2d(p, rp)}{U}\right)E_{rech} - 2E_m d(p, rp)}{(E_r + E_t)\nu' + E_{id}}, \tag{10}$$

where $\left(Xt^{j-1} - \frac{2d(p, rp)}{U}\right)E_{rech} > 2E_m d(p, rp), \forall\, j \in \mathbb{N}^*$.

4   CoverMe

"CoverMe" is a localized algorithm that takes into consideration the ability of nodes to move towards a recharging point. It describes the basic steps a relay node can follow to be replaced by nodes with higher remaining energy. CoverMe is not a routing protocol but a trade-off mechanism which sacrifices part of the coverage to extend network lifetime.

The replacement process is divided in three steps. During the first step all the nodes of the network compute a threshold as it has been described in Formula (3). This is a critical threshold that the nodes use to avoid running out of energy. Note that this is a time threshold but since the nodes consume energy with a constant rate, this time threshold can be easily transformed to an energy threshold.

At this moment, the relay nodes do not take into account the time of Formula (4) since they have not chosen yet a replacement node. This is done in the next step where the relay nodes communicate with non-relay nodes and select their substitutes. In case of multiple candidates, a relay node chooses the node that is placed closer to it. In case where there exist multiple relays and multiple candidates, the replacement nodes are chosen in a first-come, first-served manner and no evaluation is done between them. It is worth pointing out that depending on how long is the communication range and how many hops the messages travel away, a relay node may find one, many or no candidates.

Once the replacement node has been chosen, the nodes recompute the threshold considering the time the candidate needs to travel towards the relay node. During the last step, when the relay's energy is close to the threshold, the corresponding substitute starts moving. After the replacement the sensing node becomes the new relay node, while the old one is driven towards the recharging point. It is worth mentioning that a partially charged node is considered to be a candidate node and it can be used for future replacements unless Criterion (8) does not hold true. CoverMe prefers choosing partially recharged candidates since they are usually closer to the 1-hop relay nodes and, moreover, the remainder of the sensing nodes keep their positions prolonging the coverage time.

This three-step process continues until all the sensing nodes have depleted their energy or none of the sensing nodes can reach the sink. A relay node that cannot support any sensing node due to a network partition is considered isolated and goes recharging even if its energy threshold is placed much later. Fully recharged nodes take the initial position of the node that they were replaced by during the last replacement. This means that a relay node which was replaced by a sensing node, it will take the initial position of the sensing node.

5   Evaluation and Discussion of the Results

In this section we present theoretical results based on the analysis done in Section 3 and we discuss the feasibility of node replacement using rechargeable mobile nodes based on real values. At the same time, we simulate "CoverMe" and we compare its performance to other approaches. The results are divided in two parts. In the first part, we present results related to 2-hop networks and we compare the performance of CoverMe to theoretical results and to the approach where no replacement is performed. In the second part, we assess CoverMe in multi-hop networks and we compare its performance to the approach where no replacement is done and to the approach where rerouting is chosen to reconnect non-connected nodes to the sink. The simulations were performed on a custom simulator[1] using an ideal MAC layer. The highest density of the network was one node per 400 square meters.

For the evaluation purposes we used the following values concerning the energy consumption parameters and the speed of the nodes: $E_r = 2.5J/s$, $E_t = 5J/s$, $E_s = 2.5J/s$, $E_i = 8J/s$, $E_m = 25J$, $U = 0.9m/s$, $E_{rech} = 27J/s$. E_r, E_i, E_m and U were experimentally found using Wifibots[2]. E_t was computed considering the first-order radio model [25], a communication range of 50m, a packet size of 1KB, and transmission rate of 1 packet/s.

A simple 2-hop simulation scenario was considered for the first part of the simulation with one relay node and ν sensing nodes with equal distances from the relay. The sink as well as the recharging point were located in the middle of the left side of the 10K m^2 terrain. Only the relay node had direct communication with the sink. Each simulation has been executed 50 times and the average results are presented. We must mention that using the energy consumption values presented in the previous paragraph Condition (7) cannot be achieved. It actually means that the network will die before the recharging node gets fully recharged. Partial recharge has been used in that case.

Concerning the second part of simulations, we present the average results of 50 instances per scenario as well as the 95% confidence intervals. We assume two deployment types; a random uniform and a non-uniform based on the Gaussian distribution. The sink is located in the middle of the left side of the terrain and the terrain size is enlarged to 40K m^2. For fair comparison reasons the normalized network lifetime is presented instead of the actual one. It is given by the sum, $\sum_{i=1}^{\tau} \frac{\#\_of\_sensors\_active_i}{total\_\#\_of\_sensors}$, where τ is the time where no active sensing node exists in the network.

5.1   Two-hop Networks

The first figure depicts the theoretical lifetime extension when no recharging is applied (see Figure 1). Equation (6) was used to create this figure with the previously mentioned values regarding the energy consumption model. The figure shows how lifetime changes with the increase of initial node energy for different sensor populations and distances. Concerning the distance between the nodes, we assume that $d(i', i)$ is equal to $d(i, rp)$. The theoretical results show that the higher the initial energy, the more the lifetime can be improved. However, it is worth observing that for all populations of ν the improvement converges to a maximum value which is higher when the number of successors is high. When the distance between the nodes is high, at least 4 KJ of initial energy are needed since the nodes consume more energy for the movement.

The corresponding simulation results are presented in Figure 2. It can be observed that CoverMe presents similar behavior to the theoretical measurements extending the lifetime up to 95%. For low initial energy, the trends seem to be slightly different. This happens because in CoverMe the time is divided in

---

[1] http://autonomous-tree.gforge.inria.fr/
[2] Wifibots mobile robots, http://www.wifibot.com/

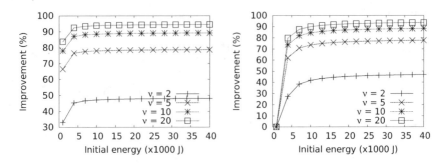

Fig. 1. Theoretical lifetime improvement (%) for different values of initial energy, ν, and 10m distance (left figure) or 50m distance (right figure)

rounds. In each round every node checks if its remaining energy will fall below the threshold during the next round according to the current energy consumption. This process may lead to an early departure of the node, and thus to slightly different results compared to the theoretical absolute values when the initial energy is low.

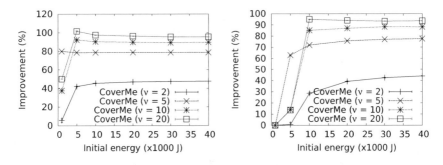

Fig. 2. Simulated lifetime improvement (%) for different values of initial energy, ν, and 10m distance (left figure) or 50m distance (right figure)

Figure 3 presents the theoretical lifetime improvement that can be achieved when recharging is taken into account. Formula (10) was used for drawing the graphs. First of all, we can observe that the improvement is high even when the nodes have low initial energy (5-10 KJ for low distance), and it gradually converges to a maximum value for all values of ν. Second, the maximum improvement is achieved when the number of successor sensing nodes is 5 for both low and long distances. It is impressive that in this case the lifetime can be improved up to 200% when the initial energy is high.

Finally, Figure 4 illustrates the corresponding simulation results of the lifetime improvement over the approach where no replacements are done. The line trends are similar to those of the previous figure while the maximum absolute values

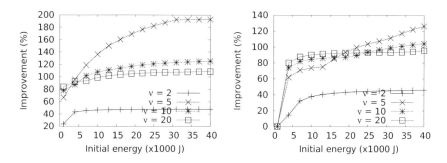

Fig. 3. Theoretical lifetime improvement (%) for different values of initial energy, ν, and 10m distance (left figure) or 50m distance (right figure). Node recharging is taken into account.

are quite the same. The results are slightly different for low energy values for the same reason explained in the first simulation. The best improvement is achieved when ν is 5 and when the initial energy is above 5 KJ and 20 KJ for the low distance and the high distance case respectively.

It is important to notice here that there exists a trade-off between the value of ν and the lifetime improvement. Indeed, the higher the value of ν, the higher the quantity of data a relay node has to forward and, thus, the more its energy consumption. However, the higher the value of ν the higher the number of nodes that can replace a dying relay. This trade-off explains the fact that when $\nu = 2$, the lifetime extension is lower than the lifetime extension when $\nu = 5$. In the same way, the lifetime extension when $\nu = 10$ is higher than the lifetime extension when $\nu = 20$. Indeed, the existence of this trade-off rises the issue of a balanced network tree construction with limited leaves.

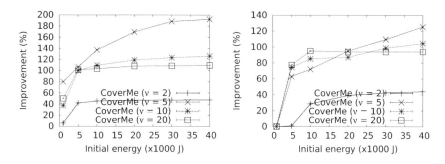

Fig. 4. Simulated lifetime improvement (%) for different values of initial energy, ν, and 10m distance (left figure) or 50m distance (right figure). Node recharging is taken into account.

5.2   Multi-hop Networks

This section evaluates the use of CoverMe in multi-hop networks where branches
may consist of multiple relay nodes and multiple other sub-branches. It actually
means that different nodes can move or be replaced at the same time while several
others may be disconnected. Due to the very large terrain size we assume that
a sensing node can connect to another relay node if no substitute is selected
by CoverMe. Figure 5 depicts the performance of the three approaches with
uniform node positions (left figure) and non-uniform positions (right figure). The
normalized network lifetime is measured for different sensing node populations,
10K Joules of energy, and when no recharging is done. The algorithms present
similar performance in uniform topologies but CoverMe yields more lifetime in
non-uniform topologies with a maximum improvement of 60%. "Rerouting" does
not perform the same since in the non-uniform scenario the probability of finding
a new path to the sink is lower than in the uniform scenario.

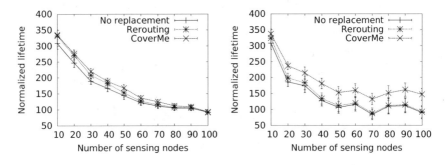

Fig. 5. Normalized network lifetime for a multi-hop scenario with no recharging, variable number of sensing nodes, 10K Joules initial energy and uniform (left figure) or
non-uniform (right figure) node deployment

In Figure 6 we measure the normalized network lifetime for a similar scenario. In this case, the initial energy varies and the number of sensing nodes is
fixed to 50. For both types of deployment CoverMe presents better performance
which increases with the initial energy. Similarly to the previous scenario, the
gap between CoverMe and the other approaches is higher for non-uniform node
deployments.

An almost identical performance is achieved when recharging is taken into
account. The corresponding results are presented in Figures 7 and 8. This almost
identical behavior appears since only a few relay nodes can take advantage of
the extra recharging energy. As explained in Section 4, the relay nodes select
the closest to them sensing nodes when multiple candidates exist. Since many
relay nodes are far from the sink (and recharging point) they will most likely
select a sensing node for the replacement than a recharging one which may be
far away. An opposite strategy which prefers recharging nodes instead of the

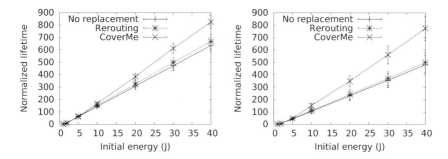

Fig. 6. Normalized network lifetime for a multi-hop scenario with no recharging, variable initial energy, 50 sensing nodes and uniform (left figure) or non-uniform (right figure) node deployment

closest candidate could also be used. However, much energy would be wasted in traveling. Nevertheless, the decision of selecting the optimal strategy is an open problem.

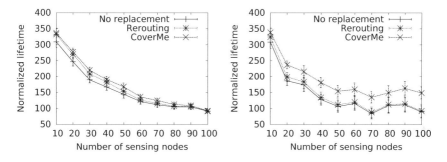

Fig. 7. Normalized network lifetime for a multi-hop scenario with recharging, variable number of sensing nodes, 10K Joules initial energy and uniform (left figure) or non-uniform (right figure) node deployment

Figure 9 depicts the number of messages sent by "CoverMe" and "Rerouting" throughout the process. The simplest (No replacement) method has been excluded from this simulation due to its negligible overhead cost. We assumed that each relay node communicates with its selected substitute every 10 iterations to ensure that it is alive. CoverMe sends less messages than Rerouting for the most of node populations since its behavior mainly depends on the activity of the relays. On the other hand, the activity of "Rerouting" depends on the number of sensing nodes. In case of a disconnection all the disconnected sensing nodes send messages to nearby relays in order to find a route towards the base station. When a few sensing nodes are placed, the number of sensing nodes is comparable to that of relays, so the two approaches produce more or less equal

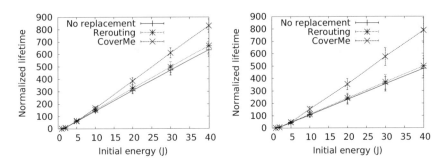

Fig. 8. Normalized network lifetime for a multi-hop scenario with recharging, variable initial energy, 50 sensing nodes and uniform (left figure) or non-uniform (right figure) node deployment

number of messages. We must mention that when the communication between the relays and the substitutes is done less frequently (every 30 iterations), the number of messages is reduced by 20%.

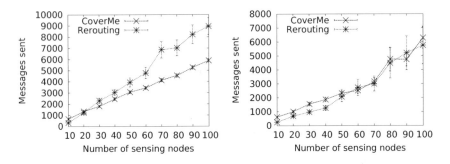

Fig. 9. Number of messages sent for a scenario with variable number of nodes and uniform (left figure) or non-uniform (right figure) node deployment

6   Conclusion and Future Work

The lifetime extension problem of energy constrained network trees was examined in this paper. In particular, we analyzed the feasibility of node replacement and recharging when nodes with high communication burden are replaced by other network members with high remaining energy. The theoretical and the simulation results showed a high performance gain in terms of lifetime for both 2-hop and multi-hop networks. The lifetime can be improved up to 200% in 2-hop networks and up to 60% for multi-hop networks, especially in the case where the nodes are not uniformly placed.

Although the proposal exhibits a high performance gain, there is room for further improvement and investigation. The selection of the best candidate and the best strategy (during the replacement process) it seems to be a critical issue since a trade-off appears between coverage time and network lifetime. On the other hand, the percentage of lifetime extension can be maximized using a certain number of predecessor nodes (sensing nodes) for each relay. This number derives from Equation (10) and can be used to construct balanced network trees with specific number of relays and leaves.

Acknowledgments. This work is partially supported by the French National Research Agency (ANR) under the VERSO RESCUE project (ANR-10-VERS-003) and by a grant from CPER Nord-Pas-de-Calais FEDER CIA.

References

1. Santi, P.: Topology control in wireless ad hoc and sensor networks. ACM Comput. Surv. 37(2), 164–194 (2005)
2. Abolhasan, M., Wysocki, T., Dutkiewicz, E.: A review of routing protocols for mobile ad hoc networks. Ad Hoc Networks 2(1), 1–22 (2004)
3. Gupta, G., Younis, M.: Load-balanced clustering of wireless sensor networks. In: IEEE International Conference on Communications, ICC 2003, vol. 3, pp. 1848–1852 (2003)
4. Dantu, K., Sukhatme, G.: Connectivity vs. control: Using directional and positional cues to stabilize routing in robot networks. In: Second International Conference on Robot Communication and Coordination, pp. 1–6 (March 2009)
5. Li, L., Halpern, J.Y., Bahl, P., Wang, Y.M., Wattenhofer, R.: A cone-based distributed topology-control algorithm for wireless multi-hop networks. IEEE/ACM Transactions on Networking 13(1), 147–159 (2005)
6. Butterfield, J., Dantu, K., Gerkey, B., Jenkins, O., Sukhatme, G.: Autonomous biconnected networks of mobile robots. In: 6th International Symposium on Modeling and Optimization in Mobile, Ad Hoc, and Wireless Networks and Workshops, WiOPT 2008, pp. 640–646 (April 2008)
7. Williams, R., Sukhatme, G.: Locally constrained connectivity control in mobile robot networks. In: 2013 IEEE International Conference on Robotics and Automation (ICRA), pp. 901–906 (2013)
8. Beer, B., Mead, R., Weinberg, J.B.: A distributed spanning tree method for extracting systems and environmental information from a network of mobile robots. In: AAAI Spring Symposium: Multirobot Systems and Physical Data Structures. AAAI (2011)
9. Erdelj, M., Razafindralambo, T.: Design and implementation of architecture for multi-robot cooperation in the context of wsn. In: Proceedings of the 10th ACM Symposium on Performance Evaluation of Wireless Ad Hoc, Sensor, & Ubiquitous Networks, PE-WASUN 2013, pp. 33–40. ACM, New York (2013)
10. Loscrí, V., Natalizio, E., Razafindralambo, T., Mitton, N.: Distributed algorithm to improve coverage for mobile swarms of sensors. In: 2013 IEEE International Conference on Distributed Computing in Sensor Systems (DCOSS), pp. 292–294 (2013)

11. Costanzo, C., Loscr, V., Natalizio, E., Razafindralambo, T.: Nodes self-deployment for coverage maximization in mobile robot networks using an evolving neural network. Computer Communications 35(9), 1047–1055 (2012); Special Issue: Wireless Sensor and Robot Networks: Algorithms and Experiments

12. Zorbas, D., Razafindralambo, T., Di Puglia Pugliese, L., Guerriero, F.: Energy efficient mobile target tracking using flying drones. Procedia Computer Science 19, 80–87 (2013); The 4th International Conference on Ambient Systems, Networks and Technologies (ANT 2013)

13. Gao, T., Jin, R., Song, J., Xu, T., Wang, L.: Energy-efficient cluster head selection scheme based on multiple criteria decision making for wireless sensor networks. Wireless Personal Communications 63(4), 871–894 (2012)

14. Zorbas, D., Razafindralambo, T.: Wireless sensor network redeployment under the target coverage constraint. In: 2012 5th International Conference on New Technologies, Mobility and Security (NTMS), pp. 1–5 (May 2012)

15. Chang, J.H., Tassiulas, L.: Maximum lifetime routing in wireless sensor networks. IEEE/ACM Trans. Netw. 12(4), 609–619 (2004)

16. Wang, Q., Zhang, T.: Bottleneck zone analysis in energy-constrained wireless sensor networks. IEEE Communications Letters 13(6), 423–425 (2009)

17. Yang, Y., Fonoage, M.I., Cardei, M.: Improving network lifetime with mobile wireless sensor networks. Computer Communications 33(4), 409–419 (2010)

18. Tong, B., Wang, G., Zhang, W., Wang, C.: Node reclamation and replacement for long-lived sensor networks. IEEE Transactions on Parallel and Distributed Systems 22(9), 1550–1563 (2011)

19. Pryyma, V., Bölöni, L., Turgut, D.: Uniform sensing protocol for autonomous rechargeable sensor networks. In: Proceedings of the 11th International Symposium on Modeling, Analysis and Simulation of Wireless and Mobile Systems, MSWiM 2008, pp. 92–99. ACM, New York (2008)

20. Mei, Y., Xian, C., Das, S., Hu, Y.C., Lu, Y.H.: Sensor replacement using mobile robots. Comput. Commun. 30(13), 2615–2626 (2007)

21. Magklara, K., Zorbas, D., Razafindralambo, T.: Node discovery and replacement using mobile robot. In: Zheng, J., Mitton, N., Li, J., Lorenz, P. (eds.) ADHOC-NETS 2012. LNICST, vol. 111, pp. 59–71. Springer, Heidelberg (2013)

22. Lee, H., Keshavarzian, A., Aghajan, H.: Near-lifetime-optimal data collection in wireless sensor networks via spatio-temporal load balancing. ACM Trans. Sen. Netw. 6(3), 26:1–26:32 (2010)

23. Uthra, R.A., Raja, S.V.K.: Qos routing in wireless sensor networks—a survey. ACM Comput. Surv. 45(1), 9:1–9:12 (2012)

24. Ishmanov, F., Malik, A.S., Kim, S.W.: Energy consumption balancing (ecb) issues and mechanisms in wireless sensor networks (wsns): a comprehensive overview. European Transactions on Telecommunications 22(4), 151–167 (2011)

25. Heinzelman, W., Chandrakasan, A., Balakrishnan, H.: Energy-efficient communication protocol for wireless microsensor networks. In: Proceedings of the 33rd Annual Hawaii International Conference on System Sciences, vol. 2, p. 10 (2000)

Energy Efficient Stable Routing
Using Adjustable Transmission Ranges
in Mobile Ad Hoc Networks

Abedalmotaleb Zadin and Thomas Fevens

Department of Computer Science and Software Engineering
Concordia University, Montréal, Québec, Canada
{a_zadin,fevens}@cse.concordia.ca

Abstract. As the demand for mobile ad hoc wireless network (MANET) applications grows, so does its use for many important services where reliability and stability of communication are of great importance. In MANETs, exchanging messages during route discovery and the actual communication over these routes come at a cost of energy expenditure by the device. In this paper, we study this problem in the context of maintaining stable connections between mobile nodes. We propose a position-based routing protocol (GBR-DTR-CNR) that is both link stable and energy efficient for MANETs. Link stability will be supported by choosing links for routing based on a link expiration time estimation and a conservative neighborhood range, while energy efficiency will improved by using an adjustable dynamic transmission range that takes into account the mobility of the nodes. Simulation results demonstrate the effectiveness of GBR-DTR-CNR compared to similar routing protocols.

Keywords: Mobile Ad Hoc Networks, Dynamic Transmission Range, Energy Efficiency, Stable Connection.

1 Introduction

Wireless networks are formed by interconnected devices communicating wirelessly within a relatively limited area. Mobile ad hoc networks (MANETs) are a type of wireless network where mobile devices are themselves responsible for communication with each other without the presence of a centralized infrastructure. Devices in MANETs can typically move in any direction they want and therefore links between them and other devices may frequently change. Link failures leading to a connection break will cause the traffic flow to be interrupted until a new route is found, which leads to packet delivery gaps which are unacceptable for real-time applications such as mobile wireless telemedicine. Moreover, each device in a MANET is not only responsible for network traffic related to itself but also has to forward unrelated traffic as an intermediary which results in added energy consumption per device. MANETs devices are typically only powered by batteries, with limited computing capability while the number

S. Guo et al. (Eds.): ADHOC-NOW 2014, LNCS 8487, pp. 237–250, 2014.

of such devices may be large. The constrained battery capacity is one of the most important limitations in developing applications and services for mobile devices [1]. As MANET systems become more widely deployed, in addition to seeking to maintain stable connections, the energy consumption that is required to communicate over these connections, consuming a large part of the available energy resources of the mobile devices, has to also be taken into consideration.

Many routing protocols have been proposed to improve the routing efficiency in MANETs. Those protocols can be broadly categorized into two approaches: Topology-based and Position-based routing. For MANETs, most recent work on routing that is stability-oriented has been for topology-based routing with exception of the position-based Greedy-based Backup Routing Protocol (GBR) [2] using multi-paths to maintain link stable paths. For position-based routing on mobile ad hoc networks, no previous work has studied stable routing in combination with energy efficiency.

In this paper, we study this problem by using position-based routing based on the combination of link stable routing with dynamic transmission ranges. The dynamic transmission ranges establish energy efficiency while maintaining the high connection throughput enabled by stable connections. We will use, in particular, an approach based on GBR [2] to ensure link stability.

The rest of this paper is organized as follows. In Section 2, we will review specifically related work including some details on position-based stable routing protocols, approaches to determine energy consumption in MANETs, and using dynamic transmission power during routing. In Section 3, we propose different energy efficient routing protocols based on stable routing protocols and discuss the motivations for each. Simulation results are given in Section 4. Finally, concluding remarks are made in Section 5.

2   Related Work

Routing in a MANET depends on many factors including network topology, the type of information available during routing, and specific underlying network characteristics that could that can be used to define a heuristic to find a path quickly and efficiently.

2.1   Network Model

A MANET can be modeled using a graph $G = (V, E)$ where V represents the set of nodes/vertices, and E represents the set of links/edges. Each edge represents a link between two nodes currently within the transmission range which, for this paper, we will assume to be the same for all nodes [3] (the resulting graph is termed a unit disk graph (UDG)). Two nodes are considered to be neighbors if they are within the transmission range of each other and an edge exists between them. We will denote the set of neighbors of a node v_i by $N(v_i)$. A path of length n between a source node S and a destination node D is denoted by $(S = v_0, v_1, v_2, \ldots, v_n = D)$ where $v_i \in V$ and $v_i \in N(v_{i-1})$. Each node in the MANET will have a unique identifier and know its geographic position. In the real world, we will assume that the location of the nodes in a MANET will be tracked using Global Positioning System (GPS) and/or Location Services (LS) [2, 4]. We will assume the nodes are arranged in a two dimensional (2D) Euclidean space such that G is a geometric graph. Since the nodes may be mobile, we will also assume that all nodes will, at regular intervals, broadcast their positions to their neighbors using HELLO messages.

2.2   Routing and Route Discovery Protocol on MANETs

There are two main categories of routing on ad hoc networks, namely, Topology-based routing and Position-based routing. Topology-based routing protocols use link information available from the network to determine a route between the nodes. Position-based routing uses the positions of nodes to determine routes [5]. In position-based routing each network node is informed about its position, its neighbors' positions, and position of the destination. Position based routing use one of, or a combination of, two types of geometric based routing which are Greedy and Face based routing. In Greedy routing each node makes a local optimum choice while selecting its next hop, such as the node geographically closest to the destination [6]. Another position-based routing algorithm is Face routing, or perimeter routing. If the network topology is a connected graph then if Face routing is performed with respect to a planar subgraph, then the routing guarantees packet delivery [7, 8]. Hybrid routing combines the benefits of both Greedy and Face routing. For example, when greedy routing fails to find closer node(s) leading to the destination, face routing helps to obtain an alternate path until a closer node is found whereupon greedy routing resumes. This routing technique is called greedy face greedy (GFG) [7]. A similar hybrid based routing algorithm is Greedy Perimeter Stateless Routing (GPSR) [8].

In this paper, to determine stable connections for MANETs we will base our proposed protocols on the basic multi-path route discovery protocol GBR originally presented by Yang et $al.$ [2]. In GBR, to discover a route from source node S to destination D, GPSR is used. The path discovered is termed the primary path. Then backup paths are determined that provide link protection for the links of the primary path (see Figure 1). Since these backup paths have to survive after the link expires we need to know the lifetimes of both individual links and paths as a whole. Following [2], we denote these lifetimes, respectively, as Link Expiration Time $LET(v_i, v_{i+1})$ for the link $v_i v_{i+1}$, and Path Expiration Time $PET(P)$ for a path P. $LET(v_{i-1}, v_i)$ is defined as

$$LET(v_{i-1}, v_i) = \frac{-(pl + qd) + \sqrt{(p^2 + q^2)R^2 - (pd - lq)^2}}{p^2 + q^2} \tag{1}$$

where $p = \tau_{i-1} \sin \theta_{i-1} - \tau_i \sin \theta_i$, $q = \tau_{i-1} \cos \theta_i - \tau_i \cos \theta_i$, $l = X_{i-1} - X_i$, $d = Y_{i-1} - Y_i$, and (X_i, Y_i) are the node coordinates, τ_{i-1} and τ_i are the node velocities, θ_{i-1} and θ_i are the direction angles (with respect to the positive X-axis), and R is the transmission range.

2.3   Stable and/or Energy Efficient Routing on MANETs

For MANETs, most recent work on routing that is stability-oriented (occasionally combined with energy efficiency) has been for reactive topology-based routing, specifically for DSR [9, 10] and AODV [11, 12]. Hamad *et al.* [13] presented a routing protocol called Line Stability and Energy Aware (LSEA) which was a modified version of AODV. For protocols based on multicasting, Zhang *et al.* [14] use multicast trees and a stability evaluation metric to propose a stability-based multicast routing protocol, while Mohamamdzadeh *et al.* [15] used multicast trees to create an energy-aware, stable routing protocol. For static ad hoc networks, energy-efficient routing has been extensively studied [16–21]. For position-based routing for static ad hoc networks, Seada *et al.* [20] used an analytical link loss model to strike a balance between shorter, high-quality links and longer lossy links. Wang *et al.* [21] based their choice of neighbors for routing on a critical transmission radius for energy efficiency combined with dynamic transmission ranges, to define their Energy-Efficient protocol (LEARN: Localized Energy Aware Restricted Neighborhood Routing for Ad Hoc Networks).

For position-based routing for mobile ad hoc networks, using the position information of the nodes, we can evaluate links in terms of Euclidean distance and Link Expiration Time (LET) between two neighborhood nodes participating in the path. To increase the stability of routing in MANETs and provides a reliable end-to-end route, one approach would be for each node to choose the most stable route from its options [22]. Alternatively, to improve the stability along the path in the presence of expiring links, another approach is to maintain multiple paths along the connection. In particular, Yang *et al.* [2] achieve a reliable connection by protecting the links between each pair of nodes participating in the primary path by maintaining local backup paths in parallel with each link in the path to be used when that link expires. The protocol presented by Yang *et al.* is called the Greedy-based Backup Routing Protocol (GBR). Zadin and Fevens [23] introduced a variation of GBR used only neighbors during routing from a conservative neighborhood range that maintained path stability without the need to determine backup paths. Also, Zadin and Fevens [24] introduced a variation of GBR that improved path stability in the presence of both node and link failure.

The backup path for the link $v_i v_{i+1}$ is determined, if it exists, using a neighboring node m_j not on the primary path with the maximum PET as given by Equation 2.

$$PET(v_i, m_j, v_{i+1}) = \min(LET(v_i, m_j), LET(m_j, v_{i+1})) \qquad (2)$$

where m_j is a neighboring node to both v_i and v_{i+1}.

Dynamic Transmission Power in MANETs. The battery lifetime of wireless devices is one of the most important issues that affect the energy stability in MANETs. Thus many protocols have been proposed to improve the energy usage in MANETs through control of the transmission power. The basic approach of

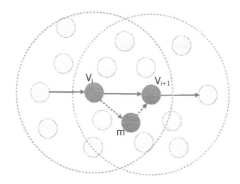

Fig. 1. Selecting a one hop backup path for link protection

assignment of different transmission powers to different nodes has been explored for static wireless devices [25] and centralized systems [26] leading to extended battery life of nodes. Kim and Eom [27] presented novel reprogramming scheme that uses dynamic transmission power control, to deal with the energy consumption of each wireless sensor node and the network load distribution. Also, Wu and Dai [28] proposed the distributed solution based on reducing energy consumption and density of the virtual backbone network using adjustable transmission range combined with clustering. In this paper, we use adjustable transmission power dynamically dependent on the location of next hop node to improve both link stability and energy efficiency for MANETs.

3   Proposed Stable, Energy Efficient Routing Protocols

Our study is focused on developing protocols that improve both energy consumption and communication stability. Furthering the work in [2, 23], we propose to reduce the energy consumption while maintaining overall communication stability of the routing by using idea of an adjustable transmission range that has been adapted to account for the mobility of the nodes. In the following, we introduce and discuss various routing protocols used to explore this idea.

3.1   GBR-STR

The GBR-STR protocol uses the stable routing protocol GBR with a Static Transmission Range (STR) which uses the fixed transmission range R. This protocol is identical to GBR as proposed by Yang *et al.* [2]. The rest of the protocols will be compared primarily to this protocol, which represents the best known position-based stable routing for MANETs.

3.2   GBR-DTR

If a node v_{i-1} wants to send a message to the node v_i, where the link has a link expiration time of $LET(v_{i-1}, v_i)$, it can conserve energy while maintaining the link by adjusting its transmission range from R to a range that is closer to the distance to v_i. For the selection of an adjustable transmission range R_a as shown in Figure 2, we assume that all packets have the same size and that the maximum transmission range is R. For energy considerations, at the beginning of each HELLO interval, we seek to use the smallest radius of transmission along the link from node v_{i-1} to node v_i in the connection while not allowing the connection to break prematurely during the HELLO interval. Let I denote the time between HELLO messages, let L in Figure 2 be the distance from node v_{i-1} to node v_i, and LET is link expiration time which calculated by Equation 1. Then we can define $W = (R - L)/\min(I, LET)$. Since $\min(I, LET)$ is the time until the distance between v_{i-1} and v_i equals R (when $I \leq LET$) or makes its closest approach to R (when $I > LET$), then the value W is the velocity of v_i approaching distance R in a radial direction. Then the adjusted transmission range R_a can be calculated as $R_a = L + r$ when W is positive, where $r = Wt$ and t is the interval of time between HELLO beacon broadcasts; or $R_a = L$ otherwise.

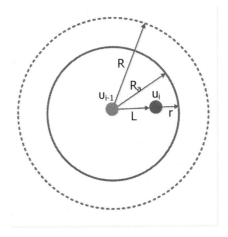

Fig. 2. Selecting a Transmission Range

The GBR-DTR protocol uses the stable routing protocol GBR with a Dynamic Transmission Range (DTR).

3.3 GBR-DTR-CNR

Sun *et al.* [11] presented a link stability based routing protocol based on AODV, utilizing the idea of a stable zone and caution zone around nodes so to initiate re-routing when a routing neighbor enters the caution zone. In [23] we proposed a similar approach for position-based routing, which we review here. Since nodes that are moving constantly with different speeds and directions, a node positioned within the transmission range of another neighboring node at certain time moment might be out of the range in another moment. In GBR, to construct the primary path, each node selects the closest node to the destination within its transmission range, if any, as its next hop. Therefore, such a node which is at the time of selection within the transmission range of the routing node may move out of range before the next HELLO beacon will broadcast. Indeed, by preferably picking nodes near the boundary of the transmission range, there is a high probability that the nodes which are picked as the next hops will no longer be within transmission range before the next HELLO beacon broadcast resulting broken links on the primary path. As we introduced in [23], with a Conservative Neighborhood Range (CNR) we take into account the possibility of nodes that can go out of the range during the interval and subsequently avoid including them in the path leading to a significant reduction in the packets losses as well as increasing the reliability of communication. The CNR depends on the interval between HELLO message broadcasts and its value is R_c as given as $R_c = R - (v_{max}t)$ where R_c is conservative neighborhood range, R the actual transmission range, v_{max} the maximum node velocity, and t is the time interval between HELLO message broadcasts. If the next hop neighbor v_{i+1} is chosen within this CNR of v_i then v_{i+1} will not go out of the transmission range of v_i during this interval, and thus no links in the primary path will break before the next HELLO beacon will be broadcast.

The GBR-DTR-CNR protocol uses the stable routing protocol GBR with a Dynamic Transmission Range (DTR) with Conservative Neighborhood Range (CNR) for neighbor selection.

3.4   LBR-STR

In order to try to develop a more energy-efficient variation of GBR, we also consider the Energy-Efficient protocol LEARN proposed by Wang *et al.* [21] for static ad hoc networks. Assume that energy required for a transmission from node u to a neighbor v is $E(\|uv\|)$. Then LEARN chooses the next node during route discovery on the neighboring node with respect to a critical transmission radius r_0 which is the distance d where $d/E(d)$ is maximum. For a node u, define the interior region of a 2-D cone CN with respect to the destination node D with its apex at u and centered on the line from u to D with cone half-angle (the angle from center line of the cone to the side of cone) of θ, and define the interior area of a 2-D torus TO which includes the region with distance between $\eta_1 r_0$ and $\eta_2 r_0$, for constant parameters θ, η_1, and η_2, as illustrated in Figure 3. Further, they define a Restricted Neighborhood Area (RNA) for a node u to be intersection of CN and TO. Then during route discovery for the next hop from u, LEARN will choose the neighbor v_i with maximum $\|uv_i\|/E(\|uv_i\|)$ in RNA, or if none, then choose the neighbor closest to D in CN, or if none still, by default, a neighbor is chosen as would be chosen by GPSR.

To maximize the energy-efficiency properties of the stable routing protocol GBR, it would appear reasonable to replace the primary greedy routing algorithm GPSR with the energy-efficient algorithm LEARN. Therefore, we propose a variation of GBR by replacing GPSR with LEARN which we will simply reference as LEARN-based Backup Routing (LBR). The LBR-STR protocol uses the stable routing protocol LBR with a Static Transmission Range (STR) which uses the fixed transmission range R.

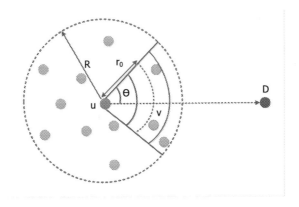

Fig. 3. Example of how LEARN chooses the next hop

3.5   LBR-DTR

The LBR-DTR protocol uses the stable routing protocol LBR with a Dynamic Transmission Range (DTR) which uses the adjustable transmission range R_a to reduce the energy expenditure during transmissions to neighboring routing nodes.

4   Performance Evaluation

This section presents simulation results comparing the algorithms GBR-STR, GBR-DTR, GBR-DTR-CNR, LBR-STR, and LBR-DTR. First we discuss the simulation setup and then give the simulation results.

4.1   Simulation Setup

For all algorithms GBR-STR, GBR-DTR, GBR-DTR-CNR, LBR-STR, and LBR-DTR, we constructed both the primary path and the backup path as described in Section 2.2. The simulation environment is modeled using network parameters that are a network area of size $2200m \times 2200m$; a varying number of nodes from $200, 250, 300, \ldots, 600$; a maximum transmission range of $R = 250m$ or adjusted transmission range. Each simulation ran for 600 seconds with enough packets and energy assigned for the simulation time. There are 20 pairs of Constant Bit Rate (CBR) data flows in the network layer, and non-identical source and destination flows were randomly selected, each flow did not change its source and destination throughout the simulations. The direction in which a node can move is given randomly at the beginning of the simulation. However, when a node reaches the boundary at angle ϕ, we reflect the node off the boundary using the formula $\phi + \pi/2 + C$ [29]. For each different node density, randomly distributed 40 connected graphs were used as a starting network topology for each run of the simulation for all algorithms. This is done to get average performance results for better analysis. The velocity was chosen to be the same for all nodes at $V = 10m/s$, and the HELLO beacon interval, I, was set to 2 seconds. At the end of the 2 second interval, if a path is determined to fail within the next 2 second interval (from the path's PET value), then at the beginning of the next HELLO interval, a new path is determined between the source and destination.

For the energy cost calculation for the power required for the transmission of a signal over a link from node v_{i-1} to node v_i, following the model of Stojmenović and colleagues [16,17], we will use the energy cost function $E(\|v_{i-1}v_i\|)$ as shown in Equation 3:

$$E(\|v_{i-1}v_i\|) = a \left(\frac{\|v_{i-1}v_i\|}{R} \right)^{\alpha} + c. \tag{3}$$

For energy calculations, we set $\alpha = 2$, $a = R^{\alpha}$, and $c = (R/2)^2$. Further, we set the receiver cost will be a constant $E_{recv} = c$, and we assume that there is no energy cost effect for idle periods or discarded packets. For the LEARN algorithm, we use the constant parameter values $\theta = \pi/3$, $\eta_1 = 1/2$, and $\eta_2 = 2$. Since $c = (R/2)^2$ then the critical transmission radius r_0 for LEARN is $\sqrt{c} = R/2$ [21].

4.2   Simulation Results

This subsection shows a comparison of the performance of the GBR-STR, GBR-DTR, GBR-DTR-CNR, LBR-STR, and LBR-DTR. In particular, we will show the experimental results regarding the effect of varying the node density. In order to compare the performance of our proposed GBR-DTR-CNR protecting scheme with the rest of the algorithms, an intensive set of experiments using Matlab were performed. The performance metrics that we are interested in are the rate at which the transmitted packets are delivered, PDR (Packet Delivery Ratio) which equals $\#P_d/\#P_s$ where $\#P_d$ is the total packets delivered during the simulation, and $\#P_s$ is the total number of packets sent during the simulation; the total energy expended, E_T, across all nodes; maximum energy expended by any node; average energy expended per node; and the average energy expended per packet delivered, E_P. If T_P is the total number of packets delivered over the entire simulation period, then E_P is calculated as $E_P = E_T/T_P$. The error bars in each graph represent 95% confidence intervals.

Figure 4 shows the relative performance of the routing protocols in terms of Packet Delivery Ratio. Out of the protocols, GBR-DTR-CNR performed consistently best over the various number of nodes, followed closely by both LEARN-based protocols LBR-DTR and LBR-STR, with the remaining two, GBR-DTR and GBR-STR, performing noticeably worse. The difference in the performance between the two groups can be understood by where the next node in a path

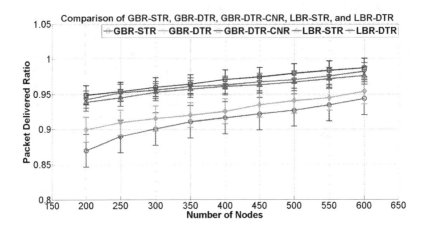

Fig. 4. Packet Delivery Ratio versus Number of Nodes

is selected by the routing algorithms. For GBR-DTR-CNR the next node is se-
lected from those neighboring nodes that will not move out of range before the
next HELLO message broadcast, while for LBR-DTR and LBR-STR, following
LEARN, the next nodes are selected closest to the critical transmission radius
$r_0 = R/2$. Therefore, the connections formed are much more stable than for
GBR-DTR and GBR-STR where the next node chosen is closest to the destination therefore more likely to move out of transmission range before the next
HELLO broadcast, more likely breaking the connection such that the path has
to be recomputed, leading to more dropped packets and a lower delivery ratio.

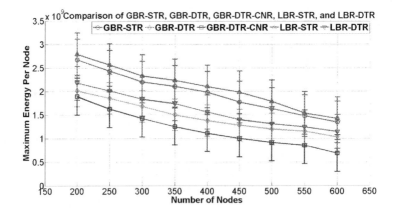

Fig. 5. Maximum Energy per Node versus Number of Nodes

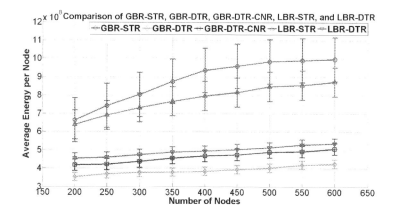

Fig. 6. Average Energy per Node versus Number of Nodes

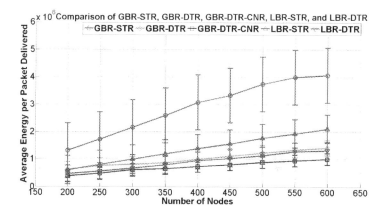

Fig. 7. Average Energy per Packet Delivered versus Number of Nodes

Figures 5, 6 and 7 show the performance of the routing protocols in terms of various energy measurements (the units of energy used are the same for all figures). In terms of maximum energy expended per node (Figure 5), as expected, the dynamic transmission range protocols fared the best with GBR-DTR-CNR again performing consistently best over the various number of nodes, followed GBR-DTR and then by LBR-DTR. For a MANET containing nodes with limited capacity batteries this means that the first battery failures are more likely to occur later for GBR-DTR-CNR than for the other protocols.

With respect to the average energy expended per node (Figure 6), GBR-DTR performed best followed by GBR-DTR-CNR, and then by LBR-DTR. Interestingly, for both these energy consumption measures per node, using LEARN in place of GPSR did not improve the energy-efficiency properties of the connections. Finally, for average energy expended per packet delivered (Figure 7), the

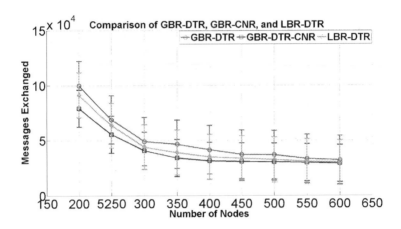

Fig. 8. Average of Overhead Routing Control Messages versus Number of Nodes

GBR-DTR-CNR again performing consistently best over the various number of nodes, followed by LBR-DTR and then by GBR-DTR.

Since MANET does not have a single centralized system to make decisions (e.g. join a network, network routing, shortest path, scheduling, queuing, priority, and fairness) for the wireless nodes, the number of message exchanges among the wireless nodes to make such decisions is larger than for a centralized system. Reducing the high number of message exchanges in the decentralized system without sacrificing performance is one of the important challenges in deploying these systems. Also, exchanging messages in wireless system comes at a cost of energy lost by the node during route discovery and actual communication. Figure 8 shows the comparison between GBR-DTR, GBR-DTR-CNR and LBR-DTR in terms of number of routing control messages (RREQ and RREP) to establish and repair connections, where GBR-DTR-CNR followed by LBR-DTR required the fewest number of messages of the three classes of protocols. The results for GBR-STR and LBR-STR are indistinguishable from their dynamic range counterparts and thus are not shown.

5   Conclusions

The overall conclusions that we reach are the following. Our proposed protocol GBR-DTR-CNR was demonstrated to maximize both the stability of connections, measured in terms of throughput, as well as the energy-efficiency of these connections, while requiring the fewest number of routing control message exchanges. The only energy expenditure measure for which GBR-DTR-CNR did not outperform the other protocols over various node densities was for the average energy expended per node where it was second only to GBR-DTR. For all the energy-efficiency measures, the use of the adjustable transmission range R_a consistently improved the energy consumption properties of the routing protocols, as expected. But the use of LEARN in place of GPSR in the stability routing protocol GBR did not lead to a more energy-efficient stable routing protocol, although a byproduct of selecting routing nodes near the critical transmission radius r_0 was a higher throughput for LBR-STR and LBR-DTR compared to GBR-STR and GBR-DTR.

References

1. Heikkinen, M.V., Nurminen, J.K.: Consumer attitudes towards energy consumption of mobile phones and services. In: 72nd Vehicular Technology Conference – Fall (VTC), pp. 1–5. IEEE (2010)
2. Yang, W., Yang, X., Yang, S., Yang, D.: A greedy-based stable multi-path routing protocol in mobile ad hoc networks. Ad Hoc Networks 9, 662–674 (2011)
3. Kuhn, F., Zollinger, A.: Ad-hoc networks beyond unit disk graphs. In: Joint Workshop on Foundations of Mobile Comp., pp. 69–78 (2003)
4. Barriére, L., Fraigniaud, P., Narayanan, L.: Robust position-based routing in wireless ad hoc networks with unstable transmission ranges. In: 5th Inter'l Workshop on Discrete Algorithms and Methods for Mobile Computing and Communications, pp. 19–27. ACM (2001)
5. Stojmenović, I.: Position-based routing in ad hoc networks. IEEE Communications Magazine 40(7), 128–134 (2002)
6. Flury, R., Pemmaraju, S., Wattenhofer, R.: Greedy routing with bounded stretch. In: INFOCOM, pp. 1737–1745. IEEE (2009)
7. Bose, P., Morin, P., Stojmenović, I., Urrutia, J.: Routing with guaranteed delivery in ad hoc wireless networks. Wireless Networks 7(6), 609–616 (2001)
8. Karp, B., Kung, H.: GPSR: Greedy perimeter stateless routing for wireless networks. In: 6th Annual International Conference on Mobile Computing and Networking (MOBICOM), pp. 243–254. ACM (2000)
9. Agarwal, S., Ahuja, A., Singh, J., Shorey, R.: Route-lifetime assessment based routing (RABR) protocol for mobile ad-hoc networks. In: IEEE International Conference on Communications (ICC), vol. 3, pp. 1697–1701 (2000)
10. Wenqing, L.: Path stability based source routing in mobile ad hoc networks. In: Cross Strait Quad-Regional Radio Science and Wireless Technology Conference (CSQRWC), vol. 1, pp. 774–778 (2011)
11. Sun, J., Liu, Y., Hu, H., Yuan, D.: Link stability based routing in mobile ad hoc networks. In: 5th IEEE Conference on Industrial Electronics and Applications (ICIEA), pp. 1821–1825 (2010)
12. Hu, X., Wang, J., Wang, C.: Stability-enhanced routing for mobile ad hoc networks. In: International Conference on Computer Design and Applications (ICCDA), vol. 4, pp. V4-553–V4-556 (2010)
13. Hamad, S., Noureddine, H., Al-Raweshidy, H.: LSEA: Link stability and energy aware for efficient routing in mobile ad hoc network. In: 14th Inter'l Symp. on Wireless Personal Multimedia Comm (WPMC), pp. 1–5 (2011)
14. Zhang, Z., Jia, Z., Xia, H.: Link stability evaluation and stability based multicast routing protocol in mobile ad hoc networks. In: IEEE 11th International Conference on Trust, Security and Privacy in Computing and Communications (TrustCom), pp. 1570–1577 (2012)
15. Mohamamdzadeh, H., Jamali, S., Begdillo, S., Kheirandish, D.: An energy-aware high network lifetime routing protocol for mobile ad hoc network. In: IEEE 3rd Inter'l Conf. on Comm. Software & Networks (ICCSN), pp. 78–82 (2011)

16. Stojmenović, I., Lin, X.: Power-aware localized routing in wireless networks. IEEE Trans. on Parallel & Distr. Systems 12(11), 1122–1133 (2001)
17. Stojmenović, I., Datta, S.: Power and cost aware localized routing with guaranteed delivery in unit graph based ad hoc networks. Wireless Communications & Mobile Computing 4(2), 175–188 (2004)
18. Kuruvila, J., Nayak, A., Stojmenović, I.: Progress and location based localized power aware routing for ad hoc and sensor wireless networks. International Journal of Distributed Sensor Networks 2(2), 147–159 (2006)
19. Feeney, L., Nilsson, M.: Investigating the energy consumption of a wireless network interface in an ad hoc networking environment. In: 20th Ann'l Joint Conf. of the IEEE Computer & Comm. Societies (INFOCOM), vol. 3, pp. 1548–1557 (2001)
20. Seada, K., Zuniga, M., Helmy, A., Krishnamachari, B.: Energy-efficient forwarding strategies for geographic routing in lossy wireless sensor networks. In: 2nd International Conference on Embedded Networked Sensor Systems (SenSys), pp. 108–121. ACM (2004)
21. Wang, Y., Li, X.-Y., Song, W.-Z., Huang, M., Dahlberg, T.A.: Energy-efficient localized routing in random multihop wireless networks. IEEE Transactions on Parallel and Distributed Systems 22(8), 1249–1257 (2011)
22. Marie, R., Molnár, M., Idoudi, H.: A simple automata based model for stable routing in dynamic ad hoc networks. In: 2nd Workshop on Perform. Monitoring & Measurement of Heter. Wireless & Wired Networks, pp. 72–79 (2007)
23. Zadin, A., Fevens, T.: Stable connections using multi-paths and conservative neighborhood ranges in mobile ad hoc networks. In: Canadian Conf. on Elec. & Comp. Eng (CCECE), pp. 1–4 (2013)
24. Zadin, A., Fevens, T.: Maintaining path stability with node failure in mobile ad hoc networks. In: Shakshuki, E., Djouani, K., Sheng, M., Younis, M., Vaz, E., Groszko, W. (eds.) ANT/SEIT. Procedia Computer Science, vol. 19, pp. 1068–1073. Elsevier (2013)
25. Ahmed, G., Khan, N.M., Masood, M.M.Y.: A dynamic transmission power control routing protocol to avoid network partitioning in wireless sensor networks. In: International Conference on Information and Communication Technologies (ICICT), pp. 1–4 (2011)
26. Jose, A.D., Wang, J., de Dieu, I.J., Lee, S.: Adjustable range routing algorithm based on position for wireless sensor networks. In: 2nd International Conference on Next Generation Information Technology (ICNIT), pp. 72–77 (2011)
27. Kim, S., Eom, D.-S.: Dynamic transmission power control for wireless sensor network reprogramming. In: 17th Asia-Pacific Conference on Communications (APCC), pp. 145–150 (2011)
28. Wu, J., Dai, F.: Virtual backbone construction in manets using adjustable transmission ranges. IEEE Transactions on Mobile Computing 5(9), 1188–1200 (2006)
29. Pazand, B., McDonald, C.: A critique of mobility models for wireless network simulation. In: Inter. Conf. on Computer and Inform. Science, pp. 141–146 (2007)

K Nearest Neighbour Query Processing in Wireless Sensor and Robot Networks

Wei Xie[1], Xu Li[2], Venkat Narasimhan[3], and Amiya Nayak[1]

[1] University of Ottawa, Ottawa, Canada
xwei082,nayak@uottawa.ca
[2] Huawei Technologies, Ottawa, Canada
easylix@gmail.com
[3] Norleaf Networks Inc., Gatineau, Canada
narasim@norleaf.ca

Abstract. In wireless sensor and robot networks (WSRNs), static sensors report event information to one of robots. In the k nearest neighbour query processing problem in WSRN, robot receiving event report needs to find k nearest robots (KNN) to react to the event, among those connected to it. In this article, we propose a new method to estimate a search boundary, which is a circle centred at query point. Two algorithms are presented to disseminate the message to robots of interest and aggregate their data (e.g. the distance to query point). *Multiple Auction Aggregation* (MAA) is an algorithm based on auction protocol, with multiple copies of query message being disseminated into the network to get the best bidding from each robot. *Partial Depth First Search* (PDFS) algorithm attempts to traverse all the robots of interest with a query message to gather the data by depth first search. In this article, we also optimize a traditional itinerary-based KNN query processing method called IKNN and compare it with our proposed algorithms. The experimental results indicate that the overall performance of MAA outweighs IKNN.

Keywords: Wireless Sensor and Robot Networks, k Nearest Neighbour Query Processing.

1 Introduction

Wireless Sensor Networks (WSNs) typically consist of a large number of resource-constrained sensing devices with tiny size, which are powered by low-energy batteries and connected via wireless communication links. The basic function of the sensors is to monitor environmental conditions (e.g. temperature, humidity, sound, etc.) and report the data to a sink, and the sink will react upon the data or forward it to a base station. Sensor-robot coordination provides information about the event sensed by sensors. The information will be forwarded to the robot if necessary. Robot-robot coordination enables solutions for robots to make a decision on how to act upon the event. We will focus on robot-robot coordination in this paper. We consider scenarios where, for robustness and reliability, multiple robots are selected as a task team.

S. Guo et al. (Eds.): ADHOC-NOW 2014, LNCS 8487, pp. 251–264, 2014.
© Springer International Publishing Switzerland 2014

k Nearest Neighbour(KNN) query processing is one of the most essential problems in WSN. This problem aims at searching for k nearest neighbours of the query point, sorting them by their Euclidean distance. Researchers have proposed multiple algorithms to solve KNN query processing problem in WSN fir different scenarios. The KNN query problem in a centralized system that stores all the information of sensors and robots in a central database has been the focus of a few articles [1] [2]. However, such a centralized scheme is not recommended in WSRN due to tremendous latency, redundant transmission, and high energy consumption.

Recently, there has been some focus on processing KNN query in a localized system in which each sensor only knows the information about its neighbour and itself. Most existing localized based KNN query processing techniques estimate a KNN boundary to restrict the message propagation within a searching region and then initiate message dissemination and aggregation [3] [4] [5] [6].

Greedy Face Greedy (GFG) algorithm has been applied in most KNN query processing approaches in WSN for message delivery. The GFG, which was first presented in 2001 by Morin et al. in [7], guarantees the message delivery from the source sensor to the destination sensor over the network. By setting the source sensor as the sensor which senses the event and the destination sensor as the event point, the query message can eventually reach the nearest sensor to the query point.

Instead of flooding the message in the whole network, several approaches have been presented to estimate a KNN boundary and ensure that all searching progress are confined within this area. Julian Winter and Wang-Chien Lee have proposed the Maximum Hop distance (MHD) algorithm [3] to calculate KNN boundary based on the data collected during the searching phase. Jayaraman et al. have proposed another approach to estimate KNN boundary in 3D sensor network [8] based on the information of the network density stored in advance.

Several itinerary based approaches have been proposed for query dissemination for KNN query processing problem in WSN. The basic idea of an itinerary-based algorithm is to enforce each sensor which receives the query message to forward the message to the next target sensor over a predetermined itinerary [9] [10] [11]. The itinerary is initiated by the Home Sensor and expands away from the query point to visit all the sensors within the KNN boundary. When the boundary is reached, the message is forwarded back to the Home Robot for processing. The drawback of itinerary-based algorithms is that the latency may significantly increase because most of the itinerary based solutions are single-threaded.

Although extensive works have been done to solve KNN query processing in WSN, very few researches are conducted to solve the KNN query processing problem in WSRNs. Multi-Robots Systems (MRS) have potential applications in forest fire detection, transport systems etc. Recently, the coordination between WSN and MRS was studied in literature. In WSRN, robots optimize the performance of WSN, such as sensor deployment [12] [13] and sensor relocation [14]. Robots can interact with sensors to receive event information from them and react. An example of WSRN is illustrated in Fig.1. In this figure, red circles denote robots and the black dots denote sensors.

Mezei et al. proposed a novel auction protocol named Auction Aggregation Protocol (AAP) in [15] to solve KNN query processing problem in WSRN. In this algorithm, query message is disseminated by tree expansion and tree construction. Query message is forwarded to every robot neighbouring the query point (the message will not be forwarded more than two hop away), and each robot that receives the message is aware of its parent robot through an identification scheme. Each robot can only choose one robot as its parent robot and thus becomes a leaf. Bid information is kept in the query message to record the best bid so far. Specifically, each node that receives the message will compare the bid stored in the message with its own bid and replace it if it offers a better bid. Eventually, the best bid so far is stored when the message reaches a leaf robot. In order to prevent the message from flooding the whole network, each robot relays only when it is within a distance of k hops from collecting robot. If the robot is more than k hops away, the message will be dropped.

1.1   Our Contribution

Propagating the query message to the whole network to search for k nearest robot to interact with certain event is inefficient and will result in tremendous latency. As a result, a search boundary should be estimated to restrict the propagation within a certain boundary. We propose an efficient and accurate approach to implement boundary estimation. This algorithm estimates the network density around the query point instead of the average value of the entire network. To improve the accuracy and robustness, multiple samples instead of one are involved in this progress. We then propose the following two algorithms to address KNN query processing.

- Multiple Auction Aggregation (MAA): This multithread algorithm constructs a tree rooted at Home Robot and disseminates multiple copies of the query message into the network to get a set of best biddings (i.e. distance with query point) from the children and then sorts them by distance.
- Partial Depth-First Search (PDFS): With a low protocol complexity, PDFS approach attempts to traverse all the robots with depth-first search within the KNN boundary and aggregates their data for processing.

Several fault-tolerant schemes for MAA and PDFS are discussed later to enhance their performance. Through extensive simulation, we evaluate the performance of MAA, PDFS, and compare that with an existing approach, IKNN [9].

2   KNN Query Processing in WSRN

We define the KNN query processing problem as follows. In a WSRN, sensors and robots are deployed randomly. An event occurs and sensed by a sensor which then tries to find k nearest robots to the query point, sorting them by their Euclidean distance. We assume that a query message aiming at searching for k nearest robots to act upon the sensed event is generated by this sensor. Afterward, the query message is forwarded over the sensor network to a robot in the vicinity (denoted as Vicinity Robot), which may not be the closest one to the query point. This robot then forwards the query message to the nearest robot (denoted as the Home Robot) over robot network. The problem is how to disseminate the query message to the robot network efficiently and find k nearest robots with high accuracy and low latency.

We assume that the sensor network and the robot network in WSRN have different responsibilities. The occurrence of an event can be only sensed in the sensor network. After the query message enters the robot network (i.e. the query message is received by the robot), subsequent process (i.e. query message dissemination and data aggregation) will be the responsibility of the robot network. The scenario that robots communicate with each other via sensor layer is expected, but we will not take this method into account for energy efficiency and stability reason. Specifically, due to the limited size, sensors operate on an extremely frugal energy budget, and limited number of message transmissions is expected before the batteries are drained. This assignment of responsibilities enhances the efficiency and sustainability of WSRN. The basic execution phases of our approach (KNN query processing for WSRN) and previously stated approach (KNN query processing for WSN) are similar. The only difference is that the carrier of the message is the robots in WSRN, corresponding the sensors in WSN. In this paper, we also assume that the robots are equipped with GPS.

To better understand the problem, let us refer to the Fig.1 as an example. In this figure, an event is sensed by a sensor, and this sensor forwards the query message to a robot in the vicinity (R7). After receiving the query message, R7 forwards the query message to the nearest robot (R6) to the query point, and R6 tries to find k nearest robots around the query point. We choose R6 as the Home Robot instead of R7 due to the fact that R6 is among k nearest robots, while R7 may not be one of nearest robots.

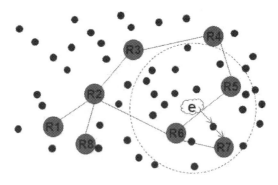

Fig. 1. Wireless Sensor and Robot Network

To search for k nearest robots to react upon query message, we break our solution down into the following phases:

- Home Robot Searching Phase: After an event being sensed, the sensor forwards the query message to a robot in its vicinity. This robot, namely *Vicinity Robot* (VR), forwards this message to the nearest robot to query point. To the best of our knowledge, GFG is the only solution to achieve this goal based on previous researches [11] [3] [4] [10]. Thus, we will continue to search the Home Robot via GFG algorithm without further discussion.
- KNN Boundary Estimation Phase: Instead of flooding the query message in the entire network, the Home Robot will first estimate a KNN boundary to restrict message propagation. We propose a partial density-based boundary estimation algorithm to improve the overall performance.
- Message Dissemination Phase: The most important and challenging part of KNN query processing is how to aggregate data from different robots of interest within the KNN boundary. In this article, we presented two new algorithms to execute message dissemination and aggregation.

In the following sections, we discuss the last two phases.

2.1   KNN Boundary Estimation

Precision of KNN boundary estimation has a significant impact on the KNN query processing. More unnecessary nodes will be involved if the boundary is overestimated which will result in large energy consumption and latency. On the contrary, an underestimated boundary can lead to the situation where less than k nodes are found at the end of the processing (i.e. $|N'| < k$). Home Robot can expand the boundary and search again, which is also not desirable from energy-saving point of view.

To deal with this situation, we propose a method to estimate the boundary in WSRN network based on partial network density around the Home Robot. The first step of our algorithm is to aggregate the hop distance information around Home Robot. After the selection of the Home Robot, a query message is generated and sent to every neighbour of the Home Robot. This message is only be forwarded to Home Robot's neighbours and not be propagated further. Each robot that receives the query message then calculates the average distance between its neighbours and itself and then adds this parameter K_i to the query message, where i indicates that this robot is the *ith* neighbour of the Home Robot. Eventually, the query message should be sent back to its source (i.e. the Home Robot). Let us denote the average distance between the Home Robot and its neighbours as K and the degree of the Home Robot is M. The Home Robot estimates average hop distance after receiving i messages from its neighbours. The partial average hop distance d is given by:

$$d = \frac{K + \sum_{i=1}^{M}(K_i)}{M + 1} \qquad (1)$$

We assume that the target number of nearest neighbour we are searching for around the query point is k and the radius of the search boundary is r. The ideal situation is when there are exactly k robots within the KNN boundary centered by query point and our target is to calculate the searching radius r. With the purpose of increasing the possibility of the occurrence of this event, we propose a network estimation algorithm. Before that, in order to better illustrate our approach, some discussions are needed to clarify and understand the relationship between r, k and the density of the network D.

We also assume that all robots are deployed randomly. A circular area centered by a random point (marked as query point in our example) with radius r is generated as the target area(denoted as A). In view that the deployment of each robot is an independent event, the possibility that a robot is encircled by this area equals to the ratio of the target area to the total area. Assuming that the total number of deployed robots is N, we denote network density D as N/A. The expected number of robots n within this area is then given by:

$$n = \frac{N\pi r^2}{A} = D\pi r^2 \qquad (2)$$

As presented in Eqn.2, for a fixed A, r is positively correlated with n. For a fixed n, r is negatively correlated with D. As a result, D and k are essential parameters to estimate KNN boundary. The average hop distance of robots neighbouring the Home Robot, to some extent, indicates the partial network density around query point. It can be verified that a sparse network should result in a larger d. In other words, d is positively correlated with D. For these concerns, after d is obtained, we propose a linear equation to approximate the radius of KNN boundary r as follows:

$$r = kdc, \qquad (3)$$

where c is the adjustable coefficient. The most essential part of this equation is the value of c, and we will optimize the value in section3 later. The calculated r is kept in the query message which is prepared to be forwarded in the next phase.

2.2   Multiple Auction Aggregation Algorithm

We extend the *Auction Aggregation Protocol* (AAP) proposed by Mezei et al. in [15] to gather the best bids of coordinated robots. Instead of keeping one bid, we keep multiple bids in the query message. In our Multiple Auction Aggregation algorithm (MAA), query message is disseminated by tree expansion and tree construction. Tree expansion starts at the Home Robot. Query message is forwarded to every robot neighbouring it, and each robot that receives the message is aware of its parent robot through the identification number. Each robot can only choose one robot as its parent robot and thus become a leaf if it does have any child robot. Bidding information is kept in the query message to record the list of best bids (i.e. the distance to query point) so far (we assume that the size of the list is L). Specifically, each node that receives the message first checks whether the list is full and adds itself to the list if the number of robots stored in the list is less than L. In the case that the list is full, the robot compares the bids stored in the list with its own bid and replaces it with the smallest bid in the list; if it offers a better bid then it forwards the query message to its children. Eventually, the best bids so far are stored when the message reaches leaf robot. In order to prevent the message from flooding the entire network, each robot transmits only when it is within KNN boundary. If the query message reaches a leaf robot or a robot outside the KNN boundary, this message will enter backtracking mode and is prepared to be sent back as backtracking message.

An example of tree expansion is illustrated in Fig.2. For simplicity, we assume the case where $k = 2$, $L = 1$, which means that every query message will keep two best bids. In Fig.2(a), an event occurs at a random point in the network and is sensed by a sensor. The sensor forwards the query message to the nearest robot C, which becomes the Home Robot. Then, robot C initiates the Boundary Estimation phase and generates the radius of KNN boundary r, and all subsequent processing should be restricted to this area. Then, query dissemination phase is initiated by forwarding the query message to all the neighbouring robots of C (i.e. D, E, and B). Meanwhile, the tree rooted at robot C is generated. Robots D, E and B mark themselves as the 1st level in this tree structure and insert this information with their respective bids (i.e. distance to the query point) to the query messages. Afterwards, the query message is forwarded to the next level in the tree structure (i.e. G, F, K, M, O and J). Please note that robot F , J, K and M will stop forwarding and enter backtracking phase to create the backtracking message, including the bidding information received so far, and forward it to their respective parents as they are out of the search boundary. The tree expansion and data aggregation progress continues until the robots which are out of KNN boundary or leaf robots are reached.

At the end of the previous phase, the query messages are kept by the leaf robots. The next challenge of MAA is how to implement data aggregation after the tree expansion is completed. For the tree structure we presented, a parent robot has multiple children and thus receives multiple backtracking messages (i.e. Node O receives the messages from node N and node Q in Fig.2). These messages should be integrated and compared to generate a new message which contains the best bids among previous messages (in our case, 2 best bids must be stored in the message). To aggregate the completed information about the child robots, a parent robot should wait for the backtracking messages from all its children before creating its own backtracking message and forwarding the message to its own parent. A fault-tolerant scheme can be added in this progress to decrease the impact of the loss of backtracking messages by children on the algorithm. Specifically, each robot will start a timer after receiving the first backtracking message from its child. After this timer expires, this robot should process the data and generate its backtracking message despite the data from its non-responsive children then forward the message to its parent.

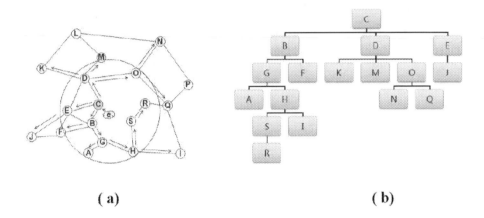

Fig. 2. (a) Tree Expansion. (b) Topology of The Tree.

Instead of simply integrating all the bids from children, the bid information from children are compared by their parent, and only k (in this case, $k=2$) best bids are selected and added to the new backtracking message. Hence, this scheme restricts the size of the backtracking message. After the Home Robot receives the backtracking message, k nearest neighbours are selected.

2.3 Partial Depth First Search Algorithm

Depth First Search is a classical algorithm extensively used to traverse all objects in a particular graph. We propose *Partial Depth First Search algorithm*(PDFS) to visit the robots within KNN boundary and aggregate the data of these robots to implement KNN search in WSRN. Specifically, we first treat the Home Robot as the root of the search tree. Then, the Home Robot chooses a neighbouring robot as its child and forwards the query message to that robot. The key of choosing child is the ID of the neighbouring robots (*lowest ID first*). The message includes the visiting sequence denoted by s. After receiving the message, current robot (denoted as t) first examines whether it is located outside of the KNN boundary (this can be implemented by comparing its coordinate and the coordinate of the query point) and returns the message to its parent robot if so (i.e. mark itself as leaf robot). If t is within the KNN boundary, t inserts the information of itself to the message and checks if there is any unvisited neighbour existing. Whether the neighbour has been visited is confirmed by a stack in the routing message. Each time a robot is been visited, the ID of this robot will be pushed into this stack and marked as "visited". If multiple unvisited neighbours exist, t chooses one of them based on the key and forwards the query message to that robot. If all the neighbours of t have been visited, forward the message back to t's parent and repeat the process. Eventually, query message is forwarded

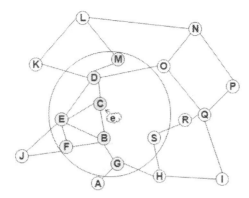

Fig. 3. PDFS for KNN Query Processing in WSRN

back to the Home Robot, and all the neighbours of the Home Robot have been visited. The query message maintains all the information of the visiting robots which enables the Home Robot to sort them by distance to the query point and to choose k nearest ones.

Fig.3 illustrates the PDFS process. The query message is first received by Home Robot C. After C's initialization, C inserts the information of itself and forwards the query message to one of its children. In our example, the sequence of traversed children is determined by key $= ID$, which means that the robot with the lowest ID will be visited first. In this case, B is selected as the next robot to be visited. Then E, D, M and L are visited in sequence. Note that L realizes that it is out of KNN boundary and thus forwards the query message directly back to M. M then realizes that it has no unvisited neighbour and forwards the message back to D. Once a robot which receives message realizes it is outside the KNN boundary, it will forward the message back to its parent immediately. This progress lasts until the message is forwarded back to C and no unvisited neighbour exists. Ultimately, C extracts the data in the query message and sorts the candidate robots by distance to choose the k nearest ones.

PDFS is fairly sensitive to packet loss which may result in unrecoverable and catastrophic consequence (i.e. the failure of the entire routing). Thus, we present a fault-tolerant scheme which can be implemented to PDFS to lower the impact of packet loss. Assume that query message m arrives at robot r for the first time. The robot r stores a copy of m in its memory before relaying the message. After that, every time r receives a new query message from its children with updated information, it replaces m with the new message. In the case that r does not receive any response from its child for a predetermined period of time (can be achieved by generating a timer after forwarding the message), r sends the backup message in its memory to its next child in the forwarding list. The non-responsive child is considered as "visited" and any subsequent message from this child will be disregarded to avoid duplicate messages in the network.

The information from the non-responsive child and its successors will be lost, but this scheme prevents the entire routing from failure when package loss occurs. The drawback of this scheme is that extra storage space for the backup message in robots is required.

3   Performance Evaluation

We conducted a wide range of simulations to evaluate the performance of MAA and PDFS and compared them with the itinerary-based algorithm, IKNN [9]. We implemented our algorithms with a Java-based simulator (JBotSim). All robots were static. System parameters and settings are summarized in Tab. 1. We used three performance metrics for comparison defined as follows:

- **Energy Consumption** (J): The total energy consumed by message transmission in one single KNN query processing round.
- **Process Latency** (ms): The average time for the result of KNN obtained by an algorithm for an event.
- **Process Accuracy** (%): The percentage ratio of correct reading of k nearest robots to the total number of robots returned to the Home Robot.

Table 1. Parameters and Values in the Experiment

Parameter	Definition	Default Value
s	Size of network	$500m \times 500m$
R	Communication range	50m
r	Radius of KNN boundary	kdc
c	Coefficient to estimate r	0.8
k	Number of target robot	5
w	Width of itinerary in IKNN	$\sqrt{3}R/2$
e	Robot energy consumption	12mw/msg
D	Message delay	20ms
N	Total number of robots	250

3.1   Assumptions

We assume that query events occur one at a time. In other words, when an query event occurs, the subsequent event will not occur until the result of the current event has been obtained. Ideal physical and MAC layer are assumed in message transmissions. Message transmission within the WSRN is also assumed to be asynchronous; therefore, all messages will eventually reach the desired destination.

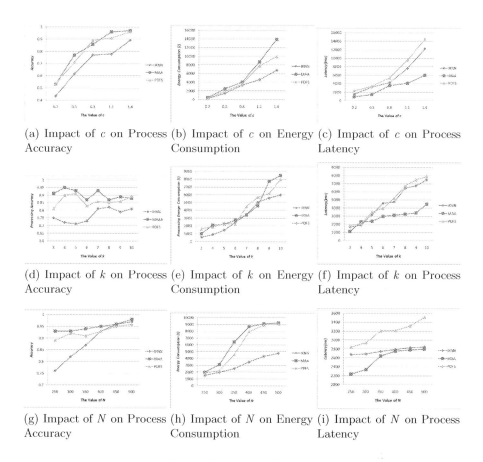

(a) Impact of c on Process Accuracy

(b) Impact of c on Energy Consumption

(c) Impact of c on Process Latency

(d) Impact of k on Process Accuracy

(e) Impact of k on Energy Consumption

(f) Impact of k on Process Latency

(g) Impact of N on Process Accuracy

(h) Impact of N on Energy Consumption

(i) Impact of N on Process Latency

Fig. 4. Experiment Results

3.2 Observations

In the following sections, we aim at studying how the various parameters affect the performance of MAA, PDFS and IKNN algorithms and try to obtain their optimal values. We only discuss the impact of one parameter in each subsection and keep the value of other parameters as default.

Impact of KNN Boundary. The radius of KNN boundary r plays an essential role in both MAA and IKNN algorithm. r is denoted by a linear equation $r = kdc$, where c is adjustable coefficients and d is the average hop distance from the Home Robot to its neighbours. To optimize the value of c, we fix $k = 5$ and then study the impact of c to the process accuracy, process latency and energy consumption. Thus, c can be determined by choosing a balance point among these three metrics.

As illustrated in Fig.4 (a), (b) and (c), both MAA and PDFS give a better accuracy compared to IKNN, but also consume more energy compared to IKNN.

This drawback tends to get worse as the KNN boundary increases. However, MAA outperforms IKNN in terms of latency as c increases. This is because the larger the KNN boundary is the less possibility that the nearest robot is missed. However, a larger KNN boundary indicates that more robots will be visited which leads to more energy consumption and larger latency. In terms of latency, MAA is a multithread algorithm and generates multiple copies of query message during the data aggregation progress to search for k nearest robots. In MAA, the data is gathered in parallel by different threads of query message but IKNN and PDFS can only aggregate the data with routing query message as a mono thread algorithm. Thus, IKNN obtains a less satisfactory performance in terms of Process Latency compared to MAA due to a long itinerary when c increases. In order to optimize the relation between accuracy, energy consumption and latency, we consider that the accuracy of both IKNN, PDFS and MAA reaches a satisfactory level (over 75%) when c is equal to 0.8. We also notice that although the accuracy of IKNN, PDFS and MAA increases when c is larger than 0.8, both MAA and PDFS suffer from dramatic energy consumption. For reasons stated above, we choose $c=0.8$ as a balance between the three parameters.

Impact of Number of Target Robots. The main purpose of our algorithm is to find k nearest robots to the query point. Thus, the impact of number of target robots (i.e. k) on the performance of the algorithm should be examined. Particularly, whether the algorithms can operate efficiently or have acceptable energy consumption and latency for different values of k is a significant evaluation criterion. For this purpose, we conduct several experiments to explore the relation between k and the performance of MAA, PDFS and IKNN.

As illustrated in Fig.4 (d), (e) and (f), IKNN, PDFS and MAA maintain a satisfactory performance as k increases. The drawback of MAA is that it continues to consume a larger amount of energy than IKNN with a fixed k. Also PDFS suffers from large latency which is larger than that of MAA for different values of k.

Impact of Network Density. We have conducted several experimental to verify the relation between network density and the performance of the proposed algorithms. To achieve this goal, the size of network area is fixed at $500m \times 500m$ and different number of robots (denoted as N) are deployed randomly within this area.

As illustrated in Fig.4 (g), (h) and (i), a dense network enhances the performance of IKNN, PDFS and MAA due to more possible routing paths around the query point. However, with more robots in search region, MAA suffers from excessive message transmission because this multiple-thread algorithm may visit multiple robots outside the search region before query message backtracking. The number these nonobject robots increases with the increase in boundary or network density which leads to extra energy consumption. However, as a single-thread algorithm, IKNN only visits one nonobject robot before backtracking. Thus, IKNN is less affected by the expansion of KNN boundary or by the increase in the network density. Particularly, IKNN gives a fairly high process accuracy which is close to that of MAA when N is over 400. MAA again consumes a relatively large amount of energy as N increases but maintains a low level of latency.

4 Conclusion

We addressed the problem of k nearest neighbour query processing with multiple solutions in wireless sensor and robot networks. We first presented a novel idea to estimate the KNN boundary based on the partial data around the query point. We then presented two viable approaches to address the message dissemination and aggregation. The first approach is called Multiple Auction Aggregation (MAA) which is an extension of an existing auction protocol. This multi-threaded algorithm constructs a tree rooted at Home Robot and disseminates multiple query messages into the network to get best bids of the children. The second approach is a depth-first based algorithm with partial search, namely Partial Depth First Search (PDFS), which traverses all robots encircled by KNN boundary and aggregates their data. We compared the performance of MAA and PDFS, and IKNN with respect to accuracy, energy consumption and latency through simulation. Based on our experiment results, MAA far outperforms IKNN in terms of accuracy in almost all occasions. However, MAA and PDFS consume more energy than IKNN. Considering the facts that the message transmission is operated in robot network layer and robots always carries much larger energy source than sensors, this drawback is not significant in most occasions. Also the latency of MAA is relatively low compared to IKNN and thus offers a quick reflection in emergency occasions. To conclude, MAA is a satisfactory algorithm to solve KNN query processing in wireless and robot networks. PDFS also provides a significant improvement in terms of accuracy compared to IKNN. As a result, PDFS can also be considered as a satisfactory approach. To conclude, with up to 30% more accuracy and 50% reduction in query response time, MAA is proved to be a satisfactory algorithm. Even the latency and the energy consumption of PDFS is generally larger than IKNN, considering the contribution on improvement of accuracy (up to 30 %), PDFS is also a satisfactory approach. Further improvement can be made to optimize the Energy Consumption of MAA and PDFS.

References

1. Roussopoulos, N., Kelley, S., Vincent, F.: Nearest neighbor queries. ACM Sigmod Record 24(2), 71–79 (1995)
2. Song, Z., Roussopoulos, N.: K-nearest neighbor search for moving query point. In: Jensen, C.S., Schneider, M., Seeger, B., Tsotras, V.J. (eds.) SSTD 2001. LNCS, vol. 2121, pp. 79–96. Springer, Heidelberg (2001)
3. Winter, J., Lee, W.C.: Kpt: a dynamic knn query processing algorithm for location-aware sensor networks. In: Proceeedings of the 1st International Workshop on Data Management for Sensor Networks: in Conjunction with VLDB 2004, pp. 119–124. ACM (2004)

4. Winter, J., Xu, Y., Lee, W.C.: Energy efficient processing of k nearest neighbor queries in location-aware sensor networks. In: The Second Annual International Conference on Mobile and Ubiquitous Systems: Networking and Services, MobiQuitous 2005, pp. 281–292. IEEE (2005)

5. Xu, Y., Lee, W.C., Xu, J., Mitchell, G.: Processingwindow queries in wireless sensor networks. In: Proceedings of the 22nd International Conference on Data Engineering, ICDE 2006, p. 70. IEEE (2006)

6. da Silva, R.I., Macedo, D.F., Nogueira, J.M.S.: Spatial query processing in wireless sensor networks-a survey. In: Information Fusion (2012)

7. Bose, P., Morin, P., Stojmenović, I., Urrutia, J.: Routing with guaranteed delivery in ad hoc wireless networks. Wireless Networks 7(6), 609–616 (2001)

8. Jayaraman, P.P., Zaslavsky, A., Delsing, J.: Cost-efficient data collection approach using k-nearest neighbors in a 3d sensor network. In: 2010 Eleventh International Conference on Mobile Data Management (MDM), pp. 183–188. IEEE (2010)

9. Xu, Y., Fu, T.Y., Lee, W.C., Winter, J.: Processing k nearest neighbor queries in location-aware sensor networks. Signal Processing 87(12), 2861–2881 (2007)

10. Wu, S.H., Chuang, K.T., Chen, C.M., Chen, M.S.: Diknn: an itinerary-based knn query processing algorithm for mobile sensor networks. In: IEEE 23rd International Conference on Data Engineering, ICDE 2007, pp. 456–465. IEEE (2007)

11. Fu, T.Y., Peng, W.C., Lee, W.C.: Parallelizing itinerary-based knn query processing in wireless sensor networks. IEEE Transactions on Knowledge and Data Engineering 22(5), 711–729 (2010)

12. Akyildiz, I.F., Su, W., Sankarasubramaniam, Y., Cayirci, E.: Wireless sensor networks: a survey. Computer Networks 38(4), 393–422 (2002)

13. Chang, C.Y., Chang, C.T., Chen, Y.C., Chang, H.R.: Obstacle-resistant deployment algorithms for wireless sensor networks. IEEE Transactions on Vehicular Technology 58(6), 2925–2941 (2009)

14. Wang, G., Cao, G., La Porta, T., Zhang, W.: Sensor relocation in mobile sensor networks. In: Proceedings of the IEEE 24th Annual Joint Conference of the IEEE Computer and Communications Societies, INFOCOM 2005, vol. 4, pp. 2302–2312. IEEE (2005)

15. Ivan, M., Veljko, M., Stojmenovic, I.: Robot to robot: communication aspects of coordination in robot wireless networks. Robotics Automation Magazine 17(4), 63–69 (2010)

Mobile Application Development with MELON

Justin Collins and Rajive Bagrodia

University of California Los Angeles, Los Angeles CA, USA
{collins,rajive}@cs.ucla.edu

Abstract. Developing distributed applications for mobile ad hoc network continues to be challenging due to the dynamic and unpredictable nature of MANETs. MELON is a general purpose coordination language designed to provide flexible communication patterns for MANET applications while remaining lightweight. Based on a distributed shared message store, MELON abstracts network communication to an asynchronous exchange of persistent messages. MELON simplifies application development by supporting read-only and remove-only messages, bulk message retrieval, and per-host ordering of messages. In this paper, we review the MELON programming model, demonstrate its utility for writing MANET applications, and quantitatively compare it to traditional distributed computing paradigms in a MANET context. For a shared whiteboard application, we find MELON achieves 100% message delivery with 95% less latency than tuple spaces while also maintaining per-host message ordering.

1 Introduction

Tiny, powerful personal computers are quickly becoming ubiquitous. In the United States, 91% of adults have cell phones, and 56% of those are smartphones[1]. Among teens, 78% have a cell phone, of which 37% are a smartphone[2]. Add smartphones to the proliferation of tablets and laptops and the ability for consumers to form mobile ad hoc networks (MANETs) is quickly becoming possible. However, applications designed for these networks remain in short supply. Developing MANET applications presents a number of challenges with the primary difficulty being the unpredictably dynamic infrastructureless wireless network itself. Unlike wired networks, failures in MANETs are commonplace instead of exceptional. Nodes may join and leave the network at any time and the network topology is in constant flux. Networked applications must be aware of and handle the challenging nature of MANETs to be effective.

Several language, middleware, and library solutions to dealing with MANET communication have been proposed to assist in developing MANET applications. The majority of these proposals are based on traditional distributed computing paradigms[3]. Of these, publish/subscribe, remote procedure calls, and tuple spaces are commonly used. However, these communication paradigms were originally designed for stable, wired networks or even interprocedural communication on a single machine. While they have been adapted to MANETs, their original designs are limiting and not well-suited to the MANET environment[4].

S. Guo et al. (Eds.): ADHOC-NOW 2014, LNCS 8487, pp. 265–278, 2014.
© Springer International Publishing Switzerland 2014

To address the need for a communication paradigm designed to operate in MANETs, we have proposed a new coordination language called MELON[5]. MELON provides a flexible, general-purpose communication abstraction which may be completely distributed. In addition to basic message exchange, MELON also implements bulk message retrieval, provides basic message access control, enforces per-host ordering of messages, and supports message streams.

In this paper we describe the MELON paradigm, compare implementing a shared whiteboard application with publish/subscribe, RPC, tuple spaces, and MELON, and finally compare the performance of each whiteboard implementation. We demonstrate MELON is well-suited for developing MANET applications, and while slower than publish/subscribe and RPC, it is the only paradigm to deliver 100% of sent messages. MELON also delivered whiteboard messages with 95% less latency and 60% more in-order messages compared to tuple spaces.

Section 2 reviews the traditional paradigms used in this paper. In Section 3 we discuss the design of MELON, the operations it provides, and our prototype implementation. We compare a whiteboard implemented each paradigm in Section 4 and then measure performance of each paradigm in Section 5 before presenting our conclusions in Section 6.

2   Related Work

This section offers a brief overview of the three paradigms we will compare with MELON. Further surveys of middleware, languages, and communication paradigms for MANET development can be found in [3] and [6].

The *publish/subscribe* paradigm divides processes into publishers and subscribers. In topic-based publish/subscribe, publishers send messages tagged with a topic. Subscribers receive the messages by subscribing to one or more topics and specifying a callback to receive the publications asynchronously and separately from the main process thread. It is also possible to handle multiple incoming publications concurrently. Publish/subscribe does not guarantee any ordering of publications nor does it specify how to deliver messages if the subscribers is not available at the time of publication. In distributed publish/subscribe such as MANETs, it is generally not expected that publishers would persist and deliver messages at a later time via brokers[7], although some implementations exist[8]. Managing an overlay network of brokers also adds considerable complexity.

Remote procedure calls (RPC) is a distributed programming paradigm which disguises remote communication as local method calls. A host can "export" an object to be accessed remotely. Remote hosts discover these remote objects by name or type and then invoke methods defined on the object. RPC is spatially coupled, since the remote object must be available in order to invoke the method. Arguments may be passed to the remote method and the return value of the method is returned to the local process. Since RPC implies code execution, failures during the remote calls can be dangerous[9].

Group RPC invokes the same method with the same arguments on all matching remote objects. In a MANET, group RPC must be performed asynchronously to be practical: the call may return multiple values but the client cannot rely on all remote hosts returning a value successfully. A timeout could be used instead, but a short timeout would cause unnecessarily lost messages, while a long timeout could cause long delays in the execution of the application.

Tuple spaces, introduced in the Linda[10] coordination language, operate on a distributed shared memory space of ordered tuples. Tuples are sent using the **out** operation then retrieved by matching templates with **rd**, which copies the tuple, or **in**, which atomically removes the tuple from the tuple space. If multiple tuples match, one is chosen nondeterministically. Tuple spaces have strict semantics for **rd** and **in**: if a matching tuple exists, it *must* be returned. **rd** and **in** are blocking operations, but typically nonblocking versions are available.

An issue particular to tuple spaces is the "multiple read problem": nondestructively retrieving all matching tuples requires repeated **rd** operations, which may return any matching tuple. One solution is to use a mutex tuple to gain exclusive access to the tuple space, remove all matching tuples using **in**, replace the tuples, and then release the mutex. However, this approach prevents concurrent access and is dangerous in MANETs where the node with the mutex may disappear. Another solution uses a counter in each tuple and each process can request tuples by the counter value in order. Multiple processes producing tuples must coordinate to generate consistent counters. A third option proposed in [11] is to introduce a **copy-collect** operation which copies all matching tuples.

Several projects have adapted tuple spaces to MANETs, including L^2imbo[12], LIME[13], TOTAM[14], and EgoSpaces[15]. LIME, a popular implementation, relies on explicit join and leave operations to federate distributed tuple spaces. This is at odds with the frequently unexpected disconnections in MANETs. [16] discusses the difficulties LIME encounters with tuple space semantics, including situations that can lead to livelocks. LIME II[17], Limone[18], and CoreLIME[19] are proposed to meet shortcomings in LIME.

3   MELON Design

MELON borrows the idea of a distributed shared message store from tuple spaces. The concept of shared message collections which may be safely manipulated by many processes fits well in a MANET context. To send a message, a process using MELON persists it in a globally shared "store". In practice these messages are stored on the client which outputs them in order maintain atomicity of removal without global state or locking. Messages are retrieved by matching against templates describing message content. This decoupling between sending and receiving is beneficial in a MANET context where maintaining connections between hosts can be challenging. With MELON and tuple spaces, disconnections do not cause lost messages or disrupt operations.

Aside from the shared message store, MELON departs significantly from typical tuple spaces in its operations and semantics. In a tuple space, any message may be read or removed by any process, and any matching message may be retrieved in any order. In contrast, MELON divides the messages into two pools:

a remove-only pool, and a read-only pool. Remove-only messages can only be retrieved once and must be removed when retrieved. Read-only messages may never be explicitly removed, only be copied from the message store. Matching messages are returned in FIFO order per host, matching the reliable FIFO-ordered multicast semantics in [9].

MELON contains some additional minor differences from tuple spaces. First, messages are not required by the paradigm to be tuples, but may be implemented as any structure which can be matched by a template (e.g., messages could be unordered tuples with named values instead). Secondly, MELON explicitly acknowledges the storage limitations of mobile devices. In any communication paradigm which persists messages, there is a limitation to how many messages may be stored. MELON attempts to mitigate this limitation by allowing messages to be automatically garbage collected. The alternatives are to not store new messages or to allow applications to exhaust available memory.

A last deviation from tuple spaces is the removal of strict semantics for returning messages. In tuple spaces, the semantics stipulate that if a matching message exists, it must be returned for a retrieval operation. In the reality of MANETs, this semantic cannot be met, so in MELON all retrieval operations are limited to best-effort.

These differences were introduced in MELON to both relieve the application developer of certain responsibilities and to allow the paradigm to operate well in a MANET. For example, read-only messages prevent a badly-behaved process from removing important messages meant to be read by many processes, and FIFO ordering is especially convenient in applications where most messages are generated by a single host and the ordering is important, such as news feeds or streaming video. MELON is also deliberately designed to avoid any global state and enable completely distributed implementations.

3.1   MELON Operations

Processes in MELON communicate by storing messages to a distributed shared message store and retrieving the messages based on templates. In this paper, we assume messages consist of an ordered list of typed values. However, as noted above, MELON does not limit how messages might be constructed. A message template is similar to a message, except it may contain both values and types. For example, a message containing [1, "hello"] could be matched by a template containing [1, String] or [Integer, "hello"] or [Integer, String]. A type will also match any subtypes.

Operations are split into read-only (**write/read/read_all**) and take-only (**store/take/take_all**) operations. Each operation is represented here as a separate function call. **store** and **write** operations return null values as soon as the saved message is available in the message store (essentially immediately). **take** and **read** operations block by default until a matching message is returned, but may be set to nonblocking on a per-call basis. If a nonblocking operation finds no matching messages, an empty set is returned.

Table 1. MELON Operations

Operation	Return Type	Action
store(*message*)	*null*	Store removable message
write(*message*)	*null*	Store read-only message
take(*template, [block = true]*)	*message* or *null*	Remove and return message
read(*template, [block = true]*)	*message* or *null*	Copy and return read-only message
take_all(*template, [block = true]*)	*array*	Bulk remove messages
read_all(*template, [block = true]*)	*array*	Bulk copy read-only messages

When called, **store** saves a copy of the message in the message store. Messages saved with **store** may only be retrieved with a **take** or **take_all** operation.

The **write** operation also stores a single message in the message store, but the message may only be copied from the storage space with a **read** operation, never explicitly removed. Messages stored with either operation may be automatically garbage collected.

A **take** operation accepts a message template as the first argument and an optional boolean indicating blocking or nonblocking for the second argument. The message template is matched against available messages in the message store which were added with a **store** operation. If a matching message is found, it will be removed from the message store and returned. Once a message has been returned by a **take** operation, it may not be returned by a subsequent operation in any process.

The **read** operation accepts the same arguments but will only return messages stored with a **write** operation which have not already been read by the current process. If a message matching the given message template is available, it will be copied and returned, but not removed from the message store. Once a message has been returned to a process, the message is considered to have been read by that process and will not be returned by any subsequent **read** or **read_all** operations in that process. A message may be **read** by any number of processes, but only once per process.

Table 2. Read from multiple processes

Process A	Process B	Process C
write([1, "hello"])	m = read([Integer, String])	m = read([Integer, String])

Table 2 illustrates one process writing a single message containing the integer 1 and the string `"hello"`. Processes B and C each perform a **read** operation with the template which matches the message stored by process A. Since **read** does not modify the storage space, the value of m for both process B and C will be a copy of the message [1, "hello"] from process A.

The **take_all** and **read_all** operations are used to retrieve a group of matching messages instead of a single message. Otherwise, the semantics match **take** and **read**: **take_all** can only remove messages from **store** operations, and **read_all** only returns unread messages from **write** operations. Table 3 demonstrates a use of **read_all**. One or more processes generate news messages containing a news category and headline. To ensure all interested parties can read the news, the server uses **write** to disallow a reader from removing a news item and preventing other readers from reading it. Any number of processes can consume the news as readers. The `fetch` method in Table 2 uses **read_all** to return all news items in a given category. Repeated calls to `fetch` will only return news items not previously read.

Table 3. News server and reader

News Server	News Reader
`def report(category, headline)`	`def fetch(category)`
`  write [category, headline]`	`  return read_all([category, String])`
`end`	`end`

By default, all retrieval operations will block the application until at least one matching message is found. The operations can also be performed in nonblocking mode, in which case **take** and **read** return null when no matching message is found, while **take_all** and **read_all** return empty collections.

3.2   MELON Implementation

We developed a prototype implementation of MELON to validate our design and obtain empirical performance data. The architecture illustrated in Figure 1 is split into five parts. The MELON API is the only interface exposed to the application and provides the six operations described in Section 3. The MELON API interacts with the distributed message storage through the storage API, which provides the same interface for both local and remote storage. The storage server proves a network interface to a local storage space and accepts connections made through the remote storage stub.

Local storage is implemented as two dynamic arrays, one for **write** / **read** messages and the other for **store**/**take** messages. For atomic updates, the **write** / **read** array uses a readers/writer lock to allow multiple **read** operations to access the array concurrently, but locks the array for **write** operations. The **store**/**take** array does not permit concurrent operations, as **store** and **take** modify the store. The two arrays may be accessed and modified independently.

Fig. 1. Paradigm Architecture

Searching for matching messages is a linear operation in the prototype implementation. Performance of local operations is explored in [5].

Network communication is handled by ZeroMQ[20], a high performance networking library. For the prototype, the networking was intentionally kept simple. For example, a **read** request queries remote hosts in random order and stops when a matching result is returned. This could possibly be improved using multicast, but it would complicate the implementation by requiring the client to handle multiple asynchronous responses, select one, request the actual message, and handle failure scenarios if the matching message cannot be returned. We traded potential performance gains for simplicity.

For **read** and **read_all** operations, it is necessary to track which messages have been read. Each process maintains its own list of read messages, which it sends with each **read** request. We use a compact sparse bit set to track message IDs and transfer this information efficiently with an average overhead of < 2 bits per ID.

4   Shared Whiteboard Example

A shared whiteboard is a digital document which may be edited and viewed by multiple users concurrently and is commonly given as an example of an application well-suited to MANETs[21]. Shared whiteboards are distributed, real-time, and interactive, which presents some interesting characteristics. Many participants must have access to update the whiteboard, but ordering of changes is very important to maintain a consistent document. Changes should be propagated quickly and reliably so that each user is working with the latest document without missing any updates.

We have implemented a shared whiteboard in MELON along with canonical versions of publish/subscribe, RPC, and tuple spaces using JRuby (a Java implementation of the Ruby language) to compare their features and performance. The programs share common code related to the actual whiteboard itself, which is implemented in the `Whiteboard` class. Changes to the shared whiteboard are encapsulated in a `Figure` object. Each version implements an `add_local_figure` method to be called when the user modifies the shared whiteboard. The MELON and tuple space versions also implement an `add_remote_figures` method which is used to retrieve updates from remote nodes.

4.1   Publish/Subscribe

The publish/subscribe whiteboard in Table 4 sets up a subscription to the "whiteboard" topic and a callback to add remotely published figures to the whiteboard. This allows the whiteboard to receive updates at any time in a separate thread, which is precisely what would be desired. To output a new figure, the whiteboard simply publishes the figure to the "whiteboard" topic.

Table 4. Publish/Subscribe Whiteboard

```
require "ps"
require "whiteboard"

class PSWhiteboard < Whiteboard
  def initialize
    @ps = PS.new

    @ps.subscribe("whiteboard") do |figure|
      add_figure(figure)
    end
  end

  def add_local_figure(figure)
    @ps.publish("whiteboard", figure)
  end
end
```

Table 5. RPC Whiteboard

```
require "rpc"
require "whiteboard"

class RPCWhiteboard < Whiteboard
  def initialize
    @rpc = RPC.new
    @rpc.export(self)
  end

  def add_local_figure(figure)
    wbs = @rpc.find_all("RPCWhiteboard")
    wbs.add_figure(figure)
  end
end
```

4.2   RPC

A shared whiteboard implementation using RPC is listed in Table 5. When the whiteboard is initialized, it exports itself as a remote object. Remotes hosts can then remotely invoke add_figure. Like publish/subscribe, this allows the white-board to accept remote figures asynchronously from the main process thread and is a natural feature of RPC. Distribution of remote figures is performed by first finding all remote instances of RPCWhiteboard, then invoking the add_figure method (defined on the parent class) directly, passing in the new figure as an argument. Since group RPC is asynchronous, it is possible that a call might complete before a prior call.

4.3   Tuple Spaces

Table 7 shows the tuple space version, which is very similar to MELON. To send an update, it outputs a tuple containing just the new figure. Unlike MELON, a misbehaving or misconfigured client could remove the messages from the tuple space, disrupting the shared whiteboard communication. Retrieval of remote messages uses a **bulk_rd** operation to read all messages containing a figure. To continuously retrieve messages asynchronously, this method can be called inside a loop in a separate thread. Once a group of figures is retrieved, each individual figure is added to the local whiteboard.

As discussed in Section 2, **copy-collect** may be used to solve the "multiple read problem". We have implemented this as the **bulk_rd** operation. However, this does not solve what might be termed the "multiple multiple read problem": since our tuple space is not static, reading all matching tuples once is not sufficient. We need to be able to perform multiple **bulk_rd**s to add all figures the whiteboard. Without *a priori* knowledge of remote hosts in the system, the only option which allows concurrent access to the tuple space is to read *all* matching tuples. Naturally, this becomes considerably expensive as the number of tuples grows large.

Table 6. MELON Whiteboard

Table 7. Tuple Space Whiteboard

```
require "melon"
require "whiteboard"

class MelonWhiteboard < Whiteboard
  def initialize
    @melon = Melon.new
  end

  def add_local_figure(figure)
    @melon.write([Figure])
  end

  def add_remote_figures
    # Returns only unread figures
    # in per-host order
    figures = @melon.read_all([Figure])

    figures.each do |figure|
      add_figure(figure[0])
    end
  end
end
```

```
require "tuplespace"
require "whiteboard"

class TSWhiteboard < Whiteboard
  def initialize
    @ts = Tuplespace.new
  end

  def add_local_figure(figure)
    @ts.out([Figure])
  end

  def add_remote_figures
    # Returns ALL figures
    # in arbitrary order
    figures = @ts.bulk_rd([Figure])

    figures.each do |figure|
      add_figure(figure[0])
    end
  end
end
```

4.4   MELON

The MELON whiteboard in Table 6 writes out each figure in a tuple containing just the new figure. It uses the **write** operation since every remote node needs to read the figures. To retrieve remote figures, MELON uses **read_all** to nondestructively read all messages containing a `Figure`. Like tuple spaces, the **add_remote_figures** method should be called in a separate thread to provide asynchronous updates. Unlike tuple spaces, MELON's **read_all** operation only retrieves unread messages, eliminating the "multiple multiple read" problem.

MELON directly provides three features which are helpful to the whiteboard application: persistent messages, reading only unread messages, and returning messages in a per-host FIFO ordering. Message persistence is crucial in MANET applications, where communication with remote nodes is often disrupted and delayed. For a shared whiteboard, every message must be delivered to keep the document synchronized between users. By managing read versus unread messages, MELON easily allows the whiteboard to efficiently fetch only newly-added figures. Finally, MELON guarantees the updates from each host will be retrieved in the order the host initiated them.

4.5   Summary

All four implementations of the whiteboard have been kept as simple and similar as possible in order to highlight the differences between the paradigms. However, while MELON appears to be as simple to use as the other paradigms, it provides more functionality and guarantees.

Publish/subscribe and RPC are push-based paradigms and allow messages to be received asynchronously by default. However, they do not provide message persistence in their canonical forms. While publish/subscribe is a multicast paradigm by default, RPC must be adapted to perform group communication. Also, RPC must explicitly discover remote objects before invoking remote methods. In both paradigms figures are sent and received singly, although multiple messages may be received concurrently.

Tuple spaces and MELON are pull-based paradigms which provide message persistence. Both tuple spaces and MELON require applications to explicitly use a separate thread to receive messages asynchronously. Tuple spaces do not provide a method to nondestructively read a subset of matching tuples, while MELON does. MELON is also the only paradigm to provide some ordering of messages by definition. Both paradigms allow bulk retrieval of messages, although tuple spaces require an extension to the canonical paradigm. This extension is necessary to allow nondestructive reads of multiple matching tuples and avoid the "multiple read" problem.

5   Quantitative Comparison

For these experiments, we implemented the shared whiteboards as described in Section 4 in the four paradigms using the same codebases as in [5]. To make the comparison as fair as possible, each paradigm shares a considerable amount of common code and utilizes ZeroMQ for network communications.The tuple space implementation uses the LighTS[22] local tuple space library from LIME.

To evaluate the implementations in a MANET environment, we used EXata[23] to provide high-fidelity wireless models and precisely repeatable scenarios while allowing us to run real applications. Our scenario uses 50 nodes with 802.11b radios using AODV moving with random waypoint with a maximum speed of 5m/s in a 500x500 meter area. The two-ray path loss model is used. To measure how the implementations fared in turbulent network conditions, we performed the experiments with increasing levels of packet loss from 1% to 30% as measured by the *ping* command. We used an experiment coordination framework written in MELON itself to manage the applications, running EXata, and collating results.

In our scenario, six real nodes are running the whiteboard application, the rest are simulated and function only as intermediate nodes. For each experiment, each of the six nodes sends 50 whiteboard updates with pauses of 5-10 seconds. This roughly models each user updating their whiteboard at a brisk pace for 4-8 minutes.

5.1   Results

For each implementation, we measured lost messages, messages received out of order, and the message latency. For out-of-order messages, we divided it into two metrics: host out-of-order and global out-of-order. Host out-of-order messages are messages from a single host which are not received in the order sent. Global out-of-order messages are those received before their preceding message. For example, node A receives a message m_1 from node B, then sends m_2. If node C receives m_2 prior to m_1, m_2 will be considered globally out of order.

As guaranteed by the paradigm, MELON maintains host ordering in all scenarios while publish/subscribe and RPC do not. For global message ordering, MELON outperforms tuples spaces by 67% and remains within 15% of RPC performance until packet loss reaches 30%. However, this is because MELON continues to achieve 100% message delivery while RPC drops nearly 12% of messages when the network conditions are poor. Reliability does decrease speed, but MELON experiences 95% less latency than tuple spaces while remaining within 3 seconds of RPC in the worst case. While we would like speed, reliability, and perfect ordering of messages for a shared whiteboard, it is better to have the document be eventually consistent than to lose information.

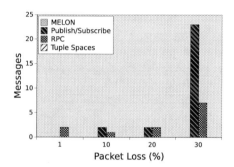
Fig. 2. Host Out-of-Order Messages

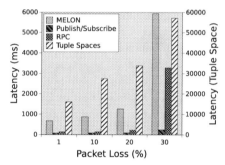
Fig. 3. Message Latency

In our experiments, messages from a single host were generally delivered in the order they were sent as shown in Figure 2. For MELON and tuple spaces, no messages were delivered out of order. Note for tuple spaces this is an accident of the implementation, whereas in MELON it is guaranteed. In LighTS, tuples are sequentially stored locally in an array in the order they are output, then returned in that same order when they are matched. Tuple spaces in general do not return matched messages in any particular order.

Asynchronous group RPC is used in this application. If one call is delayed, it is possible a subsequent call will complete before a prior one, which is why RPC delivers a small number of messages out of order. Publish/subscribe is fully asynchronous and incoming publications can be processed concurrently. However, even in the worst case publish/subscribe delivers 97.8% of the messages from a host in FIFO order. Like tuple spaces this is the result of the implementation: the RPC and publish/subscribe paradigms make no promises about the ordering of messages.

Unlike per-host ordering, many messages were delivered out of order from a global perspective as can be seen in Figure 4. This is entirely expected, since none of the paradigms provide a global ordering. Enforcing a global ordering in an unreliable network is not feasible, since nodes may become unavailable at any time while continuing to output messages. However, the global ordering remains important for a shared whiteboard.

Our results show publish/subscribe performs the best for this metric. Indeed, ordering is largely dependent on deliveries completing quickly before later messages overtake them. As shown in Figure 3, publish/subscribe is an extremely quick method for delivering messages, so it excels in ordering as well. Conversely, tuple spaces fare the worst, delivering 67% of messages out of order. Again, because tuple spaces provide no way of controlling which matches messages are returned or in what order, the whiteboard implementation must transfer large amounts of tuples in order to nondestructively read all matching messages. This is extremely slow, as reflected in Figure 3.

MELON and RPC provide about the same global ordering, although MELON is more affected when the network conditions worsen. This is likely due to MELON's reliable message delivery (Figure 5), as messages may be delayed significantly by broken routes or network partitioning. In contrast, losing messages improves ordering since a message not delivered cannot be out of order. MELON is the only paradigm to demonstrate 100% message delivery. Tuple spaces are expected to be reliable, but this application requires delivery of large numbers of messages. Since the median latency for tuple spaces reached a full minute, the experiment completed before all messages arrived.

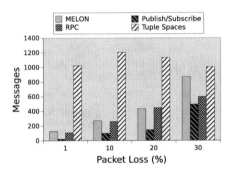

Fig. 4. Global Out-of-Order Messages

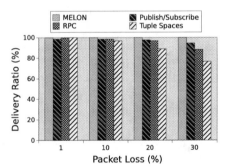

Fig. 5. Delivery Rates

While low delivery rates for publish/subscribe have been observed previously[4], here it performs well in the lossy environment due to quick delivery rates, but still dropped 1.4% of messages when the network connectivity was good. RPC never achieves 100% delivery rates and declines as the network degrades. In group RPC, clients cannot be aware of how many receivers may be available and therefore does not attempt to retry calls after timeouts. Synchronous RPC would block the process until the message is delivered, but deliveries would be considerably delayed which is unacceptable for this application.

Median time between sending and receiving a message is reported in Figure 3. Since tuple spaces are so much slower, the results are aligned with the right-hand y-axis which is an order of magnitude higher. Publish/subscribe was extremely quick, as it requires no message confirmations nor active discovery of remote hosts. RPC was also quite fast until it was disrupted by the 30% packet loss.

Logically, delivery rates and latency are directly related. With reliable delivery some messages may be very late, increasing overall latency. Since dropped messages do not count towards median latency, a lossy communication paradigm can appear very fast. MELON provides better reliability and therefore is a bit slower as the network becomes less reliable and more delivery attempts are required. Frequency of the pull attempts are another trade-off that pull-based paradigms must make. Publish/subscribe and RPC may send as soon as a message is ready, but MELON and tuple spaces must continually poll to receive messages. Faster polling results in faster delivery, but higher overall network usage, collisions and network monopolization.

6   Conclusions

Traditional distributed computing paradigms were not designed to operate in dynamic, self-organizing MANETs where disconnections and topology changes are frequent. In this paper we have introduced the MELON coordination language and compared it qualitatively and quantitatively to traditional distributed computing paradigms which have been adapted to MANETs.

In our shared whiteboard experiments, MELON was the only paradigm to deliver 100% of sent messages. It also maintained 100% host ordering of messages with 95% less latency than tuple spaces and 67% more globally in-order messages, without additional complexity for the application developer. These results indicate MELON can serve as a practical approach for distributed communication in MANET applications while providing persistent messages, reliable FIFO-ordered multicast, efficient bulk retrieval, and simple message streaming.

References

1. Smith, A.: Smartphone Ownership - 2013 Update. Pew Internet & American Life Project (June 2013)
2. Madden, M., et al.: Teens and Technology 2013. Pew Internet & American Life Project (March 2013)
3. Collins, J., Bagrodia, R.: Programming in mobile ad hoc networks. In: WICON 2008: 4th Intl. Conf. on Wireless Internet, pp. 1–9. ICST (2008)
4. Collins, J., Bagrodia, R.: A quantitative comparison of communication paradigms for manets. In: Sénac, P., Ott, M., Seneviratne, A. (eds.) MobiQuitous 2010. LNICST, vol. 73, pp. 261–272. Springer, Heidelberg (2012)
5. Collins, J., Bagrodia, R.: Melon: A persistent message-based communication paradigm for manets. In: 10th ICST Conf. on Mobile and Ubiq. Sys, Mobiquitous (2013)

6. Hadim, S., et al.: Trends in middleware for mobile ad hoc networks. Journal of Communication 1(4), 11–21 (2006)
7. Eugster, P.T., et al.: The many faces of publish/subscribe. ACM Comput. Surv. 35(2), 114–131 (2003)
8. Cugola, G., Picco, G.P.: Reds: a reconfigurable dispatching system. In: Proc. of the 6th International Workshop on Software Engineering and Middleware, pp. 9–16. ACM (2006)
9. van Steen, M., Tanenbaum, A.S.: Distributed Systems, pp. 375–381. Prentice Hall (2002)
10. Gelernter, D., Carriero, N.: Coordination languages and their significance. Commun. ACM 35(2), 97–107 (1992)
11. Rowstron, A., Wood, A.: Solving the linda multiple rd problem. In: Hankin, C., Ciancarini, P. (eds.) COORDINATION 1996. LNCS, vol. 1061, pp. 357–367. Springer, Heidelberg (1996)
12. Wade, S.P.: An investigation into the use of the tuple space paradigm in mobile computing environments. PhD thesis, Citeseer (1999)
13. Murphy, A., et al.: Lime: A coordination middleware supporting mobility of hosts and agents. ACM Trans. on Soft. Eng. and Meth. 15(3), 279–328 (2006)
14. Gonzalez, B., et al.: Programming mobile context-aware applications with totam. Journal of Systems and Software (2013)
15. Julien, C., Roman, G.-C.: Egospaces: Facilitating rapid development of context-aware mobile applications. IEEE Trans. on Soft. Eng. 32(5), 281–298 (2006)
16. Carbunar, B., Valente, M.T., Vitek, J.: Lime revisited. In: Picco, G.P. (ed.) MA 2001. LNCS, vol. 2240, pp. 54–69. Springer, Heidelberg (2001)
17. Artail, H., et al.: The design and implementation of an ad hoc network of mobile devices using the lime ii tuple-space framework. IEEE Wireless Comm. 16(3), 52–59 (2009)
18. Fok, C.-L., et al.: A lightweight coordination middleware for mobile computing. In: De Nicola, R., Ferrari, G.-L., Meredith, G. (eds.) COORDINATION 2004. LNCS, vol. 2949, pp. 135–151. Springer, Heidelberg (2004)
19. Carbunar, B., et al.: Corelime: A coordination model for mobile agents. Electronic Notes in Theoretical Computer Science 54, 17–34 (2001)
20. Hintjens, P.: ZeroMQ: Messaging for Many Applications. O'Reilly (2013)
21. Rewadkar, D.N., Karve, S.: Spontaneous wireless ad hoc networking: A review. International Journal 3(11) (2013)
22. Balzarotti, D., et al.: The lights tuple space framework and its customization for context-aware applications. Web Intelli. and Agent Sys. 5(2), 215–231 (2007)
23. Scalable Networks. Exata: An exact digital network replica for testing, training and operations of network-centric systems. Technical brief (2008)

An Analytical Model of 6LoWPAN Route-Over Forwarding Practices

Andreas Weigel and Volker Turau

Institute of Telematics
Hamburg University of Technology, Hamburg, Germany
{andreas.weigel,turau}@tuhh.de

Abstract. 6LoWPAN has been developed to bring IPv6 to even the smallest resource-constrained devices, enabling the vision of an Internet of Things. To be compliant to IPv6's minimum MTU of 1280 Bytes, its fragmentation mechanism allows transmission of datagrams the size of up to 2048 Bytes. Within low power and lossy environments, fragmentation of datagrams can lead to an increase in end-to-end loss rates and to a waste of bandwidth by propagation of fragments of an already lost datagram. We present an extension to an existing analytical, bit-error-based model, which takes into account different route-over forwarding practices in the presence of fragmentation and use it to assess their influence on the end-to-end success rate and the expected number of bits sent. While being not precisely accurate due to the neglect of effects like (self-induced) collisions, the model provides insight into the effects to be expected when using 6LoWPAN's fragmentation mechanism.

Keywords: 6LoWPAN, IPv6, fragmentation, route-over, LLN, analytical model, forwarding, Internet of Things.

1 Introduction

Several efforts have been undertaken to realize the vision of an Internet of Things, i.e., bringing IPv6 to the smallest and resource-constrained devices as well as to lossy environments. One centerpiece of those efforts is the creation of the protocol "Transmission of IPv6 Packets over IEEE 802.15.4 Networks" (RFC 4944 [1]; 6LoWPAN). It provides mechanisms to compress the comparatively large IPv6 headers to prevent the huge overhead a 40 byte header would add to 802.15.4's [2] physical payload of only 127 byte. Additionally, 6LoWPAN defines a mesh routing header and a fragmentation mechanism to realize the minimum MTU of 1280 byte.

With regard to routing within a 6LoWPAN mesh network, RFC 4944 defines two possibilities: mesh-under and route-over. With mesh-under routing, the 6LoWPAN network appears to the IPv6 layer as a single hop, routing over multiple hops is then done at the 6LoWPAN layer by some not specified routing protocol. Route-over, on the other hand, uses IPv6 routing, i.e., 6LoWPAN nodes act as IPv6 routers. Within this paper, we concentrate on the latter

S. Guo et al. (Eds.): ADHOC-NOW 2014, LNCS 8487, pp. 279–289, 2014.

scheme, which enables the use of standard protocols throughout the whole protocol stack. In the presence of fragmented IPv6 datagrams, the straight-forward realization of route-over, called "assembly mode" in the remainder of this paper, implies that each datagram is completely reassembled at each intermediate hop. During reassembly, a node has to reserve buffer space for the incoming datagram. Considering resource-constrained nodes, which usually provide reception buffers not much larger than the minimum MTU, this can easily lead to a node being forced to drop a second incoming large datagram. Additionally, – on longer multi-hop routes – the theoretical potential for pipelining fragments of a datagram is wasted.

This issue being known, the implementation guidelines for 6LoWPAN [3] suggest the use of a so-called virtual buffer, which is used to store information on the datagrams in transit while the individual fragments are forwarded directly, without waiting for complete reassembly of the datagram. We use the term "direct mode" when referring to this forwarding strategy.

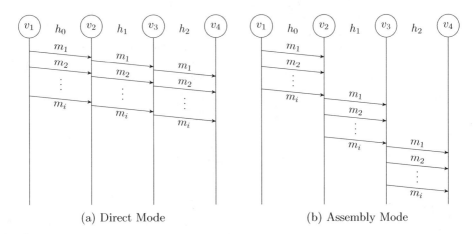

(a) Direct Mode   (b) Assembly Mode

Fig. 1. Sequence chart of messages sent in direct and assembly modes; h_k denotes the kth hop, m_x the xth fragment

Figure 1 illustrates the difference between the two modes in a sequence diagram. While at this first glance the direct mode seems much more attractive in terms of buffer space and latency, simulations and testbed experiments for some scenarios have shown a severe decrease of reliability. A major reason for this is found in self-induced collisions between consecutive fragments of a datagram due to the hidden node problem, e.g., consider fragments m_2 sent by v_1 and m_1 sent by v_3 in Figure 1a, which potentially collide at node v_2. Additionally, the direct mode may suffer from additional load caused by fragments belonging to a datagram which has long been given up on.

To better asses the actual extent of the latter effect, we extended an existing bit-error-based analytical model of a 6LoWPAN multi-hop, multi-fragment transmission and evaluated the expectation value of the number of bits sent and the actual datagram success rate. Aside from this extension of the model, the main contribution of this paper is to quantify the difference in expectation values of the number of bits sent between direct and assembly modes.

Section 2 provides an overview about existing work in this field, Section 3 introduces the analytical model and Section 4 provides an evaluation of results obtained for a given parameter set. Section 5 concludes the paper.

2   Related Work

Several research efforts have been undertaken to analyze the performance of 6LoWPAN fragmentation over 802.15.4 networks, both analytically and by simulation or testbed experiments.

The model presented in Section 3 is derived from the model of Ayadi and Thubert, who analyzed the efficiency of different TCP segment sizes ([4], [5]). One of their basic assumptions about forwarding of fragments differs from ours: They assume that a sender always will send all fragments belonging to a datagram, no matter if one already has failed, i.e., no link-layer acknowledgement has been received, while we expect a sender to give up on a datagram if it has not received an ACK for one of its fragments. In [6], the model is extended to assess the effectiveness of simple forward fragment recovery, which is described within an internet draft [7] of the IETF ROLL working group.

A whole different approach which tries to take into account collisions, but is not specifically tailored to typical 6LoWPAN scenarios, has been followed by Di Marco et al. [8], which is in turn based on [9]. Their models are based on Markov chains and model the transmission of frames over multiple hops by state transitions, taking into account clear-channel assessment, backoffs and link-layer retries. Output values of this model are latency, energy consumption (derived from a node's state) and reliability.

Evaluation of experiments in real testbeds have been carried out by Ludovici et al. [10]. Loss rate and end-to-end delay for IPv6 datagrams over 6LoWPAN were analyzed within a testbed of five TelosB nodes and one sender. Similar simulations and experiments have been carried out by Weigel et al. within a larger testbed of 13 nodes and several simulation scenarios [11]. The results of both studies show a dramatically higher reliability for the assembly mode compared to the direct mode. The assembly mode started to suffer from a drop in reliability as soon as the datagrams got bigger, so that only one at a time fits into the reassembly buffer. On the other hand, due to some potential for pipelining fragments, the observed end-to-end latency was lower for the direct mode in the presence of multiple hops.

3   Analytical Model

Our model is an extension of the one presented by Ayadi et al. and follows the same approach taken to derive probabilities and expectation values presented in their paper [4]. It calculates the number of expected bits sent depending on link bit error rates, the number of link-layer attempts and a forward error correction (FEC), i.e., a redundancy ratio which yields a number of correctable bit errors. We assume that bit errors are independent of each other. Additionally, we assume that duplicates are detected by the 6LoWPAN layer, which is realistic as it needs to keep track of the state of fragmented datagrams anyway.

The major difference between the existing model and our extension is the handling of failures and partial failures. While Ayadi et al. let a sender always send all fragments, we assume that a sender gives up on a datagram in case of a failure or partial failure. We also consider both forwarding techniques introduced in Section 1 in that context: assembly and direct mode. We implemented both our extended model and the original model of Ayadi. The latter was slightly modified to get comparable resulting quantities. The extended model for assembly and direct modes is referred to as non-persisting model, the original model is referred to as persisting model in the remainder of the paper.

Table 1. List of symbols used in this paper

Symbol	Definition
B_k	bit error rate on link k
L_F	link layer frame size (bits), including 802.15.4 headers
L_A	link layer acknowledgement frame size (bits)
c	number of correctable bit errors
h	index of last hop of the route
h_0	index of first hop of the route
r	number of link layer attempts
m	number of fragments sent
$p_{f,k}$	probability of failure for a single transmission on link k
$p_{p,k}$	probability of partial failure for a single transmission on link k
$p_{s,k}$	probability of success for a single transmission on link k
$P_{s,k}$	probability of success after r attempts on link k
$P_{p,k}$	probability of partial failure after r attempts on link k
$P_{f,k}$	probability of failure after r attempts on link k
$H_{s,k}$	expectation value of number of bits sent in case of success on link k
$H_{p,k}$	expectation value of number of bits sent in case of partial failure on link k
$H_{f,k}$	expectation value of number of bits sent in case of failure on link k
$H_{sp,k}$	exp. value of number of bits sent in case of success or part. failure on link k
Q_s	end-to-end probability of success for a whole datagram
Q_f	end-to-end probability of failure for a whole datagram
E^A	expectation value of number of bits sent (assembly)
E_s^A	cond. expectation value of number of bits sent in case of success (assembly)
E_f^A	cond. expectation value of number of bits sent in case of failure (assembly)
E^D	expectation value of number of bits sent (direct)

3.1   Link-Layer Model

At the link layer, three different outcomes of a transmission are defined:

- The transmission failed (probability $p_{f,k}$)
- The transmission partially failed, i.e., the frame arrived but the ACK was never received (probability $p_{p,k}$)
- The transmission was successful, frame and ACK arrived (probability $p_{s,k}$)

At their hearts, both models are based on the bit error rate of a link. In case of the occurrence of an uncorrectable bit error, the transmission is considered unsuccessful, yielding the introduced probabilities as

$$p_{f,k} = 1 - \sum_{i=0}^{c} \binom{L_F}{i} B_k^i (1 - B_k)^{L_F - i} \tag{1}$$

$$p_{p,k} = (1 - p_{f,k})(1 - (1 - B_k)^{L_A}) \tag{2}$$

$$p_{s,k} = (1 - p_{f,k})(1 - B_k)^{L_A} = 1 - (p_{p,k} + p_{f,k}) \tag{3}$$

with the symbols specified in Table 1. We stick with the approach from [4] to model FEC only for the data packet, not the acknowledgment. While the model itself thereby regards the possibility of a FEC mechanism, for our evaluation in this paper we consider $c = 0$ and therefore do not go into further details about the derivation of the number of fragments for a given datagram size.

Based on those basic formulas, the probabilities for success, partial failure and failure after r send attempts can be derived as

$$P_{s,k} = \sum_{j=1}^{r} p_{s,k}(1 - p_{s,k})^{j-1} = \sum_{j=1}^{r} p_{s,k} \sum_{i=0}^{j-1} \binom{j-1}{i} p_{p,k}^i p_{f,k}^{j-1-i} \tag{4}$$

$$P_{p,k} = (p_{p,k} + p_{f,k})^r - p_{f,k}^r = \sum_{j=1}^{r} \binom{r}{j} p_{p,k}^j p_{f,k}^{r-j} \tag{5}$$

$$P_{f,k} = p_{f,k}^r \tag{6}$$

and the conditional expectation value for the number of sent bits in each case can then be derived by enumerating all possible outcomes after r attempts as

$$H_{s,k} = \frac{1}{P_{s,k}} \left(\sum_{j=1}^{r} p_{s,k} \sum_{i=0}^{j-1} \binom{j-1}{i} p_{p,k}^i p_{f,k}^{j-1-i} (jL_F + (i+1)L_A) \right) \tag{7}$$

$$H_{p,k} = \frac{1}{P_{p,k}} \left(\sum_{i=1}^{r} \binom{r}{i} p_p^i p_f^{r-i} (rL_F + iL_A) \right) \tag{8}$$

$$H_{f,k} = \frac{1}{P_{f,k}} r L_F P_{f,k} = r L_F \tag{9}$$

$$H_{sp,k} = \frac{1}{1 - P_{f,k}} (P_{p,k} H_{p,k} + P_{s,k} H_{s,k}) \tag{10}$$

where $H_{y,k}$ is the expected number of bits sent in case of success, partial failure or failure on link k, with the corresponding probability $P_{y,k}$. In addition to the values for the three possible outcomes, we define as $H_{\mathrm{sp},k}$ the expected number of bits sent in case of a success or a partial failure. The formulas presented so far are basically those introduced in [4], extended by an index to identify a certain hop and a slight rearrangement due to the different results which partial failures produce in our multi-hop model.

3.2   Multi-hop Model

Moving to a multiple hop, multiple fragment scenario, we get for the end-to-end (over hops from h_0 to h) probability of success and failure of a datagram consisting of m fragments Q_{s} and Q_{f}, independent of the actual forwarding mode (direct or assembly):

$$Q_{\mathrm{s}}(h_0, h, m) = \prod_{k=h_0}^{h} (P_{\mathrm{s},k}^{m-1}(P_{\mathrm{s},k} + P_{\mathrm{p},k})), \tag{11}$$

$$Q_{\mathrm{f}}(h_0, h, m) = \sum_{k=h_0}^{h} Q_{\mathrm{s}}(h_0, k-1, m) \left(\sum_{x=1}^{m-1} P_{\mathrm{s},k}^{x-1} P_{\mathrm{p},k} + \sum_{x=1}^{m} P_{\mathrm{s},k}^{x-1} P_{\mathrm{f},k} \right), \tag{12}$$

where h_0 is the index of the starting hop, h the index of the final hop and m the number of fragments sent. For the last fragment sent, a partial failure is sufficient, because it does not matter whether the sender would give up on the datagram afterwards.

With the assembly mode, no fragments of a failed datagram are propagated any further after the first fragment failure. The expected number of bits sent in assembly mode E^{A} therefore is

$$E^{\mathrm{A}}(h_0, h, m) = Q_{\mathrm{s}}(h_0, h, m) E_{\mathrm{s}}^{\mathrm{A}}(h_0, h, m) + Q_{\mathrm{f}}(h_0, h, m) E_{\mathrm{f}}^{\mathrm{A}}(h_0, h, m) \tag{13}$$

$$E_{\mathrm{s}}^{\mathrm{A}}(h_0, h, m) = \sum_{i=h_0}^{h} ((m-1)H_{\mathrm{s},i} + H_{\mathrm{sp},i}) \tag{14}$$

$$E_{\mathrm{f}}^{\mathrm{A}}(h_0, h, m) = \frac{1}{Q_{\mathrm{f}}(h_0, h, m)} \sum_{k=h_0}^{h} Q_{\mathrm{s}}(h_0, k-1, m)$$

$$\left(\sum_{x=1}^{m-1} ((x-1)H_{\mathrm{s},k} + H_{\mathrm{p},k} + E_{\mathrm{s}}^{\mathrm{A}}(h_0, k-1, m)) P_{\mathrm{s},k}^{x-1} P_{\mathrm{p},k} \tag{15} \right.$$

$$\left. + \sum_{x=1}^{m} ((x-1)H_{\mathrm{s},k} + H_{\mathrm{f},k} + E_{\mathrm{s}}^{\mathrm{A}}(h_0, k-1, m)) P_{\mathrm{s},k}^{x-1} P_{\mathrm{f},k} \right)$$

where $E_{\mathrm{s}}^{\mathrm{A}}$ and $E_{\mathrm{f}}^{\mathrm{A}}$ are the conditional expectation values of the number of bits sent in case of success and failure, respectively.

While the conditional expectation value of the number of bits sent in case of success is the same for the direct mode, for the case of a failure we have to take all fragments into account which already have been transported to the next receiver in the route and consider their contribution to the overall number of bits sent. To include those, we define E^{D} by means of a recursive formula.

$$E^{\mathrm{D}}(h_0, h, m, H_{\mathrm{acc}}, P) = P \cdot H_{\mathrm{acc}}, \tag{16}$$

if $h < h_0$ or $m = 0$, and $E^{\mathrm{D}}(h_0, h, m, H_{\mathrm{acc}}, P) =$

$$P \cdot Q_{\mathrm{s}}(h_0, h, m) \left(E_{\mathrm{s}}^{\mathrm{A}}(h_0, h, m) + H_{\mathrm{acc}} \right) + P \sum_{k=h_0}^{h} (M_{\mathrm{p}} + M_{\mathrm{f}}), \tag{17}$$

in all other cases. In order to simplify presentation and to foster understandability, we omit the parameters h_0, h, m, k, and H_{acc} for M_{p}, M_{f}, and H_z in equations (17) to (20). M_{p} and M_{f} contain the actual recursion:

$$M_{\mathrm{p}} = \sum_{x=1}^{m-1} E^{\mathrm{D}}(k+1, h, x, H_{\mathrm{p}}, Q_{\mathrm{s}}(h_0, k-1, m) P_{\mathrm{s},k}^{x-1} P_{\mathrm{p},k}) \tag{18}$$

$$M_{\mathrm{f}} = \sum_{x=1}^{m} E^{\mathrm{D}}(k+1, h, x-1, H_{\mathrm{f}}, Q_{\mathrm{s}}(h_0, k-1, m) P_{\mathrm{s},k}^{x-1} P_{\mathrm{f},k}) \tag{19}$$

with

$$H_z = H_{\mathrm{acc}} + (x-1)H_{\mathrm{s},k} + H_{z,k} + E_{\mathrm{s}}^{\mathrm{A}}(h_0, k-1, m), z \in \{\mathrm{p},\mathrm{f}\} \tag{20}$$

As for the assembly mode, the formula sums up the expectation values for the link layer transmission multiplied by the corresponding probability for failures at a certain hop and a certain fragment. However, a failure of the xth fragment on the kth hop means a partial datagram of $x - 1$ (or x in case of partial failure) fragments will be send from hop $k + 1$ to hop h. This is captured by applying the recursion, additionally passing the accumulated expectation value of the number of bits sent so far (parameter H_{acc}) and passing the product of corresponding probabilities (parameter P). Thereby, all possible cases are enumerated. In the initial formula H_{acc} and P are set to 0 and 1, respectively. For example, the expected number of bits sent to send 10 fragments via hops 1 to 5 then can be obtained by $E^{\mathrm{D}}(1, 5, 10, 0, 1)$.

4   Evaluation

To evaluate the impact of the different forwarding techniques, we fed our model with scenarios varying bit error rate, number of link-layer attempts, number of hops and number of fragments. In this paper we do not further investigate the potential use of FEC. For all scenarios we set $L_{\mathrm{F}} = 119\,\mathrm{byte}$[1], leaving a 802.15.4

---

[1] Including 11 byte MAC header and 2 byte PHY header.

payload of 106 byte and assume a link-layer acknowledgement of 7 byte. Unless specified differently, we set the non-varying input parameters to the values shown in Table 2. Besides the overall probability of success for a transmission, our model's main output metric is the expected number of bits sent, which indicates the amount of energy used for transmission as well as the overall traffic load produced on a link. In the following, additional subscripts P and NP indicate persisting and non-persisting forwarding strategy (see Section 3), respectively.

Table 2. Default parameter values

Parameter	Value	Parameter	Value
r	5	c	0
h_0	1	h	8
m	12		

While we constructed our model in a way that each link can be assigned a different bit error rate, for the evaluation presented here we used equal rates for all links. To calculate results, the model was implemented as a Mathematica module.

Figure 2 shows the ratio of expectation values of bits sent for direct and assembly modes (E_{NP}^D/E_{NP}^A) along with the overall probability for success of the

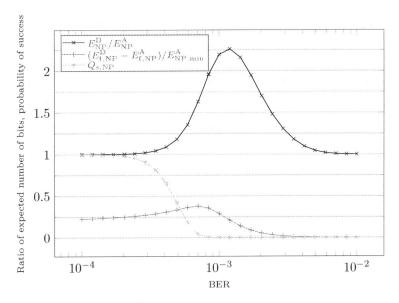

Fig. 2. Non-persisting mode; Ratio of expected number of bits sent in direct (E_{NP}^D) and assembly mode (E_{NP}^D), together with overall probability of success and the difference in bits for the number of bits sent in case of failure for both modes, normalized by $E_{NP,min}^A$, against bit error rate

whole datagram $Q_{s,NP}$, which is the same for both forwarding techniques. For success rates approaching 1 and 0, the difference between the expected number of bits sent in direct and assembly forwarding modes approaches zero. However, for a probability of success of 0.81 we get a ratio of 1.04, i.e., a 4 % increase compared to the assembly mode and even higher ratios for lower probabilities of success. This difference is exclusively caused by the direct mode's conditional expectation value of bits sent in case of a failure. To better illustrate this increased number of bits in direct mode in case of a failure, we added the difference of the corresponding conditional expectation values of direct and assembly mode to the plot, normalized by the minimum number of bits needed for a successful transmission $((E_{f,NP}^{D} - E_{f,NP}^{A})/E_{NP,min}^{A})$. The direct mode produces about one fourth of the minimum number of bits needed for successful transmission more than the assembly mode.

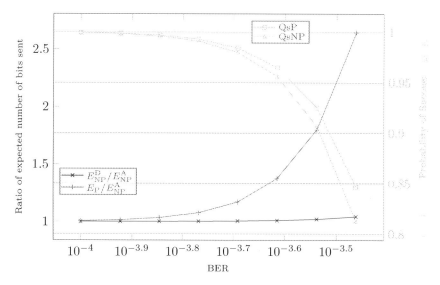

Fig. 3. Comparison of the ratios of the number of expected bits sent (left y-axis) and the probability of success (right y-axis) for persisting and non-persisting approaches

To assess the difference between persisting and non-persisting forwarding, we also compared the output from the original model with our extended version. On the one hand, we expected the probability of success to be slightly higher than for our assembly and direct non-persisting modes, as partial failures on a link are counted as an overall success with regard to the transmission of a fragment. On the other hand, the expected number of bits sent should be significantly higher for the persisting approach, as even in the case of a failure a sender continues sending all remaining fragments. Figure 3 shows a comparison of the two strategies for different values of the bit error rate. While the effect of persisting on the probability of success is comparatively small – for the shown BER values it stays below 3 % –, the impact on the expected number of bits sent is significant.

In the light of those results, the use of such a persisting strategy for the given scenario is considered inadvisable.

5   Conclusion and Future Work

We presented an analytical model based on bit error rates for 6LoWPAN multi-hop and multi-fragment transmissions. While – being solely BER-based – the model neglects various effects like collisions through (self) interference and is also ignorant of latencies, it can help to better understand results obtained through simulation or testbeds. Comparison of assembly and direct modes have shown a significant difference of the expected number of bits sent in case of transmission failures. This observation offers a possible partial explanation for the bad performance of the direct mode in recent research efforts with respect to reliability. Furthermore, the evaluation shows that using any kind of persistent forwarding technique increases the expected number of bits dramatically, while only slightly improving the probability of success and can therefore be considered inferior when using fragmentation without the usage of any fragment recovery mechanism.

5.1   Future Work

Our implementation of the model in Mathematica currently is done in a straight-forward way, given the formulas presented in Section 3. Especially the computing time of the recursion part of the direct mode grows exponentially with m and h, already limiting the values of these parameters for which the model can be calculated in reasonable time. A more efficient implementation or a transformation of the given formula is needed to make working with the model more comfortable. While being basically supported by the model, the use of forward error correction for the data frames was not evaluated in this paper and is left for future work. Furthermore, we want to assess the impact of the presented effects with simulations and testbed experiments.

References

1. Montenegro, G., Kushalnagar, N., Hui, J., Culler, D.: Transmission of IPv6 Packets over IEEE 802.15.4 Networks. RFC 4944 (Proposed Standard) (September 2007)
2. IEEE Standard for Local and Metropolitan Area Networks— Part 15.4: Low-Rate Wireless Personal Area Networks (2011)
3. Bormann, C.: 6LoWPAN Roadmap and Implementation Guide (draft) (April 2013), http://tools.ietf.org/pdf/draft-bormann-6lowpan-roadmap-04.pdf (accessed: February 07, 2014)
4. Ayadi, A., Maillé, P., Ros, D.: Tcp over low-power and lossy networks: tuning the segment size to minimize energy consumption. CoRR, abs/1010.5128 (2010)
5. Ayadi, A., Maille, P., Ros, D.: Tcp over low-power and lossy networks: Tuning the segment size to minimize energy consumption. In: 2011 4th IFIP International Conference on New Technologies, Mobility and Security (NTMS), pp. 1–5 (February 2011)

6. Ayadi, A., Maille, P., Ros, D., Toutain, L., Thubert, P.: Energy-efficient fragment recovery techniques for low-power and lossy networks. In: 2011 7th International Wireless Communications and Mobile Computing Conference (IWCMC), pp. 601–606 (July 2011)
7. Thubert, P., Hui, J.: LLN Fragment Forwarding and Recovery (draft) (February 2013), http://tools.ietf.org/html/draft-thubert-6lo-forwarding-fragments (accessed: February 07, 2014)
8. Di Marco, P., Park, P., Fischione, C., Johansson, K.H.: Analytical modeling of multi-hop ieee 802.15.4 networks. IEEE Transactions on Vehicular Technology 61(7), 3191–3208 (2012)
9. Bianchi, G.: Performance analysis of the ieee 802.11 distributed coordination function. IEEE J. Sel. A. Commun. 18(3), 535–547 (2006)
10. Ludovici, A., Calveras, A., Casademont, J.: Forwarding Techniques for IP Fragmented Packets in a Real 6LoWPAN Network. Sensors (Basel) 11(1), 992–1008 (2011)
11. Weigel, A., Ringwelski, M., Turau, V., Timm-Giel, A.: Route-over forwarding techniques in a 6lowpan. In: Pesch, D., Timm-Giel, A., Calvo, R.A., Wenning, B.-L., Pentikousis, K. (eds.) Monami 2013. LNICST, vol. 125, pp. 122–135. Springer, Heidelberg (2013)

A Traffic-Based Local Gradient Maintenance Protocol: Making Gradient Broadcast More Robust⋆

Alexandre Mouradian and Isabelle Augé-Blum

Université de Lyon, INRIA, INSA Lyon, CITI, F-69621, France
`firstname.lastname@insa-lyon.fr`

Abstract. Gradient broadcast routing is a robust scheme for data gathering in Wireless Sensor Networks (WSNs). At each hop, the sender broadcasts the packet to its neighbors and one or more nodes among its neighbors closer to the sink forward it. As long as a node has at least one neighbor with a smaller hop-count, it can route packets. Nevertheless, nodes can disappear because of energy depletion, hardware failure, etc. In this case, it cannot be ensured that a packet reaches the sink. Usually this issue is addressed by updating the gradient with a periodical flooding. Nevertheless, it consumes an important amount of energy, moreover, parts of the network may not need to be updated. In this paper, we propose GRABUP (GRAdient Broadcast UPdate), a traffic-based gradient maintenance algorithm which updates the gradient thanks to the data packets. We simulate the proposition and compare it with the classic gradient broadcast routing.

1 Introduction

After two decades of scientific research, WSNs are becoming a mature technology with many applications (smart metering, anomaly detection, environment monitoring, etc). While tremendous efforts have been made to reduce energy consumption and enhance self-organization schemes of WSNs protocols, new applications are emerging with high Quality of Service (QoS) constraints [13] [9] that cannot be met by existing protocols [2]. Critical applications require timeliness (bounded end-to-end delays) and a high degree of reliability (high delivery ratio) while keeping energy consumption as low as possible.

At the routing level, in order to increase the delivery ratio, opportunistic routing have been proposed [4] [8] [12]. In classical routing, a node with a packet to forward looks in its routing table, finds the most suited neighbor according to the routing metric and sends the packet to this neighbor. With opportunistic routing, a node with a packet to forward, broadcasts it to its neighbors, each neighbor then decides if it forwards the packet or not according to the routing metric (usually, an election mechanism is used). This kind of routing creates link

---

⋆ This work has been partially founded by French Agence Nationale de la Recherche under contract VERSO 2009-017.

S. Guo et al. (Eds.): ADHOC-NOW 2014, LNCS 8487, pp. 290–303, 2014.
© Springer International Publishing Switzerland 2014

diversity which makes it robust to packet loss [4] [7]. Moreover, the knowledge of the neighborhood is not mandatory in order to route the packets because the routing decision is not taken by the sender. This aspect is interesting for two reasons in WSNs:

- it avoids to store a neighborhood table which is memory costly (sensors usually have few KB of RAM);
- it thus also allows to avoid neighborhood maintenance with hello messages which is energy costly.

GRAdient Broadcast (GRAB) [11] is an opportunistic gradient routing protocol. It implements opportunistic routing with a metric called cost. Each node is given a cost to reach the sink in number of hops, a node can relay a packet if it has a smaller hop-count than the sender, so packets flow toward the sink (like water in a funnel). Changes in the environment or node deaths can lead to a not up to date gradient field (holes in the funnel). So the packets may not end up at the sink, but in a topology void.

In order to prevent the voids, in this type of protocols, the whole gradient field is refreshed periodically or the refresh is triggered by an event. The update of the gradient is done by flooding the network with initialization packets. Whereas, in many cases, only a part of the network need to be updated. Moreover, in highly dynamic networks, updates take place very often even if no application traffic is exchanged, which leads to a high energy consumption.

In this paper, we propose a scheme to update the gradient field only if needed and during packet exchanges. The principle is to detect when a node's hop-count is not up to date (no neighbors with smaller hop-count, so there is a void) and let nodes with the same or greater hop-count deal with the packet to avoid the void. Once the void is avoided, a local initialization is triggered to update the nodes around it. As observed by simulation, this scheme increases the delivery ratio and reduces energy consumption compared to periodical gradient refresh.

The paper is organized as follows: Section 2 presents related gradient routing and gradient maintenance schemes. Section 3 describes the details of the proposed protocol, gives an example on a simple topology and provides discussions on practical aspects for implementing such a scheme. In section 4, we comment the simulation results of the comparison between our proposition and GRAB on large scale random topologies. In section 5, we conclude and present future works.

2   Related Work

Gradient maintenance has received little attention in the literature. Apart from the works presented in this section, articles on gradient routing give very little insight on how the hop-count is updated.

In [11], the authors propose GRAB, an opportunistic gradient routing scheme. First the paper defines how the cost field is initialized: the sink triggers the construction of a cost field, the network is flooded with an initialization packet containing the cost to reach the sink from the sender (the cost is updated by each sender). Each node, on the reception of an initialization packet, computes its new cost (the cost of the sender plus the cost of the link). If it is smaller than the previous one, it keeps it and rebroadcasts the initialization packet. We can notice that, if the cost of all links is one, then the cost corresponds to the hop-count to reach the sink. In the remainder of the paper, we take the hop-count for cost, because it is highly robust to topology changes: if a node has several neighbors with a smaller hop-count, its cost does not change when one of them dies. When the cost field is initialized, packets are routed using a robust opportunistic gradient routing scheme. With this protocol, the packet must be forwarded by a node closer to the sink in terms of hop-count than the sender. The GRAB scheme allows packet replication in order to increase reliability. Each packet has a budget which allows to elect multiple forwarders among nodes closer to the sink than the sender. We can notice that packet replication also increase energy consumption. The authors of [11] claim that the maintenance of the cost field is event triggered, which reduces energy consumption compared periodical maintenance, because the initialization flooding is performed only when needed. To our understanding, the initialization is triggered when the delivery ratio goes under a predefined threshold. To compute the delivery ratio, the authors assume that the packets are numbered and ordered when received by the sink. In our opinion, these are too strong assumptions which do not hold for many WSNs applications. Typically with anomaly detection applications, it is not possible to know in advance which node will send packets and in which order. So the initialization phase has to be rerun periodically with a period that depends on the dynamics of the topology changes.

In [14], [5] and [10], the authors propose that the neighbors periodically exchange hello packets in order to maintain the hop-count field. In [5] and [10] the period changes according to the network dynamics (there are less packet exchanges if the network is very stable). Nevertheless, in this case again, energy is spent to maintain the field even in parts of the network which do not change. Moreover, this scheme implies to keep track of the neighbors of each node in the network which is something that opportunistic routing aims at avoiding.

SGF [6] extends the GRAB cost field by taking into account the energy needed to reach the sink. It also proposes to update the cost of nodes at each transmission when the forwarder of a node changes. We can note that with the hop-count cost this kind of update is not necessary because the cost does not change with the forwarder. With the SGF maintenance scheme, if no node closer to sink relays a packet, it is sent back to the predecessor node until one node can route the packet. This can lead to go several hops back unnecessarily in some cases, and in other cases, it can lead to lose the packet even if there exists a path to the sink. For example, Fig. 1 depicts a topology where the maintenance fails. A packet coming from A or B arrives at C after the death of node D. In this case, the packet will go back to A or B leading to a dead-end. Whereas it is still possible to reach the sink through H.

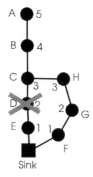

Fig. 1. Example of maintenance failure with SGF. The numbers correspond to hop-counts of the nodes.

In the remainder of this paper we present GRABUP, an extension of GRAB which allows to maintain the gradient field only when and where needed. The maintenance is triggered by traffic packets and only affects nodes which need to be updated in areas where there is traffic.

3 GRABUP

In this section we present GRABUP which extends GRAB with a traffic-based and localized gradient update scheme. We first describe the proposition in details. Then, we illustrate it with an example. Finally, we discuss further crucial aspects of the proposed scheme.

3.1 Protocol Description

GRABUP implements the same routing scheme as GRAB:

- it uses a hop-count field, each node knows its hop-count h;
- the sender broadcasts its packet to its neighbors;
- a node closer to the sink than the sender can forward the packet.

In addition, nodes with equal or greater hop-counts can forward the packet if no node with a smaller hop-count received it. This scheme allows to get around voids in the gradient field. Once the void is avoided (the packet has reached a node with an up-to-date hop-count), an update packet is sent back to update the gradient (suppress the voids).

In order to prevent all nodes closer to the sink to forward the packet (which leads to a high energy consumption), an election mechanism allows to choose only a subset of nodes as forwarders (as in [4] [11] [7]). In GRABUP we choose to limit the number of forwarders to one. This is achieved thanks to the mechanism depicted on Fig. 2. When a node has a packet to send, it broadcasts the packet to its neighbors with its hop-count h contained in the header. Each receiver of

Fig. 2. GRABUP acknowledgment period is divided in 3, one period for nodes with an hop-count smaller than h, equal to h and greater than h with h the hop-count of the sender

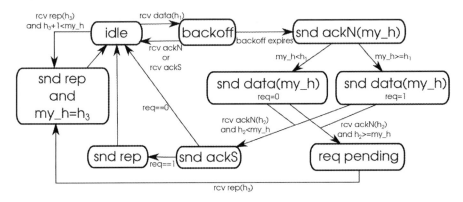

Fig. 3. State machine representation of GRABUP

the packet picks a time at random uniformly according to its hop-count. Nodes with a smaller hop-count than h pick a time between t_1 and t_2, nodes with a hop-count equal to h pick a time between t_2 and t_3 and nodes with a hop-count greater than h pick a time between t_3 and t_4. The node initializes a backoff timer with this value and waits from t_1 until either it receives an acknowledgment (another node is the forwarder) or the timer expires (the node is the forwarder). If the timer expires, the node sends an acknowledgment. The acknowledgment is repeated by the source of the packet so every neighbor of the source is aware that a forwarder has been chosen. We can notice that the value of the backoff timer can be picked according to other parameters such as energy level, receive signal strength of the packet, etc. We can also notice that with such a scheme there is still a probability that more than one forwarder is elected (if two or more nodes pick the same backoff timer). As in [4] and [11], we consider this probability small enough not to induce energy consumption issues (it does not reduce significantly the network lifetime).

The global behavior of a node is represented on Fig. 3. The protocol is depicted with a state diagram where the labels on the transitions are conditions to go from one state to another. The hop-count of the node executing the protocol is noted *my_h*. On Fig. 3, we do the distinction between an acknowledgment from a neighbor of the emitter of the packet (noted *ackN*) and one repeated by the emitter (noted *ackS*).

After the initialization (a flooding similar to the GRAB initialization, described in section 2), the nodes are in the *idle* state. If a node receives a data packet containing the hop-count h_1 in the header, it goes to the *backoff* state and pick a backoff time which depends on its hop-count as previously commented. On the reception of an *ackN* or *ackS* the node goes back to the *idle* state and stops (and reset) its backoff timer (it is not the forwarder so it drops the data packet). If its backoff timer expires, it sends an *ackN*, meaning that it is elected forwarder of the data packet. Then, if $h_1 > my\_h$ the node forwards the data packet with its hop-count my_h in the header and keep its *req* variable to 0. The *req* variable is put to one when the $h_1 \leq my\_h$, it means that the hop-count h_1 of the first sender is not up to date because no one with a hop-count less than h_1 answered. In this case, an update of the hop-count will have to be triggered. After the emission of the data packet, the node waits for an *ackN*. Depending on the hop-count h_2 of the node which sends the *ackN*, the node either sends an *ackS* (if $h_2 < my\_h$) or goes to the *request pending* state ($h_2 \geq my\_h$). In the former case, the node then sends a *rep* packet if *req* is equal to one and goes back to the *idle* state. In the latter case, the node waits for a *rep* packet in order to get an up to date hop-count (it does not take part in the forwarding scheme until its hop-count is updated), when it receives a *rep*, it updates its hop-count and forwards the *rep* packet with its new hop-count. In this case, an *ackS* is not sent, it allows potential forwarders in opposite directions to forward the packet and thus explore the topology as described in the example of section 3.2. If a node is in *idle* state and receives a *rep* packet it will update its hop-count and send a new *rep* only if the new hop-count is smaller than the previous.

With this protocol, if there is a void in the topology (a node with no neighbor with a smaller hop-count), the packet goes around the void until it reaches a node with neighbors at smaller hop-count. From this point a *rep* message is sent back to update all nodes around the hole.

We can notice that the hop-count update process is not separated from the routing. This allows to do the update exactly when and where needed and not in unused or already up to date parts of the network.

3.2 Example

In this section we give an example of the behavior of GRABUP on the topology depicted by Fig. 4. The letters represent node identities and the numbers their hop-counts before the execution of GRABUP. Only F has a packet to send, it broadcasts it to its neighbors. Only E receives the data packet, it backs off and acknowledges as described previously. E then forwards the packet. At this point there is no neighbor of E with a smaller hop-count than E. Node G, which picked a shorter backoff than D, sends an *ackN* first but as explained in the previous section, E does not send an *ackS*. As a results D, which does not receive the *ackN* from G also send one. Thus both D and G forward the packet. With this example, we see the purpose of such a behavior, indeed, if only G forwards the packet, it goes on the longest side of the void, leading to higher end-to-end delays and not minimum hop-counts. From this point, there are 2 data packets and E

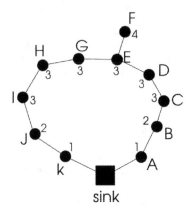

Fig. 4. Example topology, numerous node deaths lead to a not up to date hop-count field

is in the *req pending* state (because no node with a smaller hop-count relayed its packet). No more data packet duplication occurs then, because nodes E, G, D and H go to the *req pending* state after relaying the data packet (for example, it prevents E from acknowledging the packet relayed by G even if it does not receive the *ackN* from H). The data packets are thus relayed on both sides of the void. Here we assume that the packet of the left-hand side path arrives to node I before the one of the right-hand side arrives to node C. I and C have neighbors with a smaller hop-count. They thus send a *rep* packet after forwarding the data packet in order to update the nodes around the void. Since the packet arrives first at node I, the update is as if the other side did not exist. The nodes update their hop-counts and forward the *rep* packet. Then, when the data packet is forwarded by node C it triggers another update (with a *rep* packet) on the right hand side of the void. This packet allows nodes which are closer to the sink with this path (the right-hand side of the void) to update their hop-counts.

3.3   Discussion

In this section, we comment: the impact of packet duplications during the update scheme, the update decision procedure, the need for synchronization between emitter and potential forwarders, the possible MAC schemes and the impact of transient links.

Packet Duplications. As mentioned in section 3.1, we implement a version of GRAB with only one forwarder selected, in order to reduce energy consumption. This is achieved thanks to an acknowledgment of the source (*ackS*) of the packet which informs all potential forwarders that there is already a forwarder selected. Nevertheless when a sender has no neighbor with a smaller hop-count, it does not send a source acknowledgment. It is the case of node E of Fig. 4, this induces potential packet duplications because a neighbor of the sender may not receive an *ackN* from another neighbor. We argue that it is a necessary duplication because it allows to explore opposite sides of a void so the shortest one can be found. Moreover one side of the void may not be connected to the sink. We can note that duplication only comes from nodes that are neighbors of the sender but not neighbors of each others.

Updates. Another feature of GRABUP open to discussion is the way the hop-count updates are done. A node updates its hop-count on the reception of a *rep* packet either if it is in *req pending* state or if the h in the *rep* packet is smaller than its hop-count. In the example of Fig. 4 this scheme does not allow node F to update its hop-count. Indeed, the *rep* packets goes back from nodes I and C, when node E updates its hop-count to 5 and sends a *rep* packet, node F is in *idle* state. The *rep* packet sent by node E contains a greater hop-count than the one of F, it thus does not update its hop-count. This is due to the fact that nodes do not maintain neighbor tables, so F cannot know if E is its only neighbor closer to the sink.

Nevertheless, the usage of a counter which stores the number of nodes closer to the sink could be used in order to trigger the hop-count update of nodes in the position of F without the explicit knowledge of the neighborhood. A node which updates its hop-count now sends a *rep* packet with its new and former hop-counts. With this variant of GRABUP, on the reception of a *rep* packet, a node updates its hop-count in three cases: if it is in *req pending* state, if the new h in the *rep* packet is smaller than its hop-count or if old h of the *rep* packet has a smaller hop-count and the node has only one neighbor left closer to the sink.

Node Addition. In this paper we focus on topology changes which consist in nodes and links disappearance. Nevertheless, the addition of nodes in the network is not an issue for GRABUP. The addition procedure goes as follows: first, a node which is added to the network listens to packets in its neighborhood, after having overheard a packet forwarding, it knows the hop-count of at least one of its neighbors. It assigns the smallest value plus one to its hop-count and sends a *rep* packet in order to update the part of the network affected by its apparition.

Synchronization. In WSNs a duty-cycle mechanism is used to reduce energy consumption. The nodes alternately turn on (awake period) and off (sleep period) their radios. In order to communicate, they have to agree on their wakeup dates. In GRAB and GRABUP a node sends its packet to all its neighbors, this means that all its neighbors have to be awake at time t_0 represented in Fig. 2. This can be achieved either by a preamble sampling techniques or thanks to global synchronization protocols. The use of one technique or the other depends on the traffic load in the network: global synchronization may be preferred for high periodical traffic loads and preamble sampling for lower and more sporadic traffic loads [3].

Medium Access. It is possible that at time t_0, several nodes in the same neighborhood have a packet to send. If they send it at the same time it will create a collision. In order to organize medium accesses, either the nodes can use a random access scheme as recommended in GRAB or a TDMA slot can be reserved for each interfering sender. The former case is more reactive to the traffic because a node tries to access to the medium as soon as it has a packet to send, but it can lead to high collision rates for high traffic load [3]. In the latter case, a node must wait for its time slot, so it can lead to higher end-to-end delays.

Transient Links. Dynamic environment and interferences can cause packet losses in WSNs. The opportunistic nature of GRAB and GRABUP alleviate the effects of transient links by creating potential forwarder diversity [4] [7]. Nevertheless, in the case of GRABUP, the absence of acknowledgement from a node closer to the sink may not be because of a void, but because the packet was lost due to interferences. Therefore, before deciding that there is a void and that the packet can be forwarded by nodes at equal distance or farther than the sender, the packet should be retransmitted. The number of retransmissions must be determined in function of the dynamic of the links. We can notice that GRAB and other opportunistic routing schemes are affected by packet loss in the same way so retransmissions should also be implemented.

4   Performances Evaluation

In this section we compare the performance of GRAB and GRABUP and their respective gradient update mechanisms by simulation.

4.1   Simulation Environment

The simulations have been performed with the WSNet [1] simulator, an event-driven simulator for large scale wireless networks. The simulation parameters and their values are listed in Table 1.

The energy consumption parameters are taken from [7]. We only consider energy consumption coming from packet exchanges, the energy spent in order to send or receive a packet is given by $E = P \times D$ with P equal respectively to the TX or RX power consumption and D the duration of the packet. On every energy consumption figure, the unit is the Joule and the plots represent the sum of the consumption of all the nodes of the network.

We generated random uniform topologies of 200, 300 and 400 nodes. For each number of nodes, 20 topologies have been generated. This allows to observe the behavior of GRAB and GRABUP with different network densities (because the size of the simulation area remains unchanged).

We choose to evaluate GRAB and GRABUP in topologies with random node deaths: after the initialization of the gradient, every 5 seconds, a node is selected among all active nodes and killed (so at the end of the simulation about 200 nodes are killed). This allows to model random hardware failures for instance.

Table 1. Simulation parameters

Parameter	Value
Bitrate	500kbps
Data packet size	100 bytes
Ack packet size	10 bytes
Number of nodes	200 to 400
Simulation area	50×50 units
Radio range	10 units
TX power consumption	65.7mW
RX power consumption	53.7mW
Data packet traffic (for the whole network)	1 packet/5s
Node deaths rate	1 node/5s
Simulation duration	1000s

4.2   Simulation Results

Fig. 5(a) depicts a comparison of the delivery ratio between GRAB and GRABUP for topologies of 200, 300 and 400 nodes. The error bars indicate the minimum and maximum delivery ratio values of 20 different topologies. We observe that GRABUP has a higher delivery ratio in every cases. Nevertheless the difference of values decreases when the number of nodes increases. This is due to the fact that when the number of nodes increases, the network density increases as well (the simulation area stays constant) so nodes' deaths create less voids. Thus GRAB performs better in high density networks. We observe, in the simulations traces, that the packets losses of GRABUP are only due to disconnected topologies. We conclude that the routing and gradient maintenance mechanisms of GRABUP are highly reliable.

On Fig. 5(b) we compare the energy consumption of GRAB and GRABUP for topologies of 200, 300 and 400 nodes. The error bars indicate the minimum and maximum values. We observe that the energy consumption of GRABUP is greater than the one of GRAB. This is due to the energy consumption of the gradient maintenance part of the algorithm and the delivery of more packets than GRAB. The consumption difference decreases when the network density increases because less gradient updates are needed.

Fig. 6(a) depicts a comparison of end-to-end delays between GRAB and GRABUP for topologies of 200, 300 and 400 nodes. The error bars indicate the minimum and maximum values. On this figure, we notice that the maximum values of end-to-end delays of GRABUP are very high. By looking at simulation traces we observe that long delays are actually very rare and are due to cases similar to the one depicted by Fig. 7. If node D, on Fig. 7, has a packet to send just after node A's death, the packet goes to C and B and then backward until it reaches the sink through the other side. In this case, the packet is delivered in

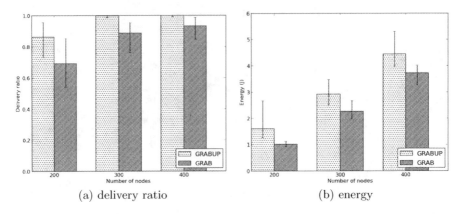

(a) delivery ratio   (b) energy

Fig. 5. Delivery ratio and energy consumption results for topologies of 200, 300 and 400 nodes: average values (the error bars represent min and max values)

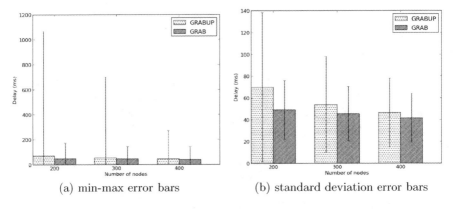

(a) min-max error bars   (b) standard deviation error bars

Fig. 6. End to end delay results for topologies of 200, 300 and 400 nodes

9 hops instead of 5 (case where the hop-counts are up to date) so the end-to-end delay is much higher. Nevertheless, it is still better than in the case of GRAB for which the packet cannot reach the sink.

Fig. 6(b) is another plot of the end-to-end delay but with error bars representing the standard deviation. It allows to better observe the difference between the average delays of GRAB and GRABUP. The difference decreases with the increase of network density because less "long delay" cases as the one of Fig. 7 appear.

As described in section 2, GRAB can update the hop-count field by periodically flooding the network. We know investigate the impact of the refresh period on the delivery ratio and energy consumption and compare the results to the ones obtained for GRABUP. The simulations have been performed on the 200 nodes topologies with gradient refresh periods ranging from 10 (the highest refresh frequency) to 50 seconds.

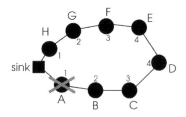

Fig. 7. Example of topology case which induces long end-to-end delays

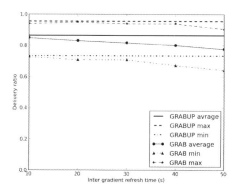

Fig. 8. Delivery ratio in function of refresh period for 200 nodes topologies

Fig. 8 depicts the average, maximum and minimum delivery ratios of GRAB and GRABUP for different refresh periods. We observe that for a refresh period of 10 seconds, the values of the delivery ratio of GRAB are very close to those of GRABUP. When the refresh period increases, it degrades the performances of GRAB as expected.

On Fig. 9 we observe the average, maximum and minimum energy consumption of GRAB and GRABUP in function of the refresh period. As expected the energy consumption of GRAB is higher for short refresh periods. It decreases exponentially when the period increases. For periods smaller than 40 seconds, the energy consumption of GRAB is higher than the one of GRABUP. For periods greater than 40 seconds, the energy consumption of GRAB is equal or smaller than the one of GRABUP. Nevertheless, the delivery ratio is also smaller for these values as can be seen on Fig. 8.

Under the conditions of traffic and node deaths simulated in this paper, GRABUP achieves a better energy/delivery ratio trade-off than GRAB. This is due to the fact that GRABUP updates the gradient only where and when needed, leaving unused or unchanged parts of the network untouched. This allows to spend energy only where needed and to keep the delivery ratio very high.

We can notice that the uniform random deaths of the nodes disadvantage GRABUP, because in this case, many different parts of the network are affected

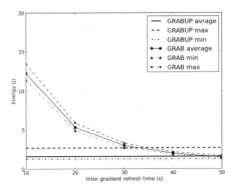

Fig. 9. Energy consumption in function of refresh period for 200 nodes topologies

by topology changes. The global refresh scheme of GRAB is thus more efficient than in cases where nodes failures occur in the same area (because the other parts of the network are refreshed for nothing).

5   Conclusion and Future Works

In this paper we present GRABUP an extension of the opportunistic gradient routing protocol GRAB. The aim of GRABUP is to update the gradient field in order to get around voids. GRABUP is a traffic-based and local algorithm, it updates the gradient only while routing the data packets and only in parts of the network affected by topology changes. We describe the protocol and illustrate its functioning with an example. We then discuss crucial aspects of the proposed solution. A comparison by simulation with the periodical flooding technique of GRAB allows to conclude that the traffic-based update scheme of GRABUP improves the energy/delivery ratio trade-off.

In the future we plan to implement GRABUP on a real testbed in order to evaluate its performances under real world conditions. We also plan to test different data packet retransmission tunings in order to find the best in function the dynamics of the links.

References

1. http://wsnet.gforge.inria.fr/
2. Anastasi, G., Conti, M., Di Francesco, M.: A Comprehensive Analysis of the MAC Unreliability Problem in IEEE 802.15.4 Wireless Sensor Networks. IEEE Transactions on Industrial Informatics 7(1), 52–65 (2011)
3. Bachir, A., Dohler, M., Watteyne, T., Leung, K.K.: MAC essentials for wireless sensor networks. IEEE Communications Surveys & Tutorials 12(2), 222–248 (2010)
4. Biswas, S., Morris, R.: ExOR: opportunistic multi-hop routing for wireless networks. ACM SIGCOMM Computer Communication Review 35(4), 133–144 (2005)

5. Gnawali, O., Fonseca, R., Jamieson, K., Levis, P.: Ctp: Robust and efficient collection through control and data plane integration. Tech. rep., Univ. of Southern California, UC Berkeley, MIT CSAIL, Stanford Univ (2008)
6. Huang, P., Chen, H., Xing, G., Tan, Y.: SGF: a state-free gradient-based forwarding protocol for wireless sensor networks. ACM Transactions on Sensor Networks 5(2), 14 (2009)
7. Lampin, Q., Barthel, D., Augé-Blum, I., Valois, F.: QoS oriented Opportunistic Routing protocol for Wireless Sensor Networks. In: Wireless Days, Dublin, Ireland (2012)
8. Schaefer, G., Ingelrest, F., Vetterli, M.: Potentials of opportunistic routing in energy-constrained wireless sensor networks. In: Roedig, U., Sreenan, C.J. (eds.) EWSN 2009. LNCS, vol. 5432, pp. 118–133. Springer, Heidelberg (2009)
9. Tan, R., Xing, G., Chen, J., Song, W.Z., Huang, R.: Quality-driven volcanic earthquake detection using wireless sensor networks. In: RTSS 2010, San Diego, CA, USA, pp. 271–280 (2010)
10. Watteyne, T., Pister, K., Barthel, D., Dohler, M., Auge-Blum, I.: Implementation of gradient routing in wireless sensor networks. In: IEEE GLOBECOM, Honolulu, USA, pp. 1–6 (2009)
11. Ye, F., Zhong, G., Lu, S., Zhang, L.: GRAdient broadcast: a robust data delivery protocol for large scale sensor networks. Wireless Networks 11, 285–298 (2005)
12. Zeng, K., Lou, W., Zhai, H.: On end-to-end throughput of opportunistic routing in multirate and multihop wireless networks. In: IEEE INFOCOM, Phoenix, USA, pp. 816–824 (2008)
13. Zhang, J., Li, W., Han, N., Kan, J.: Forest fire detection system based on a zigbee wireless sensor network. Journal of Beijing Forestry University 29(4), 369–374 (2007)
14. Zhao, Y., Chen, Y., Li, B., Zhang, Q.: Hop ID: a virtual coordinate based routing for sparse mobile ad hoc networks. IEEE Transactions on Mobile Computing 6(9), 1075–1089 (2007)

Efficient Energy-Aware Mechanisms for Real-Time Routing in Wireless Sensor Networks

Mohamed Aissani, Sofiane Bouznad, Badis Djamaa, and Ibrahim Tsabet

Computer Science Unit, Ecole Militaire Polytechnique (EMP)
P.O. Box 17, Bordj-El-Bahri, 16111, Algiers, Algeria
{maissani,bouznad.sofiane,badis.djamaa,utopia.ibrahim}@gmail.com

Abstract. We propose three mechanisms to manage nodes energy and improve the efficiency of real-time routing protocols in sensor networks. To preserve nodes' resources and to improve network fluidity, the first mechanism removes each useless packet due to its insufficient deadline in reaching the sink. To reinforce the packet real-time aspect, the second mechanism selects from the current-node queue the most urgent packet to be forwarded first. For a better node energy balancing, the third mechanism uses both the residual energy and the relay speed of the forwarding candidate neighbour to select the next forwarder of the current packet. These mechanisms are simple to implement, require very little states and rely only on local primitives. In addition they can be easily integrated in any geographic routing protocol. Associated with the real-time routing protocol SPEED in TinyOS and evaluated in the simulator TOSSIM, our proposals achieved good performance in terms of node energy balancing, packet loss ratio and energy consumption.

Keywords: Wireless sensor networks, real-time routing, energy-aware routing, node energy balancing.

1 Introduction

Wireless sensor networks (WSNs) are often characterized by a dense and large scale deployment of battery-powered sensors with limited processing, storage and communication resources. It is widely recognized that conserving energy is a key requirement in the design of WSNs, because of the strict constraints that imposes on network operations [1, 2, 3]. In fact, the energy consumed by sensor nodes has an important impact on the network lifetime that has become the dominant performance criterion. Extending lifetime of a WSN is a shared objective by WSN designers and researchers across the whole network stack. At the routing layer, for instance, it is necessary that routing algorithms use less energy-consuming paths; this challenge may be aggravated in real-time routing protocols where the real-time quality of service (QoS) imposes more constraints on these paths.

To address this issue, we propose in this paper, three energy-aware mechanisms for real-time geographic routing protocols in WSNs. While strictly respecting the real-time constraint of the routed packets, the proposed mechanisms address both network energy consumption and node energy balancing. The first mechanism, called UPR (Useless Packet Removing), removes any packet considered unnecessary since it has no chance

S. Guo et al. (Eds.): ADHOC-NOW 2014, LNCS 8487, pp. 304–317, 2014.

to reach its destination before the expiration of its deadline. The second mechanism, called UPS (Urgent Packet Selecting), selects, from the current-node queue, the most urgent packet to be forwarded first. The third mechanism, called CAB (Cost-Aware energy Balancing), uses both the residual energy and the relay speed of a forwarding candidate neighbor to select the next forwarder of the most urgent selected packet.

The rest of the paper is organized as follows. Section 2 summarizes the related work. Section 3 describes the proposed energy-aware mechanisms (UPR, UPS and CAB) that improve the performance of real-time geographic routing protocols in WSNs. Section 4 presents evaluations and discussions of the performance of our proposals. Section 5 concludes the paper and discusses future research directions.

2   Related Work

Some existing real-time routing protocols [4, 5, 6, 7] define explicit mechanisms to save energy of sensor nodes and/or maximize the network lifetime. PATH [4] improves real-time routing performance by means of reducing packets' dropping in routing decisions. It is based on the concept of using two-hop neighbor information and power-control mechanism. The former is used for routing decisions while the latter is deployed to improve link quality as well as reducing the delay. The protocol dynamically adjusts transmitting power in order to reduce the probability of packet dropping and addresses practical issue like network holes, scalability and loss links in WSNs. EARTOR [5] is designed to route requests with specified end-to-end latency constraints, which strikes the elegant balance between the energy consumption and the end-to-end latency and aims to maximize the number of the requests realized in the network. The core techniques adopted include the cross-layer design that incorporates the duty cycle, a bidding mechanism for each relay candidate that takes its residual energy, location information, and relay priority into consideration. EEOR [6] improves the sensor network throughput by allowing nodes that overhear the transmission and are closer to the sink to participate in forwarding the packet, i.e., in forwarder list. The nodes in the forwarder list are prioritized and the lower priority forwarder will discard the packet if the packet has been forwarded by a higher priority forwarder. One challenging issue is to select and prioritize forwarder list such that the energy consumption by all nodes is optimized. Extensive simulations show that this protocol performs well in terms of energy consumption, packet loss ratio and average delivery delay. TREE [7] is a routing strategy with guaranty of QoS for industrial wireless sensor networks by considering the real-time routing performance, transmission reliability, and energy efficiency. By using the two-hop information, the real-time data routes with lower energy cost and better transmission reliability are used in the proposed routing strategy.

Although the previous works contribute to improving network performance, design of energy-aware real-time routing protocols in WSNs is still a challenging issue. Consequently, three efficient energy-aware mechanisms (UPR, UPS and CAB) are proposed in this paper and then associated with the well-known real-time routing protocol SPEED [8], which gives birth to the CA-SPEED (Cost Aware SPEED) protocol. Note that SPEED [8] is designed to be a stateless, location-based routing protocol with minimal control overhead. It achieves an end-to-end soft real-time communication by maintaining a desired delivery speed across the sensor network through a novel combination of feedback control and stateless non-deterministic geographic forwarding.

3   Proposed Energy-Aware Mechanisms

We consider a WSN comprising a set of homogenous nodes N, arbitrary located in the plane. A node in the system is a device which has a processing unit, a power unit and a communication unit allowing wireless communication. For the communication model, we take the protocol model, in which all nodes located within the transmission range r of a given node i, form the set of i's potential neighbors, thus the links are assumed to be bi-directional. The combination of the nodes N and the neighborhood relations E form a WSN, which can be represented by an undirected graph $G = (N, E)$. In our simulations (Section 4), a more realistic physical model which includes a realistic signal propagation model, signal interference, distortions, background noise, etc. was used.

Since our problem consists of proposing efficient energy managing and latency improving mechanisms for real-time geographic routing protocols in WSNs, two other principal assumptions made in this category are considered. First, it is assumed that every node knows its own and its network neighbors' positions. Second, the source of a message is assumed to be informed about the position of the destination. The Knowledge of nodes' and sinks' positions is assumed to be reasonable [9]. By regards, to the neighbors' positions, this is realized by a neighbor discovery mechanism (known as location beacons) implemented in the majority of geographical routing protocols including SPEED, in which nodes periodically broadcast location beacons packets containing their positions to their neighbors. Neighbors receiving such a packet store the contained information in their caches for a specific time period. Given a data packet p to be transmitted from a node u to a node v, in one- or multi-hop way, we use the following notations: $D_{uv}(p)$ is the Euclidean distance between the nodes u and v; $T_{uv}(p)$ is the delay that the packet realized from node u to node v or the expected delay the packet should register to get from u to v; $Deadline(p)$ is the packet deadline; $AD(p)$ is the packet advance in distance from its source u w.r.t the destination node v; $AT(p)$ is the packet advance in time at current-node u. It will be measured based on the packet deadline and the expected delay $(T_{uv}(p))$ that the packet should register to get from current-node u to destination v; $H_{uv}(p)$ is the number of hops the packet travelled or ought to travel from node u to node v.

3.1   UPR Mechanism

Many existing real-time routing protocols, such those summarized in Section 2, forward, often over long distances, data packets that may have no chance to reach their destination node because of their insufficient residual deadlines resulting in network resource wasting (throughput and energy). This is because those protocols do not exploit the packet deadline information to decide on the utility of a packet. Starting from this point and in order to save nodes precious resources, the proposed UPR mechanism ensures an early removal of any useless packet that will not reach its destination. Indeed, only packets with sufficient residual deadline to reach their destinations are forwarded in the network. Thus, UPR saves nodes energy and increases the network fluidity which reinforces the real-time aspects of the associated routing protocol. To do so, UPR estimates the expected end-to-end delay allowing the current packet to reach its destination node (Section 3.1.1), then applies a decision rule, based on both the expected end-to-end delay and the application time/energy requirements, expressed by an α parameter, to decide whether to remove or forward the packet (Section 3.1.2).

3.1.1 Expected End-to-End Delay

Since the end-to-end delay in a multi-hop networks depends on the distance that a packet travels, many real-time geographic routing protocols such as SPEED, and its derived protocols, route packets according to the packet's maximum delivery speed, defined as the rate at which the packet should travel along a straight line to the destination [10].

Exploiting this feature, we propose a mechanism allowing a node in the network to estimate the expected end-to-end delay for a current packet, based on the already known past registered delay tailored to the already travelled end-to-end distance and an expected end-to-end distance which the packet should travel along a straight line from the current node to the destination. For coherency reasons, the already travelled end-to-end distance is taken as the straight distance realized by the packet.

Formally, we propose in this section a mechanism, depicted in Formula (1), which allows a current-node i to estimate the expected end-to-end delay for a packet p to reach its destination d, which is denoted by $T_{id}(p)$ in Fig. 1(a). The proposed mechanism estimates $T_{id}(p)$ as a function of the already known delay, registered by p from its source node s (dubbed $T_{si}(p)$ in Fig. 1(a) and Formula (1)), tailored to a ratio between the end-to-end distance that the packet p travelled from its source node s to current-node i (dubbed $D_{si}(p)$) and the remaining end-to-end distance for the packet to reach its destination (dubbed $D_{id}(p)$).

$$T_{id}(p) = \frac{D_{id}(p)}{D_{si}(p)} * T_{si}(p) \qquad (1)$$

In addition to its simplicity, the estimator proposed in Formula (1) bases its calculus only on local information either in current packet p or in current-node i. To calculate the Euclidian distances, the current node needs only to know, in addition to its position, the positions of the source and destination nodes which are incorporated in data packets transmitted by geographic routing protocols (the source position is used to indicate the area of the reported event and the destination position is used in routing decisions). The already registered delay is derived from the deadline field, which is an intrinsic propriety of real-time data packets. Being based only on local information, our estimator is robust to topology changes; network density and more importantly scales well with network size. Note that since the proposed estimator reposes on past registered delay by the packet, it takes into account all the parameters caused this delay (processing, queuing, etc.) and inject them to estimate the expected future delay.

In addition to the above characteristics, the proposed estimator in Formula (1) seems to give best-case expected end-to-end delays at the beginning of the packet transmission, which allows it to get in the network acquiring more knowledge and hence getting more accurate end-to-end delay expectations. Thus the expected end-to-end delays approach reality, when the message travels further in the network, making deletions based on it more accurate. However, in some extreme cases, the expected end-to-end delay given by our estimator may: (1) Exceed the real delay (if we allow forwarding the packet) which may compromise decisions based on our estimator or; (2) Underestimate the real delay allowing a useless packet to be forwarded when it should be deleted, which causes resources wasting.

To illustrate the former scenario, take a case where in the first travelled part, the packet registered relatively large delays which make the estimator expect a large future delay for the remaining distance to travel. However, if the remaining network portion is uncongested, the packet will register less delay (over estimation); therefore there is a big risk to delete a real-time data packet which is more probably to reach its destination. As an example of the latter, take the opposite extreme case when the first network portion is relaxed and the second one is congested, then the remaining delay may be underestimated and a packet which should be deleted earlier will do more steps, which induce more energy consumptions. This analysis is taken into account in designing the UPR packet-removing algorithm.

3.1.2   Packet-Remove Decision

Erroneous deletion of real-time data packets based on the above estimator could be devastating for time-critical applications. To avoid this problem, we propose a packet removing algorithm that tailors the expected delay given by the estimator to the application time/energy requirements specified by the α parameter. Hence, our removal algorithm automatically and dynamically calibrates the estimated delays based on the packet advance in distance towards its destination, until a specific ratio defined by the α parameter after which the energy-requirement is given more weight.

As explained above, the proposed estimator gives more realistic, hence trusted, values when the packet being acquired more past knowledge in advancing towards its destination. This advance could be presented by the distance gain realized by the packet from its originator node s to the current-node i w.r.t the total distance between the source s and the destination d. For this reason, and to give more chance to the packet, in time critical-applications, the proposed packet-removal algorithm dynamically adjusts the expected end-to-end delay by a dynamic factor, in the interval [0, 1], based on the advance in distance realized by the packet, until a given ratio imposed by the application energy-requirements, after which a delayed packet is automatically removed. This ratio is expressed and fixed by the α parameter defined by the application, depending on its energy/time requirements. Thus, the α parameter decides on the proportion in distance advancement after which the application will prefer the energy-saving requirement against the time-saving one. Before the α ratio, the time-requirement is more weighted, thus estimated delays are calibrated to give the packet more chances.

Formally, having the expected end-to-end delay $T_{id}(p)$ (see Formula (1)) and to decide whether to remove or forward packet p, current-node i applies Decision-rule (2), where $D_{sd}(p)$ denotes the distance between source node s and destination node d, $AD(p)$ shown in Fig. 1(b) and given by Formula (3) represents the advance in distance of packet p toward its destination node, and α is the energy/time parameter set in the interval [0,1] according to the real-time application requirements. The threshold α must be close to 0 in energy-critical applications maximizing the network lifetime and close to 1 in time-critical applications minimizing the packet loss ratio of the used routing protocol.

Decision-rule (2) is explained as follows. If $AD(p)$ is greater than $\alpha * D_{sd}(p)$ then node i prefers saving nodes energy and hence removes each delayed packet. Otherwise (i.e. before reaching the threshold α), $T_{id}(p)$ is calibrated and multiplied by a dynamic factor less than 1: $AD(p)/(\alpha * D_{sd}(p))$ to give more chance to the packet p to advance toward its destination d. If the result exceeds $Deadline(p)$ despite the given chance then packet p is removed to save network resources and to increase the network fluidity. In our simulations (Section 4), parameter α is set to 0.5. Thus, each delayed packet p with an advance $AD(p)$ greater than 50% of the total distance $D_{sd}(p)$ is immediately removed by current-node i in order to increase fluidity of links and to save the limited energy of sensor nodes.

IF $AD(p) > \alpha * D_{sd}(p)$ THEN
    IF $T_{id}(p) > Deadline(p)$ THEN remove packet p;
    ELSE packet p is to forward;
    ENDIF
ELSE   (2)
    IF$\frac{AD(p)}{\alpha * D_{sd}(p)} * T_{id}(p) > Deadline(p)$ THEN remove packet p;
    ELSE packet p is to forward;
    ENDIF

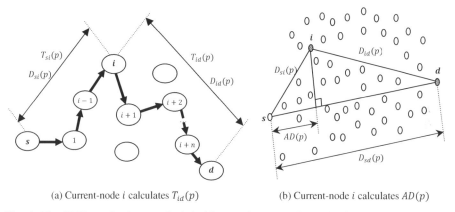

(a) Current-node i calculates $T_{id}(p)$ (b) Current-node i calculates $AD(p)$

Fig. 1. The UPR mechanism: node i decides on the removal (or the forwarding) of current packet p

To express the advancement in distance $AD(p)$ for a packet p, we apply the Pythagorean Theorem to two rectangular triangles depicted in Fig. 1(b). resulting in Equation (3) below.

$$AD(p) = \begin{cases} \dfrac{D_{si}^2(p) - D_{id}^2(p) + D_{sd}^2(p)}{2D_{sd}(p)} & \text{IF } \quad D_{si}(p) < D_{sd}(p) \\ D_{sd}(p) & \text{OTHERWISE} \end{cases} \qquad (3)$$

3.2 UPS Mechanism

Most existing real-time routing protocols use scheduling schemes based only on the residual deadline of a packet to be forwarded [11]. However, these schemes may not be effective when packets are sent to different destinations; case of a network using several sinks. In fact, these schemes prioritize forwarding a packet whose deadline is the smallest although it is very close to its destination. To illustrate this point, Fig. 2 depicts an example in which current-node i has two packets to forward: p_1 for destination $d1$ with 2 ms (milliseconds) as deadline and p_2 for destination $d2$ with 3 ms as deadline. According to existing scheduling schemes based only on the residual deadline, node i will firstly forward p_1 and then probably will removep_2 because of its distance to destination $d2$. To provide an efficient solution to this problem, the proposed UPS mechanism combines both the residual deadline and the expected end-to-end delay of a packet in a decision parameter $D(p)$ to decide on its urgency. In other words, the UPS decision parameter schedules packets based on their gained-deadline to reach their destinations, expressed as the difference between the packet residual deadline and its expected end-to-end delay to reach its destination. Formally, UPS performs as follows: (a) For a data packet p_j in the queue of current-node i, calculate the decision parameter $D(p_j)$ by Formula (4), where $T_{id}(p_j)$ is the expected end-to-end delay, calculated by Formula (2), allowing packet p_j to reach its destination d; and (b) Selects data packet p_k having the smallest decision parameter $D(p_k)$ by running Function (5).

$$D(p_j) = Deadline\ (p_j) - T_{id}(p_j) \qquad (4)$$

$$D(p_k) = Min\{D(p_j);\ \ \forall p_j \in Queue(i)\} \qquad (5)$$

Using the UPS mechanism, the two packets p_1 and p_2 (Fig. 2) will probably reach their respective destination before their deadlines expire as current-node i will forward packet p_2 before packet p_1. Note that UPS is also valid for a network using one sink (destination node).

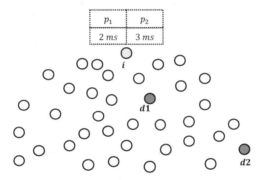

Fig. 2. Selection of the most urgent packet from the queue of current-node i

The urgent packet selection mechanisms can be performed in two ways: (a) during the packet reception by a node where its queue is scheduled according to Formula (4) or (b) during the forwarding process where the most urgent packet is timely chosen from the current-node queue. By using way (a), UPS minimizes calculations but loses reliability because the queuing time of packet p_j is not considered when estimating its expected end-to-end delay $T_{id}(p_j)$. But by adopting way (b), the current-node queue is not scheduled and selection of the most urgent packet requires extraction of all packets belonging to this queue. Since UPS is designed to achieve lower loss ratio and energy efficiency, we implemented way (b) where a current node applies both Decision-rule (2) and Function (5) on all packets in its queue, in order to remove each delayed packet and to select the most urgent packet among the not delayed ones.

3.3   CAB Mechanism

To choose the best next forwarder in terms of residual energy and relay speed, the current node uses the CAB mechanism, which is based on two efficient algorithms:

neighbor energy estimation (Section 3.3.1) and next forwarder selection (Section 3.3.2). This way, CAB aims to deliver a maximum number of real-time packets with a good node energy balancing in order to maximize the network lifetime.

3.3.1   Neighbor Energy Estimation

In order to allow nodes to be aware of their neighbors' residual energies, we propose a mechanism allowing nodes to exchange this information. In fact, each node keeps in its neighbor table a *ResidualEnergy* field representing the residual energy of each neighbor. The update of this field is discussed below. Location beacons, used in many geographical routing protocols, seem to be a good way to exchange residual energy information between nodes of a WSN. However, since location beacons are not sent frequently in static topologies, they are not sufficient to update energy information. This drives us to include the *ResidualEnergy*, in each forwarded packet. Relying on this field, the receiving node updates its own neighbors table. In this solution, the most energy updates are carried out through the location beacons that are more regular and sent in a broadcast mode to all neighbors of a node. Data packets, which are generally routed over long distances, consolidate these updates. Acknowledgments packets, sent after receiving data packets, improve the efficiency of our energy estimator.

This solution is very reliable in both static and dynamic environments. Thus, in mobile networks location beacons are sent frequently and hence updates are regular which increases routing decisions efficiency in our CAB mechanism. For lightly loaded static networks, location beacons are less frequent, which leads to less energy updates. However, in less active networks, energy consumption is reduced, so that frequency of energy updates is proportional to both network activity and energy consumption.

3.3.2   Next Forwarder Selection

We use the same definition of *FS* (Forwarding candidate neighbors Set) proposed in SPEED (Fig. 3), but we propose improved strategies to select next forwarder of packet p. In SNGF (Stateless Nondeterministic Geographic Forwarding) component of SPEED [8], the next forwarder of current packet p toward the destination node is selected by current-node i from its *FS*, which is constructed from its *NS* (Neighbors Set), according to a relay speed metric. Node n_i is a forwarding candidate if $distance(n_i, d) < distance(i, d)$.   In   Fig.   3, $FS = \{ n_1, n_2, n_3 \}$ and $NS = \{ n_1, n_2, n_3, n_4, n_5, n_6, n_7 \}$. The relay speed provided by a neighbor n_i in *FS* is given in SPEED by Formula (6), where L is the distance between current-node i and destination d, *Lnext* is the distance between neighbor n_i in *FS* and destination d, and $HopDelay(n_i)$ is the estimated delay of the link $i \rightarrow n_i$.

$$S(n_i) = \frac{L - Lnext}{HopDelay(n_i)} \quad ; \quad n_i \in FS \qquad (6)$$

Our goal is to insure a good node energy balancing without affecting the real-time aspect of the routed flows in a WSN; i.e. to achieve a lower packet loss ratio and a higher energy balancing factor. To select the next forwarder, the proposed CAB mechanism combines both the relay speed and the residual energy of all neighbors in *FS*. Neighbor n_k with the largest decision parameter $D(n_k)$ is selected by current-node i as next forwarder. Formally, node i applies Formula (7) that uses $D(n_k)$ given by Formula (8), where $S(n_i)$ is the relay speed provided by neighbor n_i, $E(n_i)$ is the residual energy of n_i, S_{max} is the maximal relay speed in FS, E_{max} is the maximal residual energy in FS, and f is the factor speed/energy varying in the interval $[0,1]$.

The weighting factor f is dynamically adjusted in a way that considers the network energy-balancing only when the message has realized an advance in its time-requirement allowing the node to choose the neighbor that better balances the energy consumption without disrupting the application real-time exigencies. Its adjustment is tightly related to the α parameter, representing the application energy/time requirement. Formally, the weighting factor f is obtained using Formula (9), where $H_{id}(p)$ denotes the remaining hops of packet p that is given by Formula (10), $H_{si}(p)$ is the number of hops traveled by p until node i (read from the packet TTL field), $AT(p)$ is the packet advance in time given by Formula (11), $AD(p)$ and $T_{id}(p)$ are given by Formulas (3) and (1) respectively.

$$NextHop(p) = n_k \; ; \quad \text{with} : D(n_k) = Max\{D(n_i) \; ; \; \forall n_i \in FS\} \tag{7}$$

$$D(n_i) = f * \frac{S(n_i) - S_{min}}{S_{max} - S_{min}} + (1-f) * \frac{E(n_i) - E_{min}}{E_{max} - E_{min}} \; ; \quad n_i \in FS \tag{8}$$

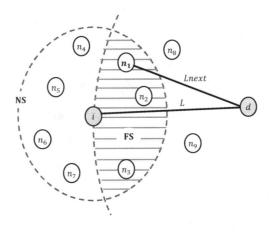

Fig. 3. Sets NS and FS of current-node i in the SPEED real-time routing protocol

Formula (9) is explained as follows: before reaching the threshold α where the expected end-to-end delay may be underestimated, the weighting factor f prioritizes the relay speed allowing the message to get forwarded and increasingly weights the residual energy in the forwarding decision, when the packet realizes an advance in distance, w.r.t to the destination, which is expressed by $AD(p)/D_{sd}(p)$ in Formula (9). After the threshold, the CAB mechanism always prioritizes the rely-speed unless the packet has gained time and has fewer hops to go. This is expressed by the $AT(p)/H_{id}(p)$ ratio in Formula (9).

$$\begin{cases} 1 - \dfrac{AT(p)}{H_{id}(p)} & \text{IF } AD(p) > \alpha * D_{sd}(p) \\[4mm] 1 - \dfrac{AD(p)}{D_{sd}(p)} & \text{OTHERWISE} \end{cases} \tag{9}$$

$$H_{id}(p) = D_{id}(p) * \frac{H_{si}(p)}{D_{si}(p)} \tag{10}$$

$$AT(p) = \frac{Deadline(p) - T_{id}(p)}{Deadline(p)} \tag{11}$$

The above behavior of the weighting factor f makes it always preferring the rely speed on the energy balancing. It only exploits the gain in time realized by the packet and the application energy/time requirements to explicitly enhance the protocol energy balancing; thus allowing the CAB mechanism to enhance network lifetime, without disturbing the application real time real-time exigencies.

4   Performance Evaluation

To evaluate the performance of the proposed mechanisms (UPR, UPS and CAB), we associated them with the well-known real-time routing protocol SPEED [8] and the resulting protocol is called CA-SPEED (Cost-Aware SPEED). We changed only the SNGF component of SPEED. In CA-SPEED, when a current node has to forward a data packet it: (1) Removes all delayed packets from its queue by executing the UPR mechanism and selects, during this queue consultation, the most urgent packet to forward first by using the UPS mechanism; and (2) Forwards the selected urgent packet to its best neighbor node which is obtained by executing the CAB mechanism.

The protocols SPEED and CA-SPEED were implemented in TinyOS [12] and evaluated in its embedded sensor network simulator TOSSIM [13]. Also, the recent proposed routing protocol EEOR [6] was evaluated in this simulator in the same conditions. Since we are interested in real-time applications, we used a scenario of detecting events that occur randomly in a field of interest. Once an event is detected, the information captured will be forwarded in a required deadline towards a sink, which is usually connected to an actuator. Our simulation scene uses a uniform random distribution of sensor nodes. We performed simulations on a 500×500 meters terrain with 625 deployed sensor nodes (an average density of 12 neighbors per node). Two destination nodes are deployed and each one receives packets concerning particular event detection. At each time period, 20 randomly source nodes, equitably distributed on each side of the network, detect an event and forward corresponding information to one destination node (sink). Each simulation runs for 230 seconds. Table 1 summarizes the parameters used in our simulations.

Table 1. Simulation parameters

Parameter	Value
MAC layer	CSMA-TinyOS
Radio layer	CC2420 radio layer
Propagation model	log-normal path loss model
Queue size	50 packets
Transmission channel	WirelessChannel
Bandwidth	200 Kilobytes per second
Packet size	32 bytes
Energy model	PowerTOSSIMz model
Node radio range	40 meters

For each run, we measure the packet loss ratio, energy consumption per delivered packet and energy balancing factor (ebf). The later represents the variance in energy consumed by all sensors having the same initial-energy quantity. Formally, we have $ebf = (1/ns) * \sum_{k=1}^{ns}(ec_k - ec_{avr})^2$, where ec_k is the energy consumed by sensor k and ns is the number of deployed sensors, ec_{avr} is the average energy consumed by all deployed sensors.

To measure the impact of packet generation rate, resulting from event detection applied in the source nodes, when generating real-time packets on each protocol performance, we set the packet deadline to 500 ms and we vary the source rate from 3 to 23 pps (packets per second). For each simulation, we measure performance achieved by the protocols SPEED [8], EEOR [6] and CA-SPEED. Obtained simulation results, shown in Fig. 4, illustrate that our energy-aware mechanisms, used in the CA-SPEED protocol, are efficient in terms of delivering real-time flows and managing energy of sensor nodes.

It can be clearly seen that CA-SPEED loses fewer packets (Fig. 4(a)) and consumes less energy (Fig. 4(b)) compared with EEOR and SPEED protocols. This is due to both the UPR mechanism which increases the network fluidity by removing each packet having less chance to reach its destination according to its residual

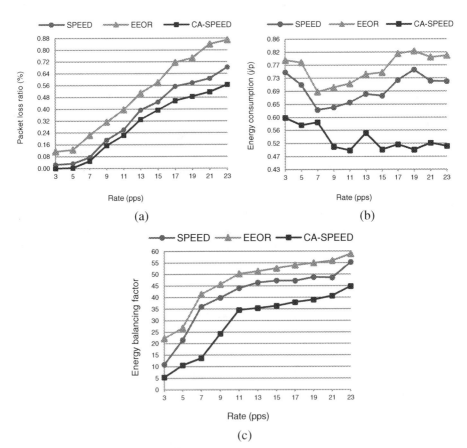

Fig. 4. Routing performance with several rates of the source nodes

deadline and the expected end-to-end delay, and the UPS mechanism which forwards first the most urgent packet extracted from the current-node queue. In applications with high rate, the EEOR and SPEED protocols lose more packets, because the deadline parameter is not considered in their routing decisions. Fig. 4(c) shows that CA-SPEED outperforms the protocols SPEED and EEOR in balancing energy of sensor nodes. This performance is due essentially to the CAB mechanism which selects as next forwarder, the neighbor with more energy when the packet is not late according to its residual deadline and expected end-to-end delay. However, it selects the neighbor providing the better relay speed when the packet is late.

To evaluate effectiveness of our energy-aware mechanisms, used by the CA-SPEED protocol, in time critical applications, we set the rate of source nodes to 1 pps and we vary the deadline of generated packets from 180 to 280 ms. For each simulation, we measure the performance of SPEED, EEOR and CA-SPEED protocols. Obtained results, shown in Fig. 5, indicate that CA-SPEED achieves good performance compared to other protocols, especially for small packet deadlines (less than 230 ms). This is due to efficiency of the associated mechanisms (UPR, UPS and CAB).

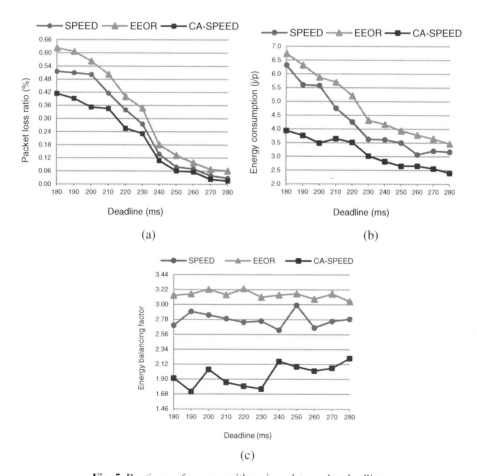

Fig. 5. Routing performance with various data packet deadlines

Fig. 5(a) shows the positive influence of the proposed energy-aware real-time routing mechanisms on the packet loss ratio in very critical applications, particularly when the packet deadline is less than 230 ms. The CA-SPEED protocol always achieves the best performance by taking advantage of each associated mechanism (UPR, UPS and CAB). CA-SPEED achieves the lower energy consumption per delivered packet (Fig. 5(b)) because firstly, it takes advantages from the UPR mechanism to delete earlier useless packets and hence liberate bandwidth for more messages to be forwarded; secondly it forwards first the most urgent message, which, with the first mechanism, minimizes the packet loss ratio and hence increases utile consumed energy and thirdly it explicitly considers energy in its forwarding decisions using CAB. Thus, the limited energy of sensor nodes is optimally managed by the proposed mechanisms in the CA-SPEED protocol. In addition, Fig. 5(c) shows that CA-SPEED clearly outperforms the existing protocols SPEED and EEOR in node energy balancing. This is mainly due to the CAB mechanism which selects as next forwarder of the current packet the neighbor realizing the best tradeoff between the packet residual energy and the neighbor relay speed.

5   Conclusion

The work carried out in this paper deals with the the energy management problem in real-time geographic routing protocols in WSNs. To contribute in this active research field, we have proposed three energy-aware real-time routing mechanisms (UPR, UPS and CAB) that aim to improve energy managing and deliver maximum number of real-time packets in WSNs. The UPR mechanism forwards only packets with sufficient deadlines to reach their destinations. The UPS mechanism chooses the most urgent packet to be forwarded first, in order to both reduce packet loss ratio of the associated routing protocol and improve the network fluidity. This urgency is calculated relying on the expected end-to-end delay allowing the current packet to reach its destination node. The CAB mechanism provides good energy balancing while minimizing the packet loss ratio by combining both residual energy and relay speed of the forwarding candidate neighbor when selecting the next forwarder in the routing path.

We have associated the proposed mechanisms with the SPEED real-time routing protocol. The resulting protocol CA-SPEED: early removes each delayed packet in network; then selects the most urgent packet to be forwarded first; and finally forwards the selected urgent packet to the next forwarder neighbor performing the best tradeoff between the residual energy of the neighbor and its relay speed. Obtained simulation results showed that CA-SPEED outperforms the two evaluated protocols SPEED and EEOR in terms of packet loss ratio, energy consumed per delivered packet and node energy balancing.

Actually, we are developing a power-aware real-time routing mechanism which combines the adjusted transmission power of the current node with the relay speed of the forwarding candidate neighbors when selecting the next forwarder of the current packet. This mechanism will, then, be combined with the CAB mechanism in a hybrid routing approach. The resulting cost-power-aware mechanism should deliver the maximum of real-time packets, save and balance more effectively the limited energy of sensor nodes. The useless delayed packets will be removed early and the most urgent packet in the node queue will be always forwarded first.

Since we base dropping decisions concerning delayed packets simply on estimated travel times towards the sink, our future work will consider any kind of weights, urgencies, fairness, or importance values of packets in order to obtain a less aggressive approach. We also plan to put our developed source codes in Imote2 sensor nodes for experimental tests in order to consolidate the simulation results presented in this paper.

References

[1] Akkaya, K., Younis, M.: A Survey on Routing Protocols for Wireless Sensor Networks. Ad Hoc Networks 3(3), 325–349 (2005)

[2] Ehsan, S., Hamdaoui, B.: A Survey on Energy-Efficient Routing Techniques with QoS Assurances for Wireless Multimedia Sensor Networks. IEEE Communications Surveys & Tutorials 14(2), 265–278 (2012)

[3] Marjan, R., Behnam, D., Kamalrulnizam, A.B., Malrey, L.: Multipath Routing in Wireless Sensor Networks: Survey and Research Challenges. Sensors Journal 12(1), 650–685 (2012)

[4] Rezayat, P., Mahdavi, M., Ghasemzadeh, M., AghaSarram, M.: A Novel Real-Time Power Aware Routing Protocol in Wireless Sensor Networks. Journal of Computer Science 10(4), 300–305 (2010)

[5] Yang, W., Liang, W., Dou, W.: Energy-Aware Real-Time Opportunistic Routing for Wireless Ad Hoc Networks. In: Proceedings of the IEEE Global Telecommunications Conference, Miami, FL, USA, pp. 1–6 (2010)

[6] Mao, X., Tang, S., Xu, X.: Energy-Efficient Opportunistic Routing in Wireless Sensor Networks. IEEE Transactions on Parallel and Distributed Systems 22(11), 1934–1942 (2011)

[7] Xue, L., Guan, X., Liu, Z., Yang, B.: TREE: Routing strategy with guarantee of QoS for industrial WSNs. International Journal of Communication Systems (IJCS), http://onlinelibrary.wiley.com/doi/10.1002/dac.2376/full (last accessed June 20, 2012)

[8] He, T., Stankovic, J.A., Lu, C., Abdelzaher, T.: A Spatiotemporal Communication Protocol for Wireless Sensor Networks. IEEE Transactions on Parallel and Distributed Systems 16(10), 995–1006 (2005)

[9] Zollinger, A.: Networking unleashed: Geographic routing and topology control in ad hoc and sensor networks. PhD thesis, ETH Zurich, Switzerland, Diss. ETH 16025 (2005)

[10] Soro, S., Heinzelman, W.: A Survey of Visual Sensor Networks. Advances in Multimedia 21, 1–21 (2009)

[11] Liu, K., Abu-Ghazaleh, N., Kang, K.D.: JiTS: Just-in-Time Scheduling for Real-Time Sensor Data Dissemination. In: Proceedings of the 4th IEEE Annual International Conference on Pervasive Computing and Communications, Pisa, Italy, pp. 42–46 (2006)

[12] Levis, P., Gay, D.: TinyOS programming. Cambridge University Press, USA (2009)

[13] Levis, P., Lee, N., Welsh, M., Culler, D.: TOSSIM: Accurate and Scalable Simulation of Entire TinyOS Applications. In: Proceedings of the 1st ACM Conference on Embedded Networked Sensor Systems, LA, California, USA, pp. 126–137 (2003)

AODV and SAODV under Attack: Performance Comparison

Mohamed A. Abdelshafy and Peter J.B. King

School of Mathematical & Computer Sciences
Heriot-Watt University, Edinburgh, UK
{ma814,P.J.B.King}@hw.ac.uk

Abstract. AODV is a reactive MANET routing protocol that does not support security of routing messages. SAODV is an extension of the AODV routing protocol that is designed to fulfil security features of the routing messages. In this paper, we study the performance of both AODV and SAODV routing protocols under the presence of blackhole, grayhole, selfish and flooding attacks. We conclude that the performance of SAODV is better than AODV in the presence of blackhole, grayhole and selfish attacks while its performance is worse than AODV in the presence of flooding attack. The blackhole and flooding attacks have a severe impact on the AODV and SAODV performance while the grayhole and selfish attacks have less significant effect on it.

Keywords: MANET, Routing protocol, AODV, SAODV, Security, Attack, Blackhole, Grayhole, Selfish, Flooding.

1 Introduction

Routing protocols for Mobile Ad Hoc Networks (MANETs) are usually designed assuming that all nodes cooperate to forward data [1,2]. However, the existence of malicious nodes cannot be ignored in MANETs because their wireless nature makes them vulnerable. A large number of attack types of varying severity are known [3].

Security mechanisms are added to existing routing protocols to resist attacks. Cryptographic techniques are used to ensure the authenticity and integrity of routing messages [4]. A major concern is the trade off between security and performance, given the limited resources available at many MANET nodes. Both symmetric and asymmetric cryptography have been used as well as hash chaining. Examples of these security enhanced protocols are Authenticated Routing for Ad-hoc Networks (ARAN) [5], Secure Link State Routing Protocol (SLSP) [6], and Secure Ad-hoc On-demand Distance Vector routing (SAODV) [7].

SAODV is an enhancement of Ad-hoc On-demand Distance Vector routing (AODV) [8]. SAODV provides an end to end authentication of the route and node by node verification of routing messages, using asymmetric cryptography and has chaining. No new message types are introduced, but routing packets are significantly larger than in AODV.

S. Guo et al. (Eds.): ADHOC-NOW 2014, LNCS 8487, pp. 318–331, 2014.

The rest of the paper is organized as follows. In section 2, an overview of the AODV and SAODV routing protocols is presented. In Section 3, the impact of some attacks on MANET is discussed. In section 4, the simulation approach and parameters is presented. In section 5, simulation results are given. In section 6, conclusions are drawn.

2   AODV and SAODV Routing Protocols

AODV [8] is a reactive routing protocol. It uses destination sequence numbers to ensure the freshness of routes and guarantee loop freedom. To find a path to a destination, a node broadcasts a route request (RREQ) packet to its neighbors using a new sequence number. Each node that receives the broadcast sets up a reverse route towards the originator of the RREQ unless it has a fresher one. When the intended destination or an intermediate node that has a fresh route to the destination receives the RREQ, it unicasts a reply by sending a route reply (RREP) packet along the reverse path established at intermediate nodes during the route discovery process. Then the source node starts sending data packets to the destination node through the neighboring node that first responded with an RREP. When an intermediate node along the route moves, its upstream neighbor will notice route breakage due to the movement and propagate a route error (RERR) packet to each of its active upstream neighbors.

SAODV [7] is an enhancement of AODV routing protocol to fulfil security feature. The protocol operates mainly by appending an extension message to each AODV message. The extension messages include a digital signature of the AODV packet using the private key of the original sender of the routing message and a hash value of the hop count. SAODV uses asymmetric cryptography to authenticate all non-mutable fields of routing messages as well as hash chain to authenticate the hop count (the only mutable) field.

Since all fields except the hop count of routing messages are non-mutable they can be authenticated by verifying the signature using the public key of the message originator. So, when a routing message is received by a node, the node verifies the signature of the received packet. If the signature is verified, the node computes the hash value of the hop count; if the routing message is RREQ or RREP; and compares it with the corresponding value in the SAODV extension. If they match, the routing message is valid and will be forwarded with an incremented hop count and a new hash value or if the destination has been reached generate the RREP. As RERR messages have a large amount of mutable information, SAODV suggests that every node (generating or forwarding a RERR message) will use digital signature to sign all fields of the routing message.

3   MANET Routing Attacks

MANETs are more vulnerable to security attacks than fixed networks due their inherent characteristics. MANET routing protocols is designed based on the assumption that all nodes cooperate without maliciously disrupting the operation of the protocol. However, the existence of malicious nodes cannot be disregarded in any system, especially in MANETs because of the wireless nature of the network. A malicious node aims to cause congestion, propagate fake routing information or disturb nodes from providing services. Attacks against MANET are classified based on modification, impersonation or fabrication of the routing messages. While there is large number of existing attacks, our paper is focused on flooding, grayhole, selfish and blackhole attacks.

3.1   AODV under Blackhole Attack

In a blackhole attack [9], a malicious node absorbs the network traffic and drops all packets. Once a malicious node receives a RREQ packet from any other node, it immediately sends a false RREP with a high sequence number and hop count equals 1 to spoof its neighbours that it has the best route to the destination. Thus, the malicious node reply will be received by the source node before any other replies and will be selected to send data packets through the route that includes the malicious node. When the data packets routed by the source node reach the blackhole node, it drops the packets rather than forwarding them to the destination node.

3.2   AODV under Grayhole Attack

In a grayhole attack [10], a malicious node behaves normally as a truthful node by replying with true RREP packets to the nodes that started RREQ packets. After the source node starts sending data through the malicious node, the malicious node starts dropping these data packets.

3.3   AODV under Selfish Attack

In a selfish attack [11], a malicious node saves its resources; such as battery, by not cooperating in the network operations. A selfish node affects the network performance as it does not correctly process routing or data packets based on the routing protocol. The selfish node drops all data and control packets even if these packets are sent to it. When a selfish node needs to send data to another node, it starts working as normal AODV operation. After it finishes sending its data, the node returns to its silent mode and the selfish behavior.

3.4   AODV under Flooding Attack

In a flooding attack [12], a malicious node floods the network with a large number of RREQs to non-existent destinations in the network. Since the destination does not exist in the network, a RREP packet cannot be generated by any node in the network. When a large number of fake RREQ packets are broadcast into the network, new routes can no longer be added and the network is unable to transmit data packets. This leads to congestion in the network and overflow of route table in the intermediate nodes so that the nodes cannot receive new RREQ packet, resulting in a DoS attack [13].

4   Simulation Approach

NS-2 simulator [14] is used to simulate grayhole, blackhole, flooding and self-ish attacks. The simulation is used to analyse the performance of AODV and SAODV routing protocols under these attacks. The parameters used are shown in Table I. Node mobility was modelled with the random waypoint method. Our simulation results are obtained from 3 different movement scenarios, 3 different traffic scenarios and 3 different node-type (malicious or non-malicious) scenarios which means that each metric value is the mean of the 27 runs. The node-type scenario is created randomly. In all cases, the 90% confidence interval was small compared with the values being reported. While we examined the effects of the attacks on both UDP and TCP traffic, in this paper we focused on their impact on the TCP traffic only. We also examined the effect of these attacks for different node speeds (0, 5, 10, 15, 20, 25 and 30 m/s). Our analysis shows that the node mobility has no significant effect on the protocol performance in the presence of malicious nodes. So, the paper results are focused only on the static network.

Our SAODV implementation is designed by modifying the original AODV source code. OpenSSL encryption library is used for digital signature creation and hash chain generation. For the purpose of securing the hop count field of the routing RREQ and RREP messages, we use SHA-1 [15] which is the most widely used secure hash algorithm, and is employed in several widely used applications and protocols. For the purpose of securing the non-mutable fields of the routing messages, we use RSA digital signature [16].

Table 1. Simulation Parameters

Simulation Time	180 s
Simulation Area	1000 m x 1000 m
Number of Nodes	100
Number of Connections	150
Number of Malicious Nodes	0 - 10
Node Speed	0 - 30 m/s
Pause Time	10 s
Traffic Type	TCP

Packet Delivery Ratio (PDR): The ratio of packets that are successfully delivered to a destination compared to the number of packets that have been sent out by the sender.

Throughput: The number of data bits delivered to the application layer of destination node in unit time measured in bps.

End-to-End Delay (EED): The average time taken for a packet to be transmitted across the network from source to destination.

Routing Overhead: The size of routing packets measured in Kbytes for route discovery and route maintenance needed to deliver the data packets from sources to destinations.

5   Simulation Results

5.1   Blackhole Attack

The effect of blackhole attack on the packet delivery ratio is shown in Figure 1.
The result shows that the PDR of SAODV is better than its value for AODV
even for a small number of malicious nodes. While PDR remains constant for
SAODV, it decreases dramatically as the number of malicious nodes increasing
under AODV.

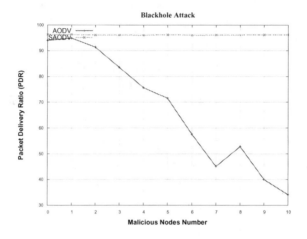

Fig. 1. PDR under Blackhole Attack

Figure 2 shows the effect of blackhole attack on the network throughput. The
result shows that while the throughput of SAODV does not change significantly
in the presence of malicious nodes, it decreases dramatically for AODV as the
number of malicious nodes increases. The first few malicious nodes have the
largest effect; beyond that increasing the number of malicious nodes has less
impact.

The effect of blackhole attack on the end-end-delay is shown in Figure 3. The
presence of malicious nodes has no effect on the delay in SAODV. While the
results show that the delay of AODV is reduced as the number of malicious
nodes increases which is slightly paradoxical as the attack improves the delay.
This is a misleading result because the delay is only measured on packets that
reach their destinations and since the blackhole nodes drop all the received data,
the number of packets that will be considered in calculating the delay decreases
as the number of malicious nodes increases. So, the routes that avoid blackhole
nodes suffer less competition, and hence reduced delay.

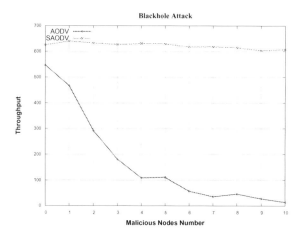

Fig. 2. Throughput under Blackhole Attack

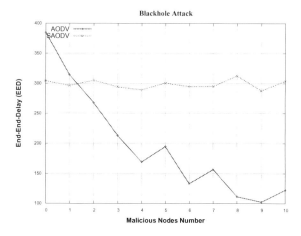

Fig. 3. EED under Blackhole Attack

Figure 4 shows the effect of blackhole attack on the routing overhead. The result shows that the routing overhead of SAODV is approximately 7 times its corresponding value in AODV. In addition, while the routing overhead of SAODV does not change significantly as a result of malicious nodes in the network, for AODV it decreases dramatically as a result of malicious nodes specially for the first two malicious nodes. These results are slightly confusing as the blackhole attack improves the routing overhead. This is because the blackhole nodes stop rebroadcasting the RREQ which decreases the number of RREQ packets, one of factors used to measure the routing overhead.

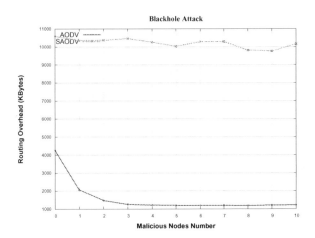

Fig. 4. Routing Overhead under Blackhole Attack

5.2   Grayhole Attack

The effect of grayhole attack on the packet delivery ratio is shown in Figure 5. The result shows that while the number of malicious nodes does not affect so much on PRD of both AODV and SAODV, SAODV enhances slightly the PDR over AODV.

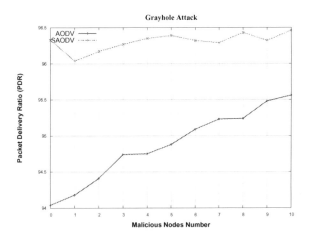

Fig. 5. PDR under Grayhole Attack

Figure 6 shows the effect of grayhole attack on the network throughput. While the malicious nodes do not introduce a significant change on the throughput of either AODV or SAODV, SAODV improves throughput by approximately 10% compared to AODV.

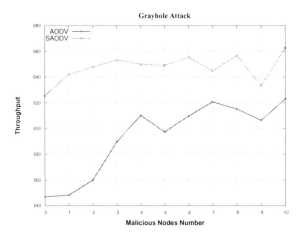

Fig. 6. Throughput under Grayhole Attack

The effect of on the grayhole attack on end-end-delay is shown in Figure 7. The result shows that the delay of SAODV does not have a significant change regardless the number of malicious nodes and that delay is better than the delay of AODV by approximately 15%. The explanation of AODV delay enhancement as the number of malicious nodes increasing is as stated in the blackhole attack because both attacks share data dropping.

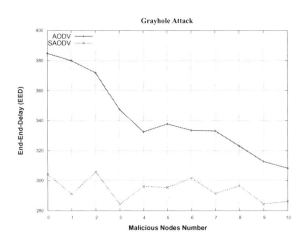

Fig. 7. EED under Grayhole Attack

Figure 8 shows the routing overhead under the grayhole attack. The results show that the routing overhead of AODV is approximately 40% of SAODV and this overhead decreases as the number of malicious nodes increases. The explanation of AODV routing overhead improvement as the number of malicious nodes increases is as stated in the blackhole attack because both attacks drop data packets.

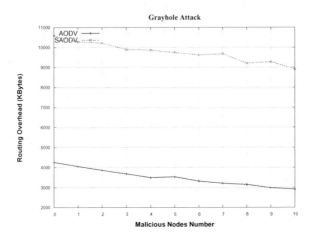

Fig. 8. Routing Overhead under Grayhole Attack

5.3   Selfish Attack

As the grayhole node drops all data packets and the selfish node drops all data and routing packets, the grayhole attack simulation produces very similar results to the selfish attack. This is because the metrics are calculated based on the received data packets which are very similar for both attacks.

The effect of selfish attack on the packet delivery ratio is shown in Figure 9. The result shows that while the number of malicious nodes does not have much effect on PRD of both AODV and SAODV, SAODV enhances slightly the PDR over AODV.

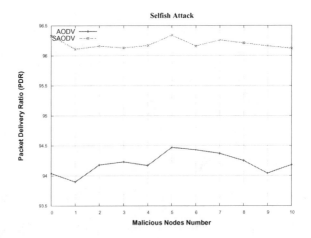

Fig. 9. PDR under Selfish Attack

Figure 10 shows the effect of selfish attack on the network throughput. While the malicious nodes do not introduce a significant change on the throughput of either AODV or SAODV, SAODV's throughput exceeds AODV by approximately 15%.

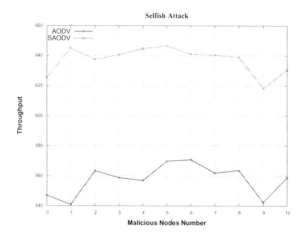

Fig. 10. Throughput under Selfish Attack

The effect of on the selfish attack on end-end-delay is shown in Figure 11. The result shows that the delay of SAODV is enhanced over AODV by approximately 20%.

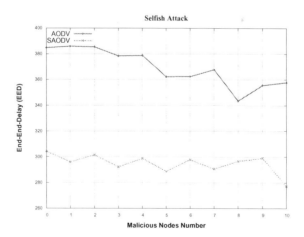

Fig. 11. EED under Selfish Attack

Figure 12 shows the routing overhead under the selfish attack. The results show that the routing overhead of AODV is approximately 40% of SAODV and this overhead is slightly decreases as the number of malicious nodes increases.

The enhancement of routing overhead under the selfish attack is real because as the number of malicious nodes increases, the number of dropped routing packets increases which reduces the routing overhead.

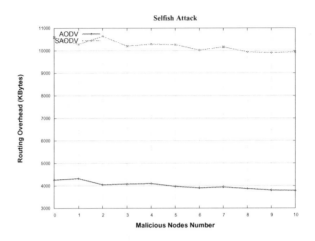

Fig. 12. Routing Overhead under Selfish Attack

5.4   Flooding Attack

The effect of flooding attack on the packet delivery ratio is shown in Figure 13. While the flooding attack has small impact on the PDR of AODV, its effect is severe on the PDR of SAODV specially for large number of malicious nodes. PDR of SAODV is slightly better than AODV for small number of malicious nodes.

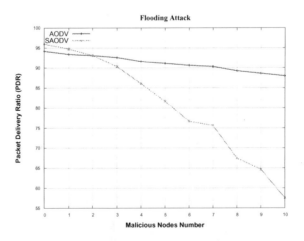

Fig. 13. PDR under Flooding Attack

Figure 14 shows the effect of flooding attack on the network throughput. Throughput of SAODV is slightly better than AODV if the number of malicious nodes is less than 2 and becomes worse for higher numbers of malicious nodes.

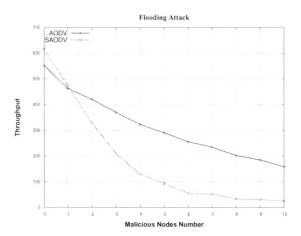

Fig. 14. Throughput under Flooding Attack

The effect of flooding attack on the end-end-delay is shown in Figure 15. The result shows that there is no significant change between the delay of both AODV and SAODV specially for small number of malicious nodes while the difference increases as the number of malicious nodes increasing.

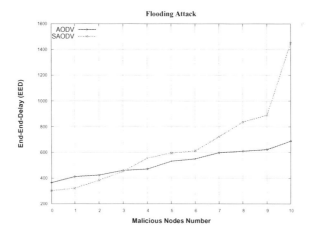

Fig. 15. EED under Flooding Attack

Figure 16 shows the effect of flooding attack on the routing overhead. The result shows that while the routing overhead of AODV slightly increases as the number of malicious nodes increases, it increases dramatically as the number of malicious nodes increases.

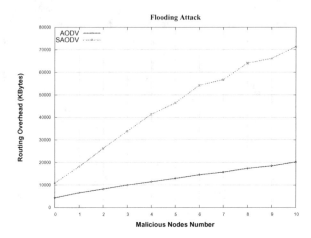

Fig. 16. Routing Overhead under Flooding Attack

6 Conclusions

In this paper, we analyse the performance of both AODV and SAODV routing protocols under the blackhole, grayhole, selfish and flooding attacks. We conclude that the performance of SAODV is better than AODV in the presence of blackhole, grayhole and selfish attacks because SAODV does not forward the routing packets without ensuring authenticity and integrity which reduces the routing packets that may cause congestion. On the other hand, the performance of SAODV is worse than AODV in the presence of flooding attack because of the malicious nodes impersonating non-existent nodes which cannot be discovered by other non-malicious nodes.

We conclude as well that the blackhole and flooding attacks have dramatic impact on the network performance. The blackhole introduces a fake RREP which affects the network performance and the flooding attack introduces a fake RREQ which affects the network performance as well. As most of the performance metrics depend on the number of received data packets, little change is observed in these metrics under grayhole and selfish attack because the malicious nodes drop data packets in these attacks.

References

1. Boukerche, A., Turgut, B., Aydin, N., Ahmad, M., Bölöni, L., Turgut, D.: Routing protocols in ad hoc networks: a survey. Computer Networks 55(13), 3032–3080 (2011)
2. Abdelshafy, M.A., King, P.J.: Analysis of security attacks on AODV routing. In: 8th International Conference for Internet Technology and Secured Transactions (ICITST), London, UK, pp. 290–295 (2013)
3. Singh, M., Singh, A., Tanwar, R., Chauhan, R.: Security attacks in mobile adhoc networks. In: IJCA Proceedings on National Workshop-Cum-Conference on Recent Trends in Mathematics and Computing 2011, vol. RTMC(11) (2012)

4. Joshi, P.: Security issues in routing protocols in MANETs at network layer. Procedia CS 3, 954–960 (2011)
5. Sanzgiri, K., Laflamme, D., Dahill, B., Neil, B., Clay, L., Elizabeth, S., Belding-royer, M.: Authenticated routing for ad hoc networks. IEEE Journal on Selected Areas In Communications 23, 598–610 (2005)
6. Papadimitratos, P., Haas, Z.J.: Secure link state routing for mobile ad hoc networks. In: Symposium on Applications and the Internet Workshops, pp. 379–383. IEEE Computer Society (2003)
7. Zapata, M.G.: Secure ad hoc on-demand distance vector routing. SIGMOBILE Mob. Comput. Commun. Rev. 6(3), 106–107 (2002)
8. Perkins, C.E., Royer, E.M.: Ad-hoc on-demand distance vector routing. In: Proceedings of the 2nd IEEE Workshop on Mobile Computing Systems and Applications, pp. 90–100 (1997)
9. Sharma, N., Sharma, A.: The black-hole node attack in MANET. In: Proceedings of the 2012 Second International Conference on Advanced Computing & Communication Technologies, ACCT 2012, pp. 546–550. IEEE Computer Society, Washington, DC (2012)
10. Manikandan, K., Satyaprasad, R., Rajasekhararao, K.: A survey on attacks and defense metrics of routing mechanism in mobile ad hoc networks. IJACSA - International Journal of Advanced Computer Science and Applications 2(3), 7–12 (2011)
11. Goyal, P., Batra, S., Singh, A.: A literature review of security attack in mobile ad-hoc networks. International Journal of Computer Applications 9(12), 11–15 (2010)
12. Guo, Y., Perreau, S.: Detect DDoS flooding attacks in mobile ad hoc networks. Int. J. Secur. Netw. 5(4), 259–269 (2010)
13. Bandyopadhyay, A., Vuppala, S., Choudhury, P.: A simulation analysis of flooding attack in MANET using ns-3. In: 2011 2nd International Conference on Wireless Communication, Vehicular Technology, Information Theory and Aerospace Electronic Systems Technology (Wireless VITAE), pp. 1–5 (2011)
14. The Network Simulator NS-2, http://www.isi.edu/nsnam/ns/
15. US Department of Commerce: Secure hash standard. Technical Report FIPS PUB 180-4, National Institute of Standards and Technology (2012)
16. Rivest, R., Shamir, A., Adleman, L.: A method for obtaining digital signatures and public-key cryptosystems. Communications of the ACM 21, 120–126 (1978)

SMART: Secure Multi-pAths Routing for wireless sensor neTworks

Noureddine Lasla[1], Abdelouahid Derhab[2], Abdelraouf Ouadjaout[1],
Miloud Bagaa[1], and Yacine Challal[3]

[1] Department of Theories and Computer Engineering, CERIST, Algiers, Algeria
[2] Center of Excellence in Information Assurance (CoEIA), King Saud University,
Riyadh, Saudi Arabia
[3] Laboratoire de Méthodes de Conception des Systèmes (LMCS),
Ecole nationale Supérieure d'Informatique, Algiers, Algeria

Abstract. In this paper, we propose a novel secure routing protocol named Secure Multi-pAths Routing for wireless sensor neTworks
(SMART) as well as its underlying key management scheme named *Extended Two-hop Keys Establishment* (ETKE). The proposed framework
keeps consistent routing topology by protecting the hop count information from being forged. It also ensures a fast detection of inconsistent routing information without referring to the sink node. We analyze
the security of the proposed scheme as well as its resilience probability
against the forged hop count attack. We have demonstrated through simulations that SMART outperforms a comparative solution in literature,
i.e., SeRINS, in terms of energy consumption.

1 Introduction

Wireless sensor networks (WSNs) [1] are defined as a large collection of tiny sensor nodes, which have scarce resources regarding energy, bandwidth, processing
capacity and storage. Such networks are designed to gather data in inhospitable
places and might be involved in critical applications meant for civil and military
use. The main task of a wireless sensor network is to collect/aggregate data from
the sensor nodes and transmit them towards the sink node using a hop-by-hop
communication. In these critical applications, establishing a reliable path free of
compromised nodes is an important security concern.

The single-path routing is not resilient to attacks as it is sufficient to compromise one node along the path to cause path failure. To deal with this failure,
a path maintenance process is initiated to find a new path, which is costly in
terms of time, control overhead, and energy consumption. The use of multipath routing can be a good solution against attacks that target the reliability of
the network. As data are transmitted redundantly through multiple paths, the
packets are likely to reach the sink even in the presence of some compromised
nodes.

The attacks against wireless sensor networks can be either *insider* or *outsider*
according to whether or not the adversary retrieves the information stored in

S. Guo et al. (Eds.): ADHOC-NOW 2014, LNCS 8487, pp. 332–345, 2014.

the sensor nodes. Using cryptography mechanism, the outsider attacks can be avoided as there is no way for an attacker to inject or read information from the network. The insider attack, however, is more powerful as by compromising a set of sensor nodes, an adversary can get access to the security materials of these nodes, change their running codes, and inject false information. For example, in routing construction protocols, an adversary can succeed at launching the Sinkhole attack by simply injecting a faked shortest path RREQ message using the cryptographic materials of the compromised nodes.

To construct the routing topology, different metrics are employed. Among them, we can find the hop count, sequence number, path identifier, etc. The hop count for example is used to select the shortest path leading to the sink and avoid routing loops, where each node should increment it by one before relaying it to its next hop. However, these metrics are mutable information, meaning that every node could manipulate it during the relay. In a security context, this mutable information is attractive for adversaries who want to compromise the network and can be exploited by many attacks like the sinkhole and the wormhole attacks. Using only cryptography techniques to ensure the integrity of route construction information (hop count, sequence number, etc.) is not sufficient especially when the adversary compromise a set of nodes in the network, as mentioned earlier. For these reasons, detecting compromised nodes is an important security concern that should be considered when designing a secure communication protocol.

In the literature, some solutions [2–5] have been designed to build a reliable routing topology. Authors in [5] propose a protocol to secure tree construction, based on braodcast key to authenticate neighboring nodes. However, although this protocol is resilient to node replication attack, the protocol cannot protect the transmitted routing information from being altered or forged when an adversary compromise a node. SEIF [3] allows the construction of more alternative disjoint paths belonging to different sub-branches. Each path from different sub-branches can be only intersected at nodes that are at one hop from the sink, and each sub-branch is tagged with a unique identifier that guarantees the construction of such topology. Furthermore, to ensure the security of the sub-branch identifiers, authors use a set of one-way hash chains to authenticate messages from the sink and the sub-branches origin. Therefore, any attempt of injecting faked sub-branch identifier can be immediately detected even if the adversary make an inside attack. However, SEIF cannot detect a Wormhole attack, when an adversary captures a valid hash chain of sub-branches from one end and replay them at the other end [6]. In addition, SEIF does not consider any metric to carry out the routing decision.

SeRINS [2] is a semi-distributed solution based on the hop count metric to select routes. This protocol provides a mechanism to protect the hop count information at the sensor node level with the help of the sink. Each sensor, first, chooses its first parent that will be used as a reference to verify the correctness of any received alternate route. After that, when a node suspects on one alternate route, it sends an alert to the sink, which makes a decision about whether the suspected or the alert sender node is malicious. However, involving the sink in the verification process of suspected nodes, causes considerable communication overhead and affects negatively the network scalability.

In this paper we propose a novel multi-path routing protocol called *Secure Multi-pAths Routing for wireless sensor neTworks* (SMART), relying on a extended version of our previous key management scheme EPKE [7], we call it *Extended Two-hop Keys Establishment* (ETKE). The main contributions of the paper are the following: Firstly, the establishment of two-hop broadcast keys enables the authentication of two-hop neighbors, which allows applying the watchdog mechanism to check the well-behaving of one-hop neighbors during the relay of RREQ messages. Secondly, we show that if two consecutive nodes along the path are not compromised, SMART ensures an immediate detection of inconsistent routing information without referring to the sink, using a two-hop verification mechanism. Thirdly, if the two consecutive nodes are compromised, an analysis of detection evasion is provided.

The remaining of this paper is organized as follows: Section 2 provides system model and basic idea. In Section 3, description of ETKE is presented. Section 4 describes SMART. Security analysis and simulation results are given in Section 5. Finally, Section 6 concludes the paper.

2   System Model and Basic Idea

2.1   Attack Model

We consider the node capture attack, in which an attacker can capture a legitimate node and turn it into a malicious one by extracting cryptographic keys from the captured node and makes it run a malicious code. The compromised node then broadcasts a fake RREQ with false hop count in order to attract the traffic that is destined to the sink. This mechanism is used by some attacks such as : Sinkhole and Wormhole.

2.2   Basic Idea

In the tree-based construction process, each node forwards a RREQ message initiated by the sink node. The RREQ message contains a mutable information (hop count) that should be incremented by one at each relay level. To secure the routing topology construction, we should ensure a correct alteration of the hop count value during the relay.

The idea behind our solution is to provide a secret that is shared between each node and its two-hop neighbors. This secret is unknown for the one-hop neighbors and allow to verify their behaviour during the relay by the two-hop neighbors.

For instance, let us consider the network in Fig. 1(a), where the dashed lines represent the communication links between neighboring nodes and the solid lines represent the selected routing paths. The number besides each node represents the hop count information relayed by that node. In Fig. 1(b), if c (the child node

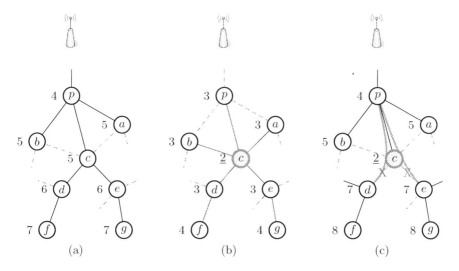

Fig. 1. The two-hop verification mechanism used in SMART

of parent p) is a compromised node, without any security mechanism, it can forge inconsistent hop count with value 2, in order to make most of the network traffic pass through it without being detected. Our idea, as shown in Fig. 1(c), is to share a secret (virtual tunnel) between node p and its two-hop neighbors d and e. This secret is used by node p to generate a proof (encryption of the current hop count vlaue). This proof is forwarded by node c without being able to decrypt it, and allows node d and e to verify if c has correctly incremented the hop count value received by its parent p. Note that each node, during the relay, (i) forwards the received proof to verify its correct relay of the received RREQ and also (ii) generates its own proof to ensure that the next forwarder node correctly relays its RREQ message.

In the next section, we give a detail description of how nodes in the network share a secret keys with their one-hop and two-hop neighbors, allowing then to check the correct relay of the RREQ messages.

2.3   Notations

The following notations in Table 1 are used throughout the paper.

Table 1. Notations

Notation	Description
$E(K, m)$	Encryption of m using key K
$A \rightarrow B : m$	A sends m to B
$A \rightarrow * : m$	A broadcasts m
$K_{A,B}$	Secret pairwise key between A and B
BK_A	Broadcast key of A
$\|\|$	Concatenation operator

3   ETKE Description

In this section we describe how to establish the required key materials between communicating nodes to secure the tree-based routing construction in SMART. ETKE extends EPKE [7] by adding the two-hop pairwise key and the two-hop broadcast key. Mainly, the required keys are the two-hop and one-hop broadcast keys, shared between each node and its one-hop and two neighbors, respectively.

To achieve key agreement between communicated nodes, the pre-distribution method is more suitable for WSN [8]. In this method, nodes are preloaded, prior to deployment, with secret information that will be used to establish secure links between neighboring nodes. In this section, we propose a solution to establish the one hop pair-wise key, one hop broadcast key, two hops pair-wise key and two hops broadcast key, based on the Transitory Initial Key setup scheme of LEAP [9] and OTMK [10]. The transitory initial key K_{IN} is used to establish keys between neighboring nodes during a key setup phase (trust period). To secure nodes against capture attacks, the K_{IN} is erased from the node's memory at the end of the trust period. The trust period represents the minimum time (T_{min}) needed by an adversary to compromise a legitimate node.

3.1   Key Setup Phase

Each node u is preloaded with a transitory initial key K_{IN} and a random number N_u. Node u compute its master key MK_u as follows: $MKu = G(K_{IN}, ID_u)$, where G is a pseudo random function. After T_{min}, each node u erases K_{IN} and N_u from its memory. Node u discovers its neighbors by broadcasting the following message

$$u \rightarrow * : Join1, u, E(K_{IN}, u||N_u) \tag{1}$$

When u's neighbors receive this message, they relay it to u's two-hop neighborhood as follows

$$Relay\_node \rightarrow * : Join2, u, E(K_{IN}, u||N_u) \tag{2}$$

Depending on the state of node u's neighbors, i.e., still keeping the initial key K_{IN} or already erased it, we distinguish two cases to establish one-hop pair-wise key, two-hop pair-wise key, one-hop broadcast key and two-hop broadcast key.

3.2   Case 1: K_{IN} Is Available

Creation of One-Hop Pair-Wise Key: To create the symmetric one-hop key $K_{u,v}$ between two direct neighbor nodes u and v, after they receive message (1) from each other, the following formula is used:

$K_{u,v} = G(MK_{min(u,v)}, ID_{max(u,v)}||N_{max(u,v)})$

Note that u and v can compute $MK_{min(u,v)}$ because each one knows K_{IN} and can generate the master key of any other node.

Creation of Two-Hop Pair-Wise Key: When two-hop neighbor nodes u and v receive message (2) from their relay nodes, the following formula is used to create the symmetric key between them:

$$K_{u,v} = G(MK_{min(u,v)}, ID_{max(u,v)}||N_{max(u,v)})$$

Creation of One-Hop Broadcast Key: Each node u creates one-hop broadcast key, which it shares with its direct neighbors. This key is not used to authenticate the node but to encrypt the message content. When node u's neighbor receives message (1), it uses K_{IN} to generate u's master key MK_u, and computes u's one-hop broadcast key as the following:

$$BK_u = G(MK_u, N_u)$$

Creation of Two-Hop Broadcast Key: After the reception of message (2), each u's two-hop neighbor node uses K_{IN} to compute u's master key MK_u, and computes u's two-hop broadcast key as follow:

$$BK2_u = G(MK_u, N_u||N_u)$$

3.3 Case 2: K_{IN} Is not Available

After T_{min}, each node v erases the transitory initial key K_{IN}, and will not be able to generate other master keys. It can only use its master key to calculate a symmetric key with a new deployed node u (since u can generate any master key). So, the following messages must be generated in order to establish a par-wise key between them: If node v is a direct neighbor of node u, then, the following message, $Reply1$, is sent to u:

$$v \rightarrow u : Reply1, v, E(MK_v, v||N_v) \tag{3}$$

Otherwise, the following two messages $Reply2$ and $Reply3$ are sent to relay nodes and to node u respectively.

$$v \rightarrow Relay\_node : Reply2, v, E(MK_v, v||N_v) \tag{4}$$

$$Relay\_node \rightarrow u : Reply3, v, E(MK_v, v||Nv) \tag{5}$$

Creation of One-Hop Pair-Wise Key: The one-hop pair-wise key $K_{u,v}$ between two direct neighbor nodes u and v is computed using the following formula:

$$K_{u,v} = G(MK_v, ID_u||N_v)$$

Creation of Two-Hop Pair-Wise Key: To calculate a two-hop pair-wise key $K_{u,v}$ between two-hop neighbor nodes u and v, the following formula is used:

$$K_{u,v} = G(MK_v, ID_u||N_v)$$

Creation of One-Hop Broadcast Key: Because u's neighbor (i.e., node v) has erased the transitory initial key K_{IN}, it cannot calculate the one-hop broadcast key BK_u of node u using the previous formula. The only way to do so is that u sends BK_u via a unicast encrypted packet using the shared one-hop pair-wise key with node v.

$u \rightarrow v : u, E(K_{u,v}, u||BK_u)$. This message is sent by node u when it receives $Reply1$ from node v.

Creation of Two-Hop Broadcast Key: The only way for v to get the two-hop broadcast key $BK2u$ of node u, is that u sends the $BK2u$ via a unicast encrypted packet using the shared two-hop pair-wise key with node v.

$u \rightarrow v : u, E(K_{u,v}, u||BK2_u)$. This message is sent by node u when it receives $Reply3$ from node v.

3.4   Special Case

When node v is newly deployed in the network, it can be inserted between two nodes i and j that was not neighbors and become two-hop neighbors through node v. In this case, node i and j should exchange the two-hop broadcast keys of each other. The only way to do so is to transmit the two-hop broadcast through node v during the trust period, using their master key, as follow.

$$i \rightarrow v : Join4, i, E(MK_i, i||BK2_i) \tag{6}$$

$$j \rightarrow v : Join4, j, E(MK_j, j||BK2_j) \tag{7}$$

Node v then forwards $Join4$ to i and j using their master keys to ensure that node v is a trust node as only node that has not yet erased its initial key, can generate the master keys of other nodes. The following $Join5$ message is then sent to j and j.

$$v \rightarrow j : Join5, v, E(MK_j, v, i||BK2_i) \tag{8}$$

$$v \rightarrow i : Join5, v, E(MK_i, v, j||BK2_j) \tag{9}$$

4   SMART Description

4.1   Initialization

Global One-Way Hash Chain (GOHC): Prior the deployment, a GOHC is generated $(S_0, S_1, S_2, \cdots, S_n)$ and stored in the sink node. Each sensor node is preloaded with the last value (S_n) of the GOHC. After the deployment, the sink node, at each round d, includes the last unused GOHC Value (S_{n-d}) in the RREQ message. Each node that receives RREQ of round d $(d = 1, \cdots, n-1)$ can check if RREQ is generated by the sink or not. This is achieved by applying the one-way hash function on the received value S_{n-d} and verifying whether the result is equal to the pre-loaded GHOC value $S_{n-(d-1)}$; $F(S_{n-d}) = S_{n-(d-1)}$.

Local One-Way Hash Chain (LOHC): To allow a one hop authentication of the broadcasted RREQ message, each node i generates a LOHC $(L_i^0, L_i^1, L_i^2, \cdots, L_i^n)$ and reveals the last LOHC value (L_i^n) to its reachable neighbors using its one-hop broadcast key to encrypt it L_i^n. At round d $(d = 1, \cdots, n-1)$, each node can check if RREQ is generated by its one-hop neighbor by applying the function F on the received value L_i^{n-d} and verifying whether the result is equal to the pre-loaded LOHC value $L_i^{n-(d-1)}$; $F(L_i^{n-d}) = L_i^{n-(d-1)}$.

4.2   Route Construction Process

Route Construction Initialisation: To start the route construction process, the sink node broadcasts the below RREQ packet, which includes the following fields as shown in Table 2

$$Sink \rightarrow * : ID_{sink}, Seq = S_{n-d}, OWC = L_{sink}^d,$$
$$h = 0, MAC2_{sink}, \emptyset, \emptyset \qquad (10)$$

Table 2. RREQ message Fields

Field	Description
ID_{Src}	Packet sender identifier
Seq	The first unused GOHC value
OWC	First unused One Way Chain value of a LOHC to authenticate the source
h	Hop count value
$MAC2_{src}$	$MAC(BK2_{src}, h)$; the MAC of the hop count value, generated with the two-hop broadcast key
ID_{Parent}	Primary (or main) parent identifier
$MAC2_{parent}$	The MAC received from the primary parent, generated with the parent's two-hop broadcast key

Primary Parent Selection: When a node i receives the first RREQ message, which indicates a new round d from a node j, it waits for a random time then it selects the primary parent node with the lowest hop count h' and relays the RREQ message as follows:

$$i \rightarrow * : ID_i, Seq = S_d, OWC = L_i^d, h = h' + 1,$$
$$MAC2_i, ID_{parent}, MAC2_{parent} \qquad (11)$$

After relaying the RREQ message, node i might keep receiving RREQ messages from other neighbors.

Alternative Parent Selection: When node i with a hop count h receives a RREQ message from node j with a hop count h', the following conditions must hold true so that j is accepted as an alternative parent:

- $h' \leq h$, in order to avoid routing loops.

– The grand parent of the received route must not exist as a grand parent or a parent of an already selected accepted route, in the routing table, except the case when the grand parent is the Sink. The accepted routes from different grand parent nodes guarantee that all paths are two-hop disjoint.

For each accepted RREQ message, node i adds a new entry in the routing table, which contains the following information: $<Parent\text{-}id, Hop\text{-}count, Grand\_Parent\text{-}id>$.

5   Analysis of SMART Protocol

5.1   Security Mechanisms

In SMART, the RREQ message encapsulates three security mechanisms that ensure sink authentication, source authentication and hop count integrity.

– The *Seq* field, containing the first unused GOHC, allows sensor nodes to verify that the RREQ message is initiated by the sink node and is not a re-injection of an old RREQ message of a previous round.
– To forbid an intruder from spoofing source identities, each node i, when receiving a RREQ from node j, can authenticates the message source by checking if the stored $L_j^{n-(d-1)}$ is equal to the hash value of the received OWC, $F(L_j^{n-d})$.
– The integrity control of the hop count is ensured through the $MAC2_{Grand\_parent}$ field. This MAC ensures that the received hop count value is a successor of the hop count of the grand parent. By this way, an adversary which tries to send a RREQ with lower hop count to perform a Sinkhole attack, is then prevented.

5.2   Forged Hop Count Detection

For an intruder to inject forged routing information, he should first get the cryptographic key materials by compromising some nodes in the network. However, even by compromising some set of nodes, our scheme can detect, in most of cases, the injection attempt of faked RREQ messages. In the following we give three possible cases of attack scenarios by an intruder and show how SMART can protect the routing construction by detecting these attacks.

1. When a compromised node c wants to inject forged routing information, it relay a RREQ of its parent p with forged hop count h' instead of h (see Fig. 1(b)). In this case, any node receiving or overhearing this RREQ can check that h' is not a successor of $h-1$, by simply comparing if $MAC2(BK2_p, h') = MAC2_p$.
2. A more intelligent attack consists in compromising two or more consecutive nodes along the path simultaneously. The first possible scenario is that nodes p, and c are compromised and node c try to inject a RREQ with a forged

hop count h'. In this case, node a and b (i.e., any common neighbor of both c and p) using the watch-dog mechanism (i.e., the watchdog consists in monitoring the neighboring nodes to check whether the latter are correctly forwarding the RREQ messages or not), detect that the injected RREQ is not consistent with the previous received RREQ from node p, and they will report a detection message to forbid the concerned nodes d and e to relay this RREQ. The adversary can succeeded at launching such an attack only if all the common neighbors of p and c are compromised. We will present in the next section the probability of success of such an attack under different network densities.

3. In the third possible scenario, both nodes p and c inject a RREQ with a fake hop count. Node a and b, in this case, cannot detect the inconsistency of the received RREQ message. However the parent node of p and any common neighbor between them can detect the forged hop count and then send a detection message to the concerned nodes to detect and reject this RREQ.

5.3 Resilience Probability against Forged Hop Count

The fake RREQ can propagate to the lower levels of the network without being detected if the following conditions hold true:

1. Two neighboring nodes in the network, the parent and the child in the tree, are compromised.
2. The child node injects a fake RREQ with false hop count.
3. There are no common legitimate neighbors between the parent and the child node.

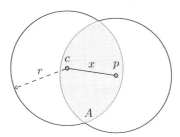

Fig. 2. The intersection region A that contains the common neighbors of p and c

The region where the common neigbors of the parent p and child c can reside, represents the intersection of two circular communication areas with radius r and centered at p and c . As shown in Fig. 2, The distance x, between the two nodes p and c, ranges from 0 to r. The area A of the common region is calculated as follows:

$$A(x) = 2r^2 cos^{-1}(\frac{x}{2r}) - x\sqrt{r^2 - \frac{x^2}{4}} \qquad (1)$$

The expected area of A, as shown in [11], is calculated as follow:

$$E[A] = \int_{x=0}^{r} A(x)f(x)dx \tag{2}$$

The probability distribution function of x, when node p and c are uniformly distributed in the deployment area, is as follow:

$$F(x) = P(distance < x) = \frac{\pi x^2}{\pi r^2} = \frac{x^2}{r^2} \tag{3}$$

Then, the probability density function of x is given by:

$$f(x) = F'(x) = \frac{2x}{r^2} \tag{4}$$

Then:

$$E[A] = \int_{x=0}^{r} \left(2r^2 cos^{-1}(\frac{x}{2r}) - x\sqrt{r^2 - \frac{x^2}{4}} \right) \frac{2x}{r^2} \, dx \tag{5}$$

$$E[A] = \left(\pi - \frac{3\sqrt{3}}{4} \right) r^2 = 1.84255 \ r^2 \tag{6}$$

The average number of common neighbors C_N, within the region A, is given by $C_N = E[A] \times d$, where d is the network density.

Let P_{comp} be the probability of compromising a node, and we define P_{attack} as the probability that an attacker forges a fake RREQ without being detected. The adversary needs to compromise all the common neighbors in A tu succeed at launching its attack, then:

$$P_{attack} = (P_{comp})^{C_N} \tag{7}$$

As depicted in Fig. 3, our protocol is effective when the network density increases. The probability for an attacker to succeed at launching a forged hop count is low, especially when the compromising probability is low. For example, for an average network density of 5, the probability of successful attack does not exceed 0.15, even if the compromising probability is hight (0.8).

5.4   Wormhole Immunity

A wormhole attack occurs when a malicious node forwards incoming packets to a distant point of the network by means of a fast link longer than a normal node's range. Wormhole attacks can cause severe damage to hop-count-based routing protocols, since the attacker can present to distant nodes an attractive path, which other uninfected paths cannot compete. If such a scenario occurs, the attacker will be advantageous compared to other nodes as a large amount of packets will pass through him.

These attacks are challenging because attackers may relay the packets without any malicious modification. Existing secure routing protocols for WSN, such as

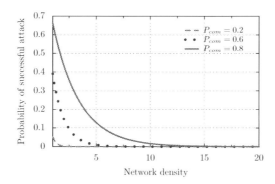

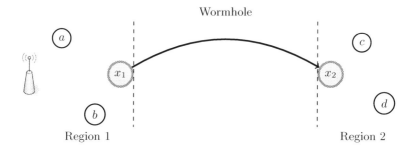

Fig. 3. The probability of successful hop count forgery attack

Fig. 4. Defense against a wormhole attack

SeRINS and SEIF, are vulnerable to wormholes. In contrast, SMART can in some cases defeat this attack thanks to its two-hop key distribution scheme. Basically, this is due to the fact that these keys represent *locality proofs* that can be used to detect false neighborhood links introduced by a wormhole.

To clarify how these locality proofs can be used to detect wormholes, let us consider the example in Fig. 4. Attacker x_1 tunnels its RREQ messages to x_2. To propagate this message, the attackers must ensure that nodes at different endpoints of the tunnel share two-hops broadcast keys to make the wormhole looks like a normal single hop. Here, we can distinguish two cases:

1. If the nodes were deployed before the compromise of x_1 and x_2, we can show that such keys cannot be established. Indeed, only a node with K_{IN} can create an authentic two-hop link between already deployed nodes. Therefore, when x_1 and x_2 become compromised, they are necessary unaware of K_{IN} and thus the keys between nodes in Regions 1 and 2 cannot be established through the tunnel. Consequently, exisitng keys will play the role of locality proofs to detect the far distance between Region 1 and 2, and hence preventing the tunnel to be viewed as a normal single hop.

2. If a node is newly deployed in the vicinity of the attackers, keys between nodes in different tunnel regions can be maliciously created. For example, suppose that node b is deployed after wormhole creation. Attacker x_1 can tunnel the key exchange messages of b to Region 2. Therefore, nodes at this endpoint will consider b as a two-hop neighbor through the wormhole link. The attackers can therefore relay consistent RREQ to connect the two regions and without being detected.

5.5   Energy Consumption

In this section, we evaluate the performance of SMART and compare it to SeRINS as both aim to secure the hop count information. Also, in a survey about secure multi-path routing in WSNs [6], SeRINS is the only one so far, which ensures the following security properties: authentication, integrity, confidentiality, freshness, and accountability. It is easy to show that these security properties are also ensured by SMART. Both protocols SMART and SeRINs have been implemented in TinyOS. To evaluate energy consumption, we have used Avrora [12] that emulates and analyzes programs written for AVR microcontroller, which is produced by Atmel and used in MICAz sensor mote.

The energy consumption determines the network lifetime and must be considered when designing protocols for WSNs. For this reason, we have measured the average energy consumption of SeRINS and SMART during one round while varying the number of intruders in the network, as shown in Fig. 5. In our simulation scenario, the total number of nodes in the network is set to 300 nodes. We can notice that SMART does not consume an additional energy when increasing the number of intruders in the network. However, the energy consumption in SeRINS increases as the number of intruders increases. This is due to the number of alerts sent to the sink for each intruder and from each intruder's neighbor.

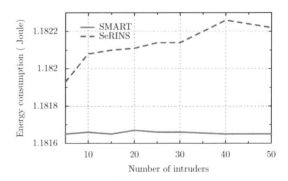

Fig. 5. Energy consumption vs. number of intruders in the network

6   Conclusion

In this paper we have proposed a security framework composed of multi-path routing protocol (SMART) and two-hop key management scheme (ETKE). The proposed framework keeps consistent routing topology by protecting the hop count information from being forged. The two-hop verification and the watch-dog mechanisms ensure a fast detection of inconsistent routing information without referring to the Sink node. The security analysis have shown that the probability of successful attack is very low under medium and high network densities. In addition, simulation experiments have shown that SMART is more energy-efficient compared to SeRINS.

References

1. Akyildiz, I., Su, W., Sankarasubramaniam, Y., Cayirci, E.: Wireless sensor networks: a survey. Computer Networks 38(4), 393–422 (2002)
2. Lee, S., Choi, Y.: A secure alternate path routing in sensor networks. Computer Communications 30(1), 153–165 (2006)
3. Challal, Y., Ouadjaout, A., Lasla, N., Bagaa, M., Hadjidj, A.: Secure and efficient disjoint multipath construction for fault tolerant routing in wireless sensor networks. Journal of Network and Computer Applications 34(4), 1380–1397 (2011)
4. Ghosal, A., Halder, S.: Intrusion detection in wireless sensor networks: Issues, challenges and approaches. In: Wireless Networks and Security. Signals and Communication Technology, pp. 329–367 (2013)
5. Dimitriou, T.D.: Securing communication trees in sensor networks. In: Nikoletseas, S.E., Rolim, J.D.P. (eds.) ALGOSENSORS 2006. LNCS, vol. 4240, pp. 47–58. Springer, Heidelberg (2006)
6. Stavrou, E., Pitsillides, A.: A survey on secure multipath routing protocols in wsns. Computer Networks 54(13), 2215–2238 (2010)
7. Bagaa, M., Challal, Y., Ouadjaout, A., Lasla, N., Badache, N.: Efficient data aggregation with in-network integrity control for wsn. J. Parallel Distrib. Comput. 72(10), 1157–1170 (2012)
8. Chen, C.Y., Chao, H.C.: A survey of key distribution in wireless sensor networks. In: Security and Communication Networks (2011)
9. Cheng, Y., Agrawal, D.P.: An improved key distribution mechanism for large-scale hierarchical wireless sensor networks. Ad Hoc Networks 5(1), 35–48 (2007)
10. Deng, J., Hartung, C., Han, R., Mishra, S.: A practical study of transitory master key establishment for wireless sensor networks. In: Proc First IEEE Int'l Conf Security and Privacy for Emerging Areas in Comm. Networks, SecureComm 2005 (2005)
11. Hai, T.H., Nam Huh, E., Jo, M.: A lightweight intrusion detection framework for wireless sensor networks. Wireless Communications and Mobile Computing 10(4), 559–572 (2010)
12. Titzer, B., Lee, D.K., Palsberg, J.: Avrora: scalable sensor network simulation with precise timing. In: Proceedings of the 4th International Symposium on Information Processing in Sensor Networks (IPSN), pp. 477–482 (2005)

A Robust Method for Indoor Localization Using Wi-Fi and SURF Based Image Fingerprint Registration

Jianwei Niu[1], Kopparapu Venkata Ramana[1], Bowei Wang[1],
and Joel J.P.C. Rodrigues[2]

[1] School of Computer Science and Engineering, Beihang University, Beijing, China
niujianwei@buaa.edu.cn, {ramana.kopparapu,wangbowei1219}@gmail.com
[2] Instituto de Telecomunicações, University of Beira Interior, Covilhã, Portugal
joeljr@ieee.org

Abstract. This paper introduces a method for the accurate indoor localization for mobile users when they are surrounded by unknown environments in places like airports, hospitals, libraries, museums, and supermarkets. Our system makes use of the combined data comprising two kinds: indoor Wi-Fi signals and the images of surroundings taken by users. We use Wi-Fi registration based on IEEE 802.11 to determine Access Point location according to the Received Signal Strength (RSS) as a distance function. Our fingerprinting method gives probability of signal strengths histogram at a given location. We use the Received Signal Strength Indicator (RSSI) data in to data collection to determine the overage area estimation and the mode of RSSI in localization. Next, we utilize the Speed Up Robust Features (SURF) descriptor to match the user-captured images with the image repository containing pre-captured images of the environment. Our method is accurate and less time consuming as compared to different approaches.

Keywords: Wi-Fi fingerprints, SIFT, SURF, image registration, indoor positioning.

1 Introduction

Obtaining user location is one of the important role for indoor locations. Current mobile phone technology has evolved to smart-phones having a variety of sensors. Wireless Local Area Networks (WLAN) are growing in popularity over wired networks for installation in offices and homes. Almost all portable devices these days come equipped with 802.11 wireless cards, which allow instant wireless connection to gadgets available within wireless range. Applications using WLAN technology, like location awareness, are also increasingly gaining popularity. The ongoing research has particularly focused in this area for it being cheap and available off the shelf. According to [1], for indoor positioning, many large Internet e-commerce businesses and research universities offer a variety

S. Guo et al. (Eds.): ADHOC-NOW 2014, LNCS 8487, pp. 346–359, 2014.

of research programs. Big companies like Google and Microsoft are also offering latest products in this area [2]. The need for localization is increasing and so is the range of related possibilities. There are multiple ways to track people in a building environment. The work in [3] combines Wi-Fi localization and static camera tracking. Wireless Local Area Networks using Wi-Fi is becoming more and more ubiquitous. As such, this approach provides a potential pre-built infrastructure for small area localization. The two main options for doing Wi-Fi localization are triangulation and fingerprinting. [18] Triangulation involves mapping signal strength as a function of distance while fingerprinting creates a probability distribution of signal strengths at a given location and uses a map of these distributions to predict a location given signal strength samples. In Wi - Foto [4] fingerprints are obtained when the photos are taken, and the photos are stored together with the fingerprints, so that the system can provide a photo of the place closest to the user-friendly device control system, which has device controllers overlaid on a current room photo.

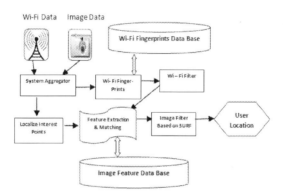

Fig. 1. Our Improved Localization Process of Fingerprints Registration

Generally, template matching methods become slow when the number of candidate photos increases, and the accuracy decreases when there are many similar candidate photos. By utilizing Wi-Fi positioning to roughly narrow down the candidate photos, Wi-Foto 2 [4] improves the performance of the template matching. To fuse the Wi-Fi and image fingerprints, image registration involves a lot of technology such as complex feature extraction, optimization algorithms, image segmentation, pattern recognition and matching, but it lacks of systematic theoretical guidance in many ways.

Therefore, a good deal of research can be directed to improve the automation, registration accuracy and speed of the registration algorithms [5]. Although the indoor environment is very complex, the house structure and housing decoration will not change frequently (such as ceilings, floors, windows, doors, furniture, etc.). Utilizing this fact, we can base our design on the variable structure of rooms within a building. Fig. 1 shows the design of a robust method for localization. It consist input system as a system aggregator, and two distinguish databases with image filer and image feature matching. Initially the system aggregator collects Wi-Fi fingerprints, and then matches those fingerprints with existed data base to identify access point location, Probability of Received Signal Strength indicator (RSSI) at a given location give as inputs to image registration. The server stores the Wi-Fi fingerprints and the feature vectors in the database. Locate the user by calculating similarity of Wi-Fi fingerprints and several template matching chooses a photo among many pictures by calculating the similarity of their feature vectors based on Speed Up Robust Features (SURF) algorithm. and match with image data base to using SURF Registration to locate the user location. Image based localization offers many benefits like easy data acquisition, high positioning accuracy, low user burden low, low-cost equipment, etc. Our method is easy data acquisition, low cost for user and it improves the speed and accuracy of the indoor localization.

The rest of the paper is structured as follows. Previous work related to indoor localization using Received Signal Strength (RSS) fingerprints is discussed in Section II. The proposed method based on Wi-Fi and Image fingerprinting registration and detailed experimental setup is mentioned in Section III. Our performance evaluation, followed by the results regarding the positioning accuracy is presented in Section IV. Finally, Section V provides the conclusions and discusses some ideas for future work.

2   Related Work

Indoor positioning is a research field that has been addressed by several authors and disciplines. Several types of signals (radio, images, and sound) and methods have been used to infer location. Each method has specific requirements as to what types of measurements are needed. Most of the pattern recognition methods, like fingerprinting [6], estimate locations by recognizing position-related patterns. The analysis of RSSI patterns is a technique that has been examined by several authors [7] [8], obtaining an accuracy ranging from 0.5 to 3 meters. Better results can be obtained by integrating the information captured by multiple sensors. An interesting mobile phone-based location system for indoor environments is via ambience fingerprinting of optical, acoustic and motion attributes. However, the optical recognition techniques proposed in Surround Sense are too limited, since the authors are only considering pictures of the floor in order to extract information about light and color. As compared to Scale-Invariant Feature Transform (SIFT), our work deals with SURF to avoids typical variations related to light, color or scale, and provides robustness and better accuracy.

2.1   Wi-Fi Based Positioning Technology Fingerprint

Wi-Fi localization using RSSI readings was also considered as a potential solution.

Triangulation

The goal of Wi-Fi triangulation is to map RSSI as a function of distance. This method requires a steep linear characterization curve in order to be properly implemented. Functions describing these curves are then used with live RSSI values as input to generate an (x, y) location prediction. This method was considered first due to its relatively simple implementation.

Fingerprinting

Wi-Fi Fingerprinting creates a radio map of a given area based on the RSSI data from several access points and generates a probability distribution of RSSI values for a given (x, y) location. Live RSSI values are then compared to the fingerprint to find the closest match and generate a predicted (x, y) location.

The traditional methods of angle of arrival positioning (Angle Of Arrival, AOA) [9] [10], arrival time positioning (Time Of Arrival, TOA) [11] and the signal strength analysis [12] [13] in complex indoor environments make it difficult to achieve precise positioning of the results. By comparing the position of fingerprint classification and matching, fingerprint signal characteristics required to locate the target position are obtained. So the availability and accuracy have a greater advantage.

2.2   RSS Fingerprint Registration

RSS location fingerprinting has two phases: training and positioning. In training phase, the system initially creates an RSS fingerprinting database which keeps entries of correlation between each physical location and its signal values from various access points (RSS fingerprint). All interested locations are kept inside this database. In positioning phase, a device measures RSS fingerprint from a location, then the measured RSS fingerprint is compared with all entry locations in the RSS fingerprinting database. With appropriate search algorithm, the system returns the outcome as an estimated location whose RSS fingerprint is the likeliest one to the currently measured one from the device. There are many search algorithms for computing location of a device. The basic algorithm is the K-nearest neighbor algorithm (KNN) [7]. The K-nearest neighbor (KNN) method is a deterministic approach that uses fingerprint from a specific location and from the database to estimate the mobile device location. Firstly, it finds the Euclidean distance (D) of the current measured fingerprint to the pre-stored fingerprint in the database.

2.3   Euclidean Distance

Euclidean distance, also known as the Euclidean space, in n-dimensional space is used to calculate the distance between two points. In an n-dimensional space the distance D between two points $A(a[1], a[2], ..., a[n])$ and $B(b[1], b[2], ..., b[n])$ is:

$$D(A, B) = \sqrt{\sum_{n}^{i=1}(a[i] - b[i])^2} \tag{1}$$

We can use the Euclidean distance calculation test data sample and fingerprint sample collected data distance between test data and determine the degree of similarity with fingerprint database data. Although the Euclidean distance is useful, but there are obvious shortcomings. Different properties of the sample (i.e., the index or variable) are equivalent to the difference between the views, which sometimes cannot meet the actual requirements. For example, in educational research, the analysis and identification of human and individual attributes to distinguish between different individuals can have different important parameters. Therefore, we sometimes need to use a different distance function to find the distance of the two points.

2.4   Image Registration

Image registration is the process of transforming different sets of data into one coordinate system. In which the image data may be at different times and different environmental conditions of the imaging device or a plurality of images obtained under. Image registration techniques are widely used in computer vision, remote sensing image processing, medical image processing and other fields. The core image registration feature point extraction step, from the images to be matched to extract important features include image feature points to perform a similarity measure between the image and the image feature points matching.

2.5   Feature Extraction Algorithm

Feature extraction and image matching represent two important tasks in computer vision, computer graphics, and all images applications. Feature extraction is the basic point for image analysis algorithms. Extraction should have repeatability as its main feature i.e., for the same scene from different angles at different times, under different environmental conditions,[16] the extracted features should be the same. In order to eliminate the shooting angle, illumination change, image blur and other effects, researchers have proposed a number of feature extraction algorithms for which computational complexity and accuracy of the results are also very different. Two widely used feature extraction algorithms are SIFT and SURF [14]. A comprehensive comparison of the traditional SIFTS and the SURF features are given in Table 1.

Table 1. Traditional SIFT and SURF Performance Comparison

Method	Time	Space	Rotate	Fuzzy	Illumination	Affined
SIFT	General	Optimal	Optimal	General	General	Better
SURF	Optimal	General	General	Better	General	General

3   Experimental Environment

The experiments were conducted on the tenth floor of campus new main building where several students, researchers and professors move around frequently. The dimension of the floor was approximately 35 x 35 meters, and included 28 rooms. We collected signals from the fingerprinting system based on RSSI. We used eight distributed 802.11 access points for our experiment. During the corresponding training phase we collected RSSI observations for each room. For our database of images, we obtained a set of images, where the number of images captured for each room depends on the room type. Thus, a room in a corridor is associated to as many images as possible due to variety in direction of movements, whereas inside the rooms we build panoramic images covering the whole dependency. For larger scenarios we plan to develop some technique like the one presented by Park et al. in [15] in order to populate larger databases of images and RSSIs using the data provided by the users as they make use of the available location services. The system tests were performed at 10th floor of Campus New Main Building G Block. We used an Intel (R) Core (TM) i3 CPU M330@2.13GHz with 4 GB RAM, running the stable operating system Windows7 (32Cbit). Wi-Fi experimental data was obtained by Atheros AR9285 Wireless Network Adapter. Test images were acquired by the Nokia N97 mobile phone camera. We have collected the appropriate software client for each device in order to collect RSSIs and images and to send them to a repository. Their locations were chosen so as to provide consistent coverage throughout the entire scenario, guaranteeing that every cell is covered by, at least 3 access points.

3.1   Experimental Data Collection

Experiments were performed at Campus New Main building Block G 10th floor. 6 rooms numbered 1024, 1036, 1038, 1039, 1043 and 1045 were used, as shown in Fig. 2. We deployed 46 APs all across the experimental environment and we did not add any auxiliary components, nor for any AP did we modify the position of interference. Although the chairs present in these rooms during the experiments were subjected to a certain degree of movement and adjustment, yet the rooms were not subjected to large-scale renovation or structural repair. Wi-Fi data was collected using the Wireless Mon software, gathering AP information in a room. Mon Wireless is a wireless adapter and allows the user to monitor the state of aggregation, display neighboring wireless access point or base station real-time information. An AP gives information such as Service Set Identification (SSID), RSSI, Media Access Control (MAC) address and other relevant information.

3.2   Wi-Fi Filter Implementation on RSSI

Traditional Wi-Fi fingerprint often contains AP MAC address and the AP signal values, the RSSI value of fingerprint is included to determine the uniqueness of each room Wi-Fi fingerprint, and use the information for each room to distinguish different fingerprints, and finally use it to achieve the location. But after

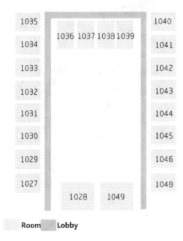

Fig. 2. Experimental Environment

test measurements and results related to continuously measuring the same AP RSSI value in the same room, the RSSI value fluctuations are very large. We collected AP information continuously for the same period in room 1045. This information included the AP MAC address and the received signal strength RSSI values. Fig. 3(a)(b)(c) shows the three histograms of occurrence of RSSI values, where each histogram corresponds to an AP placed in room 1045. We found that we could just use AP information in a message, i.e., AP MAC address, as the fingerprint data.

Fig. 4 gives the normalized frequencies of occurrence of the APs in all the 6 rooms. We scanned all the 46 AP MAC addresses (the MAC addresses of the APs were labelled from 1 to 46). Then we collected the frequency of occurrence of each AP for each room and normalized it. The Wi-Fi fingerprint data for a room was stored in a text file, which consisted of the MAC labels of all the APs, and the RSSI value for each MAC label. It can be seen that for each room, there is a unique identifiable fingerprint, which consists of the normalized frequency of occurrence of each AP in that room. This enables convenient fingerprint matching by employing the Euclidean distance between the test samples and the samples already stored in the fingerprint database, so as to decide whether there is a close correlation between the samples. For each test fingerprint, we compare its value to fingerprints of all the rooms and find out the three best matches. We then sort these matches in a descending manner to complete the Wi-Fi filter.

3.3   Wi-Fi Probabilistic Filter Model

During the testing phase, we collected Wi-Fi fingerprints of the test images and determined their Euclidean distances from the fingerprints stored in the database. Based on these distances, we appropriately attribute the test fingerprint to a certain room. However, this is estimation based on probability.

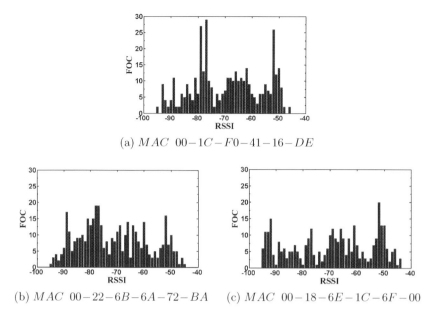

(a) $MAC\ 00-1C-F0-41-16-DE$

(b) $MAC\ 00-22-6B-6A-72-BA$ (c) $MAC\ 00-18-6E-1C-6F-00$

Fig. 3. RSSI against Frequency of Occurrence (FOC) for each AP in Room 1045

In other words, if the distance between the test fingerprint and the fingerprint for a certain room in the database is large, it implies a smaller probability of the test fingerprint belonging to that room.

According to the results, we can construct the test sample filter model. Using this model, we can find the sample most closely associated with the room position location information. Fig. 5 shows probabilistic filter results for test samples for rooms 1038 and 1039. Horizontal axis is the room number, the vertical axis is the probability of matching between the test Wi-Fi fingerprint samples and the fingerprints of these rooms. Finally, we sort descending the results of the filter, and then select the most likely value of the three rooms as the input to the image filter.

3.4   Image Filter Implementation

Analysis of SURF Algorithm

The SURF detector algorithm tries to localize the "interest points". An interest point is a point in the image which: (a) has a clear, preferably mathematically well founded, definition, (b) has a well-defined position in the image space, (c) the local image structure around it is rich in terms of local information contents, so the use of interest points simplifies further processing in the vision system, (d) is stable under local and global perturbations in the image domain, including deformations as those arising from perspective transformations such as scale changes, rotations and/or translations as well as illumination/brightness variations. Moreover, the notion of the interest point includes an attribute of scale,

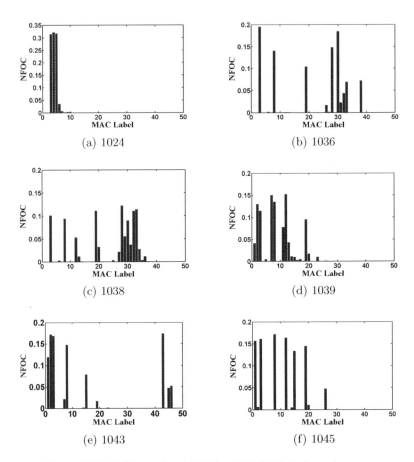

Fig. 4. NFOC (Normalized FOC) of MAC labels in each room

so as to allow it to compute interest points from real-life images as well as under scale changes.

Algorithm Step-by-step
1. Computation of the integral image of the input images.
   2. Interest points detection:

   – Computation of the Box Hessian operator [17] at several scales and sample rates using box-filters.
   – Selection of maxima responses of the determinant of the Box Hessian matrix in box space. The matrix Given a point $x = (a, b)$ in an image I, the Hessian matrix H(x, α) in x at scale α is defined as follows:

$$H(x, \alpha) = \begin{pmatrix} K_{aa}(x, \alpha) & K_{ab}(x, \alpha) \\ K_{ab}(x, \alpha) & K_{bb}(x, \alpha) \end{pmatrix} \tag{2}$$

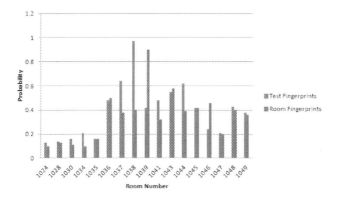

Fig. 5. Wi-Fi fingerprint filter model

Where $K_{aa}(x, \alpha)$ is the convolution of the Gaussian second order derivative $\frac{d^2}{dx^2}g(\alpha)$ with the image I in point X, and similarly for $K_{ab}(x, \alpha)$ and $K_{bb}(x, \alpha)$.

– Refinement of the corresponding interest point location by quadratic interpolation;
– Storage of the interest point with its contrast sign.

3. Local descriptors construction:

– Estimation of the dominant orientation of each interest point;
– Computation of the descriptor (16 x 4 vector) corresponding to the scaled and oriented neighborhood of the interest point.

Here we exploit integral images for speed. Moreover, only 64 dimensions (16 x 4) are used, reducing the time for feature computation and matching, and increasing simultaneously the robustness.

4. Image matching:

– Matching the SURF descriptors of both images by a nearest neighbor criterion inspired from the SIFT algorithm, speeded-up by a priori imposing that the sign of the contrast is the same for corresponding descriptors.

Image Filter Specific Implementation
Implementation of the image filter, we used SURF algorithm to achieve filters. In the implementation of SURF algorithm, in order to match the test image with the image database, by comparing the corresponding feature vectors. If the test image and the image in a database match the largest number of sub-regions, then the transfer of the image position information, as the final result is output on the graphical interface. Fig. 6(a) shows the SURF features calculated for the room 1049. Fig. 6(b) shows the matching of SURF features between 2 images of the room 1038, where the images were taken from different viewpoints. The

number of matching points is 116 with interest points 71. Fig. 6(c) shows the matching of SURF features between images of rooms 1038 and 1039. The number of matching points 116 with interest points is only 60. So we define a minimum threshold value to matching points, for a test image to qualify as belonging to a room.

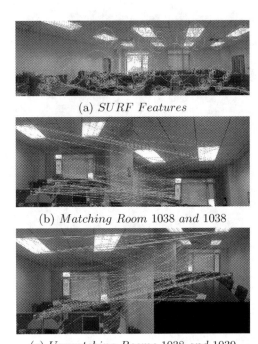

(a) *SURF Features*

(b) *Matching Room* 1038 *and* 1038

(c) *Unmatching Rooms* 1038 *and* 1039

Fig. 6. Matching images by using SURF

4   Performance Assessment and Results

We implemented our system both sequentially and in parallel. Fig. 7 gives the results of the comparison of both the approaches can be seen, for most of the rooms the accuracy remains above 0.8 of true positives. There is not a big difference between the performances of both the implementations. It is worth mentioning here that for sequential execution, Wi-Fi filter will work at correct positioning results and discard false positives, resulting in image filtering in succession Wi-Fi filter results, and generating accurate results. For room 1049, the accuracies alarmingly drop to 0.70 and 0.50, which was an unexpected outcome.

However, after further analysis, we found that due to the huge size of room 1049, our image acquisition time database does not cover all the corners of the room. So the match between the test samples and the database samples resulted in low positioning accuracy. We can conclude that whether we use sequential

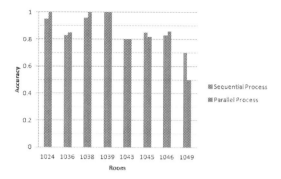

Fig. 7. Comparison of accuracy of sequential and parallel systems

or parallel implementation of the system, the accuracy of the system structure remains largely unperturbed. Fig. 8 gives a comparison of the execution times for both the implementations. It is obvious that there is a huge difference between the processing times for sequential and parallel structures.

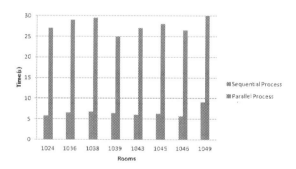

Fig. 8. Comparison of speed of sequential and parallel systems

Because the image processing time is longer, the parallel execution of two filters does not reduce the running time, but will increase the image processing filter burden. So we deduce that the running time depends entirely on the image filter implementation. When the number of rooms increases, the image filter running time also increases linearly. Hence the use of sequential structure is much more effective. In summary, although the accuracy for parallel sequence can be slightly higher than the sequential sequence for a few cases, the huge time overhead renders the parallel sequence as infeasible for achieving the positioning system needs. So we use the sequential program to efficiently obtain a more complete positioning system. In our method speed is the most essential parameter followed by illumination invariance. Hence SURF is more desirable and this makes our approach novel and robust.

5 Conclusion

The multi sensor localization system presented in this paper provides an accurate method for estimating the user location. As mentioned during the paper, the main drawback of using images is the elevated computational cost of the matching process in huge scenarios. Though our experimental scenario is relatively small, we have demonstrated that by using RSSI information we can reduce the search time up to 83%. In larger scenarios this performance improvement will be meaningfully higher since the complete tree of features will contain a higher amount of descriptors. Obtaining data from sensors to providing location estimation, in approximately three seconds. There are several augmented-reality applications which do not require a real time response. Therefore, we find our approach a valuable contribution for this kind of location-based services. As future work, we are investigating the use of different localization algorithms will be implemented in order to deal with structured and non-structured environments. The use of other algorithms to detect location with moving entities may also be incorporated to our approach in order to improve the efficiency of the dynamic objects detection.We also plan to address the case where dynamic objects move very slowly, which creates problems in the neighborhood comparison for the future work.

Acknowledgment. This work was partially supported by the National Natural Science Foundation of China under Grant No. 61170296 and 61190125, by Instituto de Telecomunicações, Next Generation Networks and Applications Group (NetGNA), Covilhã Delegation, by National Funding from the FCT V Fundação para a Ciência e Tecnologia through the Pest-OE/EEI/LA0008/2013 Project.

References

1. Pahlavan, K., Xinrong, L., Makela, J.P.: Indoor geo-location science and technology. IEEE Communications Magn. 40(2), 112–118 (2002)
2. Nisarg, K., Balajee, K., Glasgwow, E.D., Dias, M.B.: Bringing Navigation Indoors. The Way We Live Next (2008); Robust Indoor Localization on a Commercial Smart Phone. Procedia Computer Science 10, 1114–1120 (2012)
3. Van den Berghe, S., Weyn, M.: Fusing Camera and Wi-Fi Sensors for Opportunistic Localization. In: UBICOMM 2011: The Fifth International Conference on Mobile Ubiquitous Computing, Systems, Services and Technologies IARIA (2011) ISBN: 978-1-61208-171-7
4. Arai, I., Horimi, S., Nishio, N.: Wi-Foto 2: Heterogeneous device controller using Wi-Fi positioning and template matching. Adjunct Proceedigs of Pervasive (2010)
5. Zhang, Y., Ma, L., Zhang, R.: A Quick Image Registration Algorithm Based on Delaunay Triangulation. TELKOMNIKA 11(2), 761–773 (2013)
6. Chandrasekaran, G., Ergin, M.A., Yang, J., Liu, S., Chen, Y., Gruteser, M., Martin, R.P.: Empirical Evaluation of the Limits on Localization Using Signal Strength. In: Proc. SECON 2009, pp. 333–341 (2009)
7. Bahl, P., Padmanabhan, V.N.: RADAR: An In-Building RF-based User Location and Tracking System. In: Nineteenth Annual Joint Conference of the IEEE Computer and Communications Societies 2000, vol. 2, pp. 775–784 (2000)

8. Haeberlen, A., Flannery, E., Ladd, A.M., Rudys, A., Wallach, D.S., Kavraki, L.E.: Practical Robust Localization over Large Scale 802.11 Wireless Networks. In: Proc. MOBICOM 2004, pp. 70–84 (2004)

9. Niculescu, D., Nath, B.: Ad hoc positioning system (APS) using AOA. In: Niculescu, D., Nath, B. (eds.) INFOCOM 2003. Twenty-Second Annual Joint Conference of the IEEE Computer and Communications, vol. 3, pp. 1734–1743. IEEE Societies (2003)

10. Zhang, Y., Brown, A.K., Malik, W.Q., et al.: High resolution 3-D angle of arrival determination for indoor UWB multipath propagation. IEEE Transactions on Wireless Communications 7(8), 3047–3055 (2008)

11. Llombart, M., Ciurana, M., Barcelo-Arroyo, F.: On the scalability of a novel WLAN positioning system based on time of arrival measurements. In: 5th Workshop on WPNC 2008, pp. 15–21. IEEE (2008)

12. Han, D., Andersen, D.G., Kaminsky, M., Papagiannaki, K., Seshan, S.: Access point localization using local signal strength gradient. In: Moon, S.B., Teixeira, R., Uhlig, S., et al. (eds.) PAM 2009. LNCS, vol. 5448, pp. 99–108. Springer, Heidelberg (2009)

13. Gmskaya, H., Hakkoymaz, H.: WiPoD wireless positioning system based on 802.11 WLAN infrastructure. Proceedings of the Enformatika 9, 126–130 (2005)

14. Bouris, D., Nikitakis, A.: Fast and Efficient FPGA-based Feature Detection employing the SURF Algorithm. In: 18th IEEE Annual International Symposium on Field-Programmable Custom Computing Machines (2010)

15. Park, J.G., Charrow, B., Curtis, D., Battat, J., Minkov, E., Hicks, J., Teller, S., Ledlie, J.: Growing an organic indoor location system. In: Proc. MobiSys 2010, pp. 271–284 (2010)

16. El gayar, M.M., Soliman, H., meky, N.: A comparative study of image low level feature extraction algorithms. Proc. Egyptian Informatics Journal 14, 175–181 (2013)

17. Bay, H., Tuytelaars, T., Van Gool, L.: SURF: Speeded Up Robust Features. In: Leonardis, A., Bischof, H., Pinz, A. (eds.) ECCV 2006, Part I. LNCS, vol. 3951, pp. 404–417. Springer, Heidelberg (2006)

18. Quan, M., Navarro, E., Peuker, B.: Wi-Fi Localization Using RSSI Fingerprinting, Decertation from California Polytechnic State University (January 2010), http://digitalcommons.calpoly.edu/cpesp/17/

A Robust Approach for Maintenance and Refactoring of Indoor Radio Maps*

Prasanth Krishnan, Sowmyanarayanan Krishnakumar,
Raghav Seshadri, and Vidhya Balasubramanian

Department of Computer Science and Engineering,
Amrita School of Engineering, Coimbatore,
Amrita Vishwa Vidyapeetham (University)

Abstract. WiFi Fingerprinting techniques are widely used for indoor localization needs, due to better accuracy guarantees. However, the accuracy is limited by the freshness of the radio map, which is used for localization. Over time this radio map might be incompatible due to the changes in signal strength, a consequence of the dynamic nature of the environment. Therefore, repeated radio map calibration becomes a necessity. Currently radio maps are either manually calibrated, use additional infrastructure or complex algorithms to account for the radio map errors. There is therefore a need for a methodology to update the radio map with minimal additional overhead. This paper proposes a crowdsourcing approach which uses data collected from users of the localization system to maintain the freshness of the radio map. This approach leverages inertial sensors present in commonly used handheld devices, like mobile phones and tablets using which, a trajectory of the path of each user is computed. This trajectory is then coupled with the knowledge of the physical map under consideration to get the real time Received Signal Strength Indicator ($RSSI$) values at the reference points (RPs). A novel cluster based $RSSI$ propagation policy is proposed where the real time $RSSI$ values obtained are propagated within the clusters. Extensive experiments of our localization system implemented in a real indoor environment shows that this approach maintains radio map freshness while keeping the cost of update low and without need for extra infrastructure.

1 Introduction

Location Based Services are being widely deployed as part of indoor pervasive systems. RF-based localization systems, which employ trilateration or fingerprinting based techniques are commonly used in the indoor space, due to wide deployment of wireless networks in buildings.Owing to their better accuracy, fingerprinting based techniques are often adopted in indoor location-estimation systems. However, a major limitation of fingerprint-based methods is that the radio maps are static. This poses a serious problem to the effectiveness of location

---

* This work has been funded in part by DST(India) grant DyNo. 100/IFD/2764/2012-2013.

S. Guo et al. (Eds.): ADHOC-NOW 2014, LNCS 8487, pp. 360–373, 2014.

estimation. In dynamic indoor environments, radio signal is affected by factors like opening and closing of windows or doors, shifting the location of any large object, and human activity. This change in Received Signal Strength Indicator ($RSSI$) results in an error in localization if a radio map that does not reflect the change, is used as the training map for localization. To understand the impact of using a non-updated radio map, a localization experiment was conducted in an office environment using a radio map that was not updated. The localization results showed an average increase in error of about 23.22 percent when the radio map calibrated before 20 hours was used. This demonstrates a need for keeping a fingerprinting map as up-to-date as possible.

There have been some attempts at addressing the errors due to non-updated radio maps [6] and these either use extra infrastructure or additional manual process. A detailed survey of current work is done in Section 2. Manually measuring and updating the radio map for a large indoor region is tedious and not a feasible solution for real time update of radio maps. A solution that incorporates extra hardware like robots or additional infrastructure will significantly add to costs and may not always be possible. Thus, a method of update which does not require manual labour or external hardware is necessary.

In this paper, our approach is to employ the $RSSI$ data gathered from users who use the localization services, in a crowdsourced fashion to keep the radio map up-to-date. We design a comprehensive approach to solve the problem of rebuilding the radio map, which consists of the following three parts:

Trajectory Estimation and Extraction of Data from Map: Using the inertial sensors (accelerometer, orientation sensor etc) present in handheld devices, the trajectory of the user accessing our system is estimated. This information along with a knowledge of the map is used to extract the locations, hence the reference points (RPs) which fall on this trajectory are estimated and the $RSSI$ values corresponding to them is recorded.

Clustering of the Reference Points: Since the update process relies on the user's random movements, it is essential to be able to update larger portions of the map with updates to a few reference points. Clustering of reference points is employed to enable this, here reference points are clustered by using the two most dominant Access Points (APs). This ensures that similar points are grouped together, and when an update is received to any point within the cluster, the update can be propagated.

Propagation of Updates: The new $RSSI$ values obtained from user's trajectories, are propagated within each cluster based on $RSSI$ patterns in the clusters. In addition it is ensured that not all new $RSSI$ values are immediately propagated and only some $RSSI$ values which significantly impact the accuracy of localization are propagated. The localization which is done after the update, is done in a hierarchical fashion (cluster based) as it ensures less similarity between the $RSSI$ values from other clusters.

This paper will discuss this novel approach to update and maintain the Radio Maps for indoor localization and discuss the experimental setup and results demonstrating the performance of our system. The next section will outline the current literature and further motivate the need for our work.

2   Related Work

There are two approaches in dealing with the *RSSI*-position relationship in indoor *RSSI*-based positioning systems: the use of signal propagation models [13, 11, 12] and the location fingerprinting methods [8, 9]. The former are more easier to implement and use methods like triangulation and trilateration [14] to estimate the position. However since signal propagation in indoor environments is highly unpredictable due to reflections, obstacles and interference from other devices, the accuracy of these methods is poor. Therefore, the more robust Fingerprinting approach [14, 12] is usually preferred.

The problem in the fingerprint based approach is the maintenance of the radio map. Using a static radio map can be highly inaccurate and it requires repeated data calibration to maintain the accuracy. To cope with the dynamic environmental changes, several adaptive algorithms have been proposed in recent years [14, 11, 8]. An approach which adapts the static radio map by calibrating new *RSSI* samples at a few known locations and fitting a linear function between these samples and the old samples from the radio map is shown in [14]. During the online phase, new samples are first shifted to old samples using the estimated linear function, such that the original radio map can be used. The main assumption is that the adaptation can be performed independently of locations. However, in a real environment, *RSSI* values vary a lot from one location to another. The LANDMARC system [5] and the LEASE system [10] both utilize real time *RSSI* values at reference points to adaptively update the radio map, and they are able to adapt to environmental dynamics. However, these systems show a considerable accuracy only when the resolution of the radio map is high, and require extra infrastructure for the updation process, which adds to the cost.

In LEMT [6] they use a radio map collected at a certain time to learn the functional relationship in the *RSSI* samples between the mobile client and the reference points, and use a model tree to reduce the number of reference points needed in the map. However, to get the real time *RSSI* values they use additional RF receivers at the reference points. Even if the use of model trees reduces the number of reference points, the cost incurred due to the placement of the RF receivers is quite high. Our approach is to use data collected from mobile devices belonging to users of our system, and this helps overcome the extra infrastructure cost needed for update.

Crowdsourcing as a methodology for indoor localization has been used by the Zee [1] and Freeloc [4] systems. Zee uses crowdsourcing to generate the fingerprinting map by utilizing the inertial sensors present in handheld devices. A comprehensive approach to estimating user's trajectories is proposed which is then used for generating the fingerprinting map. The FreeLoc [4] system also uses inertial sensors and provides a zero calibration crowdsourcing method for localization. While these systems provide an effective solution for fingerprinting, the update of the fingerprint map has not been addressed in both works. Such comprehensive solutions while being effective for radio map creation, radio map update requires faster and simpler approaches. We address this major issue of update/maintenance of radio maps by coming up with simple and innovative solutions that uses crowdsourcing for updating $RSSI$ values at reference points, and use clustering and update propagation to quickly update larger portions of the map, thereby keeping the map as fresh as possible.

3   System Architecture

This section outlines the architecture of our fingerprinting update system, as shown in Figure 1, which uniquely and effectively uses crowdsourcing for getting real time updates to the radio map. A user who uses this system for localization, becomes the source of update. The $RSSI$ values received from the user's mobile device, and the accelerometer and orientation sensor readings are used for radio map maintenance. The process of update, based on the data received from the mobile devices consists of three main modules 1) The trajectory correlation module 2) The radio map refactoring module 3) and Update Manager which consists of the Update Policy Engine, and Update Propagation Engine. The localization engine works together with these modules to assist with both localization and radio map maintenance. The trajectory correlation module identifies the physical path of the user based on the received sensor readings from the user's mobile device. The path is sent to the radio map refactoring system where the clusters in which the user's path is located is identified. The reference points in the initial fingerprinting map are clustered using the dominant two routers with respect to $RSSI$. The localization algorithm is applied on the portion of radio map belonging to cluster and the estimated location is sent to the user's device.

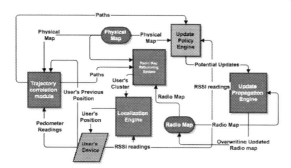

Fig. 1. System Architecture

The path matching engine sends the identified physical path to the update policy engine along with $RSSI$ values from the Wifi scans. This engine takes decisions on whether a reading has to be updated or not (based on how much the $RSSI$ values affect the overall accuracy). Information is fed to it both from the physical map and the radio map in order to achieve the same. Then the potential updates are sent to the update propagation engine for propagation. The propagation engine propagates the updates to all points inside the relevant clusters. Then, the map is re-clustered if necessary and thus, the radio map is updated.

4   Trajectory Calibration and Matching

In order to map the $RSSI$ data collected from the users to the appropriate reference points, we must first estimate the trajectory of a moving user, and translate the trajectory to the actual indoor map. Using crowdsourcing of $RSSI$ values to build maps using user trajectory has been attempted by the Zee [1] and FreeLoc [4] Systems. For real time update a simpler trajectory mapping process is sufficient. This section contains a detailed overview of the trajectory calibration and matching process.

4.1   Calibration of Trajectory

As mentioned in previous sections, the system leverages the accelerometer and the orientation sensor of the device to calibrate the trajectory. There are two components to finding the trajectory of a user, first component is estimation of distance (using the accelerometer) and the second component is the estimation of heading (direction using an orientation sensor). To estimate distance we assume that on average one step corresponds to covering half a meter. The distance is aggregate of the total number of steps taken by the user. This assumption of $0.5m$ is taken to keep the process of trajectory matching simple. To find out the user's heading direction, the orientation sensor is used. The orientation sensor functions like a compass. It gives a fixed angle as output for a particular direction e.g. East corresponds to 90 degrees. Since the distance is measured in terms of the number of steps taken by the user, the trajectory obtained from the user is stored as a series of triplets (s, θ, t). Here, the first value s of the triplet denotes the chronological order or the step i.e the 1st, 2nd, etc. step. The second value of the triplet θ denotes the heading (angle) that the step s corresponds to. The third value of the triplet t is time stamp corresponding to the time when step s was taken.

4.2   Modelling of the Physical Map

The physical map is divided into a grid of Reference Points (RPs). This grid of reference points is mapped into a graph G. Each reference point, RP_{ij} is modelled as a vertex in G. An edge connects two vertexes in G if they are connected by a path. Since the physical map is a grid, we consider an 8-point adjacency for each RP. Thus, each RP_{ij} has associated with it an an adjacency list L_{ij}, which stores the neighbors in the 8-point adjacency. This adjacency list consists of a set of triplets (N_{ij}, ω_{ij}, S). N_{ij} represents a neighbouring RP to which RP_{ij} is connected by a path. ω_{ij} represents the fixed angle at which the path to N_{ij} from RP_{ij} is oriented. S represents the number of steps taken to traverse the path between RP_{ij} and N_{ij}.

4.3   Path Matching Algorithm

The aim of the path matching is not to exactly map the user's trajectory, but to find out the reference points through which the user passes and obtain the $RSSI$ data for those reference points. A modified Depth First Search (DFS) on graph G is used for path matching. When the user is localized for the first time, the initial position of the user is obtained. Since the data regarding the placement of the reference points in the physical map is known, the reference point closest to this initial location is considered as the starting point of the DFS. The DFS will traverse through all the paths from the starting point with the number of steps walked by the user as the termination point. That is, it finds all the paths originating from the starting point whose sum of distances($\sum s$) totals the number of steps S, walked by the user. All the paths, which the DFS traverses and satisfy this termination condition, are filtered further based on the sum $\sum \theta_d$. Each $\theta_d = \theta - \omega$ is the difference between the angle θ obtained from the user for each step and ω the angle stored in the graph G. The path with the minimum $\sum \theta_d$ is selected as the user's path. This method of finding θ_d is adopted because the orientation meter from which the angles are estimated might sometimes give a spurious value as output due to jerks in the user's motion. Therefore, if the angles θ and ω were directly equated, the probability of the calculated path being wrong or of finding no matching path is high.

5   Update Policy

Once the trajectory and hence the reference points where the user has visited, has been estimated using data from the inertial sensors, the actual update of the map based on the $RSSI$ values obtained must be performed. This section explains the policies and algorithms for updating the radio map based on the crowd sourced data.

Crowdsourced $RSSI$ data from the users by recording user trajectory can be used for the update of the Radio Map, but for one set back. In order to update the radio map, $RSSI$ data is required for every point on the radio map. Since it need not be the case that the users visit all the parts of the map, it might take a long time to obtain values for all points. If the gap between the time of collection of $RSSI$ data for different points is high, it is likely that an environmental change would have taken place and the values may not be consistent. It is therefore essential to be able to update a large portion of the map with limited number of points from which $RSSI$ values are collected. This may be enabled by the propagation of updates throughout the map from $RSSI$ changes recorded at a reduced number of points. However for propagation to be effective the decisions regarding when to propagate the update and how to propagate the update are pertinent.

5.1   Update Policy and Mechanism

The $RSSI$ values collected from the user are stored as a sequence of ordered pairs $(r_i^{user}z, t_i^{user}z)$ where r is the $RSSI$ reading taken at time t from user for the Access Point(AP) z. The corresponding reference points which have been estimated as being passed by the user, are the points which are candidates for updates, and the data sent by user for each RP is accumulated in a list L_{xyz}. Here x and y are the coordinates of the RP and z is the AP from which the $RSSI$ was received.

At any time a finite set of reference points receive updated values, and it is desirable to have more points updated. To reflect a change from one point to a group of adjacent points, a pattern in how the $RSSI$ changes over time must be identified. Since $RSSI$ does not show the same pattern throughout the radio map i.e. the $RSSI$ might change differently in different areas in the radio map owing to physical blockades such as walls, entities with a non-singular state of existence such as windows and doors and to some extent, and the movement of people. Dividing the radio map into clusters with similar $RSSI$ changes based on physical proximity is one possibility, but it is not a general solution to the problem as for each location, the physical map of that location needs to be studied and experiments conducted before the map can be clustered.

To solve this problem, this system adopts a clustering model developed by Horus [8], where the points having the same q strongest Access Points are clustered together, this provides a firm and inexpensive basis to divide the radio map into clusters. For example, if there are four APs (A, B, C, D), and the 3 strongest routers are considered for the clustering process($q = 3$), then C_{ABC} is defined as the cluster in which the $RSSI$ received from the Access Points A, B, C are the strongest. Since this type of clustering is not based on physical proximity, a few points in the midst of a cluster might belong to a different cluster and this might cause some anomalies. If there exist such points, then likelihood of the $RSSI$ values recorded at these points being errant are high. If a point p of a cluster C_1 is surrounded by points from a different cluster C_2, then it can be safely assumed that p also belongs to C_2. Thus this problem is overcome by considering these points as belonging to the cluster to which the points surrounding them belong to. Therefore, the radio map is divided into such clusters as mentioned and within these clusters, the $RSSI$ values are propagated based on 'seed points' which have sufficient up-to-date $RSSI$ gathered via crowdsourcing.

The policies designed to update the radio map based on the actual data, must consider the issue of choosing the right seed point in the presence of more than one seed in a cluster, and the minimum user data required to mark a reference point as a seed point.

Policy for Choosing the Right Seed Point within Cluster. Let k seed points $m_1, m_2...m_k$ exist inside the same cluster C_{ABC}. Let the other points in the cluster excluding these points be $p_1, p_2, ..., p_{n-k}$. In order to account for the multiple seeds, cluster C_{ABC} is further divided into k sub-clusters based on the nearest neighbours of each of the k seeds. If a reference point p is closest to one of the seeds m_i than the other seeds, then it is assigned to the sub-cluster $SC_{m_i}^{ABC}$ and the $RSSI$ of the point p is increased in proportion with the increase in $RSSI$ at point m_i. Since the sub-clusters are small, it is assumed that the change in $RSSI$ over time is uniform across the sub-cluster. Let the old $RSSI$ value at m_i be r_z^{old} and the new $RSSI$ value at m_i be r_z^{new}, where z is the AP to which the $RSSI$ corresponds to. The $RSSI$ corresponding to each AP in the sub-cluster $SC_{m_i}^{ABC}$ will be increased by $\frac{r_z^{new} - r_z^{old}}{r_z^{old}}$ X 100 percent.

The update of each cluster is independent of other clusters. Once an update of a cluster is complete the list L_{xyz} corresponding to each point in the cluster is cleared for storage considerations. In addition, the $RSSI$ values of the points in the cluster are checked again to see if they still have the same strongest q APs. If the strongest q APs of some points have changed, then re-clustering takes place and those points are shifted to the cluster to which their strongest q APs correspond to. This process of update continues for the duration of the session.

Solution to the Sufficiency Problem. All along, the seed points have been defined as points with 'sufficient' user data and it is essential to define what "sufficient" represents. A frequency distribution plot of $RSSI$ at a particular point is generally a normal distribution [2]. Therefore, we define that sufficient user data is obtained for a RP when the $RSSI$ data forms a normal distribution with respect to the value with maximum frequency taken as mean.

5.2   Optimizing the Update Process

It is possible that during certain periods some access points will receive constant updates. To propagate all these updates and to recluster the map constantly is costly. Hence it is essential to analyze if all updates have to be propagated. First we characterize the threshold ϵ as the Euclidean distance value between an old $RSSI$ value and the potentially updated $RSSI$ value at any localization point whose impact on the localization accuracy is minimal. If the average difference between the old $RSSI$ values and new $RSSI$ values at the seed points within a cluster is greater than ϵ then we decide to propagate the update within the cluster. This threshold ϵ is experimentally determined as will be discussed in the next section.

5.3   Hierarchical Localization

Since the clusters are updated independently, if a usual localization algorithm which takes the entire radio map as training data for location estimation is used, some of the clusters might have been updated and others might not have been updated. This causes a dip in the accuracy of the system. Therefore, the new localization method only considers the isolated radio map of each cluster for localization. This is achieved by using the user trajectory. Using the calibration of user trajectory, a rough location of the user is obtained, hence the cluster to which the user location belongs to can be estimated approximately. Then, the a map containing only the $RSSI$ values of the RPs belonging to that cluster is isolated from the whole radio map and that new map is used for localization. This method of hierarchical is independent of the localization algorithm used and thus, any localization algorithm can be used on top of it.

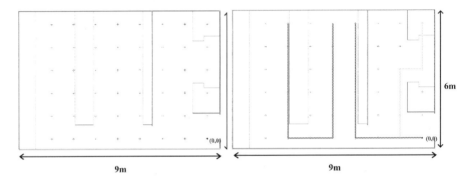

Fig. 2. a) Reference Points in Lab b) Sample Paths in Lab

6   Experiments and Evaluation

We evaluate the performance of the radio map update strategies of our system in the context of two environments 1) A small lab environment 2) A larger office environment as shown in Figures 2(a) and 3. An initial fingerprinting was done for both the environments. All experiments were conducted in both environments, and also by using both laptop and tablets as receivers to account for varying environmental conditions and devices. The $k - NN$ algorithm was used throughout the experiments. We believe that the impact of radio map freshness is highest on the $k - NN$ algorithm, and hence if our methodology can be used to demonstrate effectiveness over the same, it would be suitable for others.

As mentioned previously, it is essential to determine the update threshold ϵ so that the cost of update, specially the propagation cost is kept minimum. The ϵ is the tolerable change in $RSSI$, beyond which an update is essential. The first experiment was conducted to estimate an appropriate ϵ value. To analyze this we plotted the average localization error over different update threshold values at different localization points. Figure 5 is a sample of this plot taken in our test environments, and different devices. From this graph, it can be inferred that, when the Euclidean distance of the difference between the old and new $RSSI$ values is less than about 5 or less, the variance in error is not much and the error itself is low. However, after a threshold of 5, the average error shows many peaks and valleys, this implies that after 5, even a small increase or decrease in

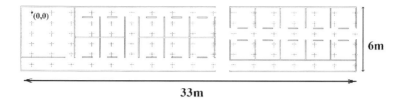

Fig. 3. Reference Points in Office

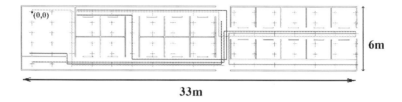

Fig. 4. Sample Paths in Office

the threshold results in steep increases or decreases in average localization error. This implies that the variance is very high and that the relationship between the average error versus update threshold is very unstable. Thus, to improve the system by not propagating unnecessary updates, the updates are only propagated if the Euclidean distance corresponding to the difference between the old and new $RSSI$ values is less than about 5 or less at a seed point. The value 5 as threshold holds for all the test environments and various devices chosen in our experiments.

Analyzing the Performance of the System. We first analyze the performance of the system, and its impact over the overall localization accuracy across both the environments. Since the lab environment is smaller, only two APs were used, and this resulted in two clusters. 4 APs were used in the other environment, resulting in 6 clusters. Initially, the radio map was calibrated manually and the $k - NN$ algorithm was run over this radio map using user localization points with $RSSI$ values measured after an interval of one hour. The average error of localization was measured. For the lab environment, the average error in the initial radio map was found to be 3.52 meters. For the office environment which is larger, for the first set of measurements recorded with a laptop, the average error in localization was observed initially was 1.90 meters. On the other hand for the same environment, the average error in localization when locating a tablet was observed to be $3.68m$. This variation in accuracy is due to the difference in devices, and environment. In order to test the system, a few of the Access Points were placed inside thick casing and further blocked physically. Then, the users' smart devices were installed with the localization application and the users were asked to move around both the environments along the paths shown in Figures 2(b) and 4. User data was collected and the calibration of trajectory was done to estimate the RPs, which the users pass by. Update was propagated to other points in the clusters using the propagation strategy. Localization for a sample set of unknown points was done using both the radio map which was updated using our framework, and using the original fingerprinting map.

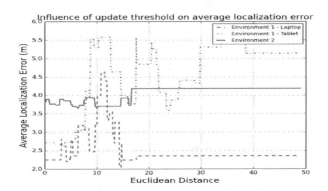

Fig. 5. Influence of update threshold on average localization error

We plot the difference in localization accuracy from initial accuracy, before and after update as shown in Figure 6. We can see that, in all three test scenarios, the update improves the performance of localization. The increase is best in the office environment, where a tablet is located. It can be observed that the tablet provided poor accuracy initially, and the update actually contributed to improving the accuracy, not just reducing the error. The experiment in the office environment locating the more stabler laptop showed the least reduction in error. This can be attributed to its high initial accuracy. This result shows that irrespective of environment or device used for localization, and as a result the initial localization accuracy, our update methodology contributes to reducing the error caused due to changes in radio maps. This update process is robus to number of updates received, as long as a normal distribution is derived. We now further analyze the update strategy in the system in the next paragraphs.

Effect on Path Estimation Error on Localization Error. Trajectory mapping, is an essential component of our update system, and any error in estimating user's location can impact the localization accuracy. Therefore we analyze the impact of path estimation errors on the localization system. If a user is initially being localized and his location is found to be $(0, 1)$ but he is actually at a point $(0, 0)$, there is an error in localization of 1 meter. Therefore, this error of 1 meter reflects on the path calibration as it initially starts with this error. Thus, since the error in path estimation will result in an inaccuracy in the points from which user $RSSI$ is collected, it will also impact the error in the next localization to be performed. As can be seen from Figure 7(a), the error in path estimation influences the average localization error to some extent. The mean standard deviation of the average localization error from all the three data sets is 0.85 meters, which is relatively small. The results shown that with a reasonable

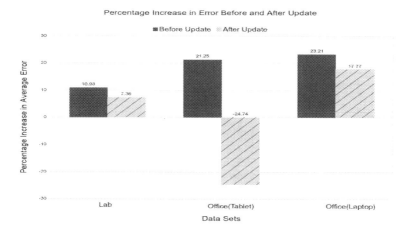

Fig. 6. Percentage increase in average localization error before and after update a) Lab, b) Office (tablet), c) Office (Laptop)

trajectory estimation, the impact on localization accuracy should be low. This, along with the overall improvement in accuracy due to updates, shows that simple trajectory estimation and potentially a knowledge of the environment can help to effectively update the radio map.

Effect of Choice of Nearest Seed Point One of the choices to be made during update policy is to determine the best seed for the propagation of update. One choice is to choose the nearest neighbor. We would like to understand if that is a good choice. Hence we plot the localization error obtained based on the seed chosen ie., the kth nearest neighbor by varying k. The graph in Figure 7(b) shows that for all three experimental setups, the nearest neighbor performs best, ie results in best accuracy. While the plot shows that a few other k^{th} nearest seeds also provide good accuracy, the accuracy does not exceed that when choosing the nearest seed for propagation, hence this demonstrates the nearest neighbor is a good option.

Analysis of the Performance of the Update Propagation Strategy. In order to reduce the cost of updates, and account for limited number of points where actual updates may be obtained, update propagation has been proposed. While previous experiments have shown that our update helps improve the localization accuracy, which is affected by changes in the radio map, it is also

Table 1. Average fluctuation in $RSSI$ from actual $RSSI$ after propagation of update (in dB)

	AP1	AP2	AP3	AP4
Lab	4.17	2.53	NA	NA
Office (Laptop)	4.73	3.58	5.35	4.38
Office (Tablet)	4.79	4.97	4.10	4.58

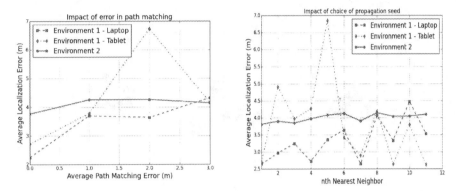

Fig. 7. a) Impact of error in path matching, b)Impact of choice of propagation seed

necessary to analyze how well the simple linear propagation policy works. To understand that we compared sample points from all the environments and analyzed the average difference between the actual update and the propagated updated. The results are tabulated in Table 1. We can see that difference from the actual value is at most 5dB, and most of the times much lower. This difference in $RSSI$ is common even when there are no interferences, and hence demonstrates that our update propagation mechanism brings the map to close to within tolerable variation from actual values, hence helps in keeping it fresh.

7   Conclusion

This paper has presented a robust system for updating and maintaining the radio map in a Fingerprint based Wifi localization system. It provides a novel method for propagating updates in a radio map using $RSSI$ information, inertial sensor measurements and the physical map of the indoor space as inputs. This data is collected from the users of this system in a crowdsourced fashion. The verification of this system through experiments on a large office environment and a small lab environment both show that there are significant reductions in the average localization errors after employing this system. The results demonstrated that by combining clustering and update propagation, the radio map can be maintained in a cost-effective manner. When there are large open spaces, the trajectory matching becomes costly, since we try to estimate the trajectory by looking at all potential paths. An efficient data structure can reduce this cost, and can be used for not only localization, but also tracking purposes.

References

[1] Rai, A., Chintalapudi, K.K., Padmanabhan, V.N., Sen, R.: Zee: Zero-Effort Crowdsourcing for Indoor Localization. In: MOBICOM 2012, Proceedings of the 18th Annual International Conference on Mobile Computing and Networking, pp. 293–304. ACM (2012)

[2] Chuan, H.F., Bose, A.: A Practical Path Loss Model For Indoor WiFi Positioning Enhancement. In: ICICS 2007, Proceedings of 6th International Conference on Information, Communications and Signal Processing, pp. 1–5. IEEE Press (2007)

[3] Radu, V., Marina, M.K.: HiMLoc: Indoor Smartphone Localization via Activity Aware Pedestrian Dead Reckoning with Selective Crowdsourced WiFi Fingerprinting. In: Proceedings of the 4th International Conference on Indoor Positioning and Indoor Navigation. IEEE Press (2013)

[4] Sungwon, Y., Dessai, P., Verma, M., Gerla, M.: FreeLoc: Calibration-Free Crowdsourced Indoor Localization. In: Proceedings of INFOCOM 2013, pp. 2481–2489. IEEE Press (2013)

[5] Ni, L.M., Yunhao, L., Yiu, C.L., Patil, A.P.: LANDMARC: indoor location sensing using active RFID. In: Proceedings of the First IEEE International Conference on Pervasive Computing And Communications, pp. 407–415. IEEE Press (2003)

[6] Jie, Y., Qiang, Y., Ni, L.M.: Learning Adaptive Temporal Radio Maps for Signal-Strength-Based Location Estimation. IEEE Transactions on Mobile Computing 7(7), 869–883 (2008)

[7] Ledlie, J., Jun-geun, P., Curtis, D., Cavalcante, A., Camara, L., Costa, A., Vieira, R.: Mol: a Scalable, User-Generated WiFi Positioning Engine. In: International Conference on Indoor Postioning and Indoor Navigation, pp. 55–80. IEEE Press (2011); Journal of Location Based Services 6(2) 2011

[8] Youssef, M., Agrawala, A.: The Horus WLAN location determination system. In: MobiSys 2005: Proceedings of the 3rd International Conference on Mobile Systems, Applications, and Services, pp. 205–218. ACM, New York (2005)

[9] Fox, D., Hightower, J., Liao, L., Borriello, G., Schulz, D.: Bayesian Filtering for Location Estimation. Pervasive Computing 2, 24–33 (2003)

[10] Krishnan, P., Krishnakumar, A.S., Ju, W.-H., Mallows, C., Gamt, S.N.: A system for LEASE: location estimation assisted by stationary emitters for indoor RF wireless networks. In: INFOMCOM 2004, Twenty-third Annual Joint Conference of the IEEE Computer and Communications Societies, vol. 2, pp. 1001–1011 (2004)

[11] Bahl, P., Padmanabhan, V.N.: RADAR: an in-building RF-based user location and tracking system. In: INFOCOM 2000 Proceedings of the IEEE Nineteenth Annual Joint Conference of the IEEE Computer and Communications Societies, vol. 2, pp. 775–784. IEEE Press (2000)

[12] Kaemarungsi, K., Krishnamurthy, P.: Properties of indoor received signal strength for wlan location fingerprinting. In: The First Annual International Conference on Mobile and Ubiquitous Systems: Networking and Services, MOBIQUITOUS 2004, pp. 14–23. ACM (August 2004)

[13] Goldsmith, A.: Wireless Communications, 1st edn. Cambridge University Press (2005)

[14] Liu, H., Darabi, H., Nanerjee, P., Liu, J.: Survey of Wireless Indoor Positioning Techniques and Systems. IEEE Transactions on Systems, Man and Cybernetics Part C: Applications and Reviews 37(6), 1067–1080 (2007)

Performance of POA-Based Sensor Nodes for Localization Purposes

Jorge Juan Robles[1], Jean-Marie Birkenmaier[2],
Xiangyi Meng[1], and Ralf Lehnert[1]

[1] Chair for Telecommunications, Technische Universität Dresden, Germany
{robles,lehnert}@ifn.et.tu-dresden.de,
Xiangyi.Meng@mailbox.tu-dresden.de
[2] TechniSat Digital GmbH
birkenmaier.j@gmail.com

Abstract. In this paper, the accuracy of the "Phase-of-Arrival" (POA) ranging method is investigated in comparison with the well-known Received Signal Strength Indicator (RSSI) method. The IEEE 802.15.4-compliant transceiver AT86RF233 [1] [2], which forms part of a low-power sensor node, was used to take measurements in different scenarios and evaluate both methods.

Additionally, a localization system based on a sensor network was built and a new measurement campaign was carried out to investigate the position accuracy of three low-complexity localization algorithms, i.e. *Multilateration, Extended-Min-Max (E-Min-Max)* and *Weighted Centroid Localization (WCL)*. The results show that it is possible to have average position errors smaller than 1m using the POA method for distance estimation and the *E-Min-Max* localization algorithm.

Index Terms: Wireless Sensor Network, Phase of Arrival, Received Signal Strength Indicator, Localization.

1 Introduction

Obtaining the location information of a sensor node is essential for many wireless applications. For instance, the positions of sensor nodes can be used in routing algorithms [3] and clustering techniques [4]. In [5] the authors introduce a location-based information delivery system. The work described in [6] shows the importance of the position information in smart home applications.

Many localization algorithms, like the algorithms investigated in this work, require the information of the distances between the mobile node (MN) and reference nodes called Anchors (ANs) to determine the MN's position. The ANs are normally fixed in the scenario and their positions are known. Therefore, they can be externally powered. On the contrary, the MNs have batteries, whose consumption has to be minimized to extend the lifetime of the MN.

To determine the distance between two nodes, the most widely used ranging methods are based on RSSI, POA and TOA (Time of Arrival). This last method measures the time that a signal requires to arrive at the receiver. In [7] and [8]

S. Guo et al. (Eds.): ADHOC-NOW 2014, LNCS 8487, pp. 374–386, 2014.

the TOA method is discussed in detail. One of the drawbacks of the methods based on TOA, is that in general they require very accurate clock signals on the transmitter and the receiver, which is difficult (or expensive) to implement in a low-power sensor node. However, there are solutions based on TOA in the market, like the sensor nodes that use the transceiver nanoLOC [9], which can achieve a relative good distance accuracy (<1m) with an acceptable energy consumption (approx. 85mW, Tx output power=0dBm). These nodes use an improved ranging protocol called Symmetric Double Sided Two Way Ranging [10], which involves a bilateral and symmetric transmission of several packets between the nodes to avoid the problem of having high-quality expensive clock oscillators.

RSSI is a measure of the signal strength. By using a path-loss model, the RSSI can be mapped to a distance value. In general, the RSSI values are dispersive and very environment dependent [11] [12] [13]. However, the use of RSSI has two important advantages: First, it does not require additional expensive hardware. Second, the reception of a short packet is enough to perform a RSSI measurement. In other methods like TOA or POA, the transmission of many packets are necessary to obtain just a distance measurement (see [10] for further information).

The POA-based approaches measure the phase of the incoming signal. By using the information of the frequency and speed of the signal, the distance between the nodes can be approximated. This method is commonly used in laser range finders [14]. In [15] a POA-based ranging technology, which detects phase-shifts in sinusoidally modulated infrared signals, is implemented to estimate the position of a mobile robot. The authors affirm that it is possible to achieve an accuracy level of about 10cm.

Recently, the company Atmel developed the transceiver AT86RF233 designed for sensor nodes, which consumes very low energy (approx. 38mW, Tx output power=4dBm) and can take POA and RSSI measurements. Thus, we were able to use the same transceiver to take distance measurements based on both methods under the same conditions and investigate their accuracy levels. In our literature review, we found a reduced number of research works that investigate the suitability of these POA-based sensor nodes for its use in localization systems. One of the first works was published in [16], where the authors evaluate different ways to exploit the POA measurements to obtain an accurate ranging estimation.

In this paper we present the results of two measurements campaigns. In the first one, we evaluated the accuracy of both distance determination methods by taking measurements with two sensor nodes located at different distances (up to 50m) in two different scenarios. In the second measurement campaign, we made use of a localization system to obtain the POA-based distance and RSSI measurements between the MN and several ANs. This information was sent to a central computer for the position estimation. In this way, it was possible to evaluate the position accuracy of the algorithms.

Fig. 1. The reference sensor node used in our experiments

2 Distance Determination Methods

2.1 Distance Estimation Using POA

Our reference sensor node consists of the module REB-CBB, which contains the microcontroller Atxmega256A3, connected with the module REB233SMAD that has the transceiver AT86RF233. Technical details about these modules can be found in [1] and [17].

The nodes are able to measure the phase shift of the incoming signal. In the ranging process, the nodes requires to perform at least two phase measurements at different frequencies. Then, the node estimates the difference between the phase measurements and with this information is able to calculate the distance between the nodes.

In order to explain the procedure, assume that two sensor nodes (A and B) are separated a certain distance d and the sensor node B receives two signals from node A at different frequencies (f_1 and f_2). Naturally, the phase measurements of both signals will be different, let say φ_1 and φ_2. The difference between these two POA measurements ($\Delta\varphi$) can be considered as:

$$\varphi_2 - \varphi_1 = 2\pi\left(\frac{d}{\lambda_2} - \left\lfloor\frac{d}{\lambda_2}\right\rfloor\right) - 2\pi\left(\frac{d}{\lambda_1} - \left\lfloor\frac{d}{\lambda_1}\right\rfloor\right) \tag{1}$$

Here, λ_1 and λ_2 are the wavelength of the signals. Note that $\left\lfloor\frac{d}{\lambda}\right\rfloor$ represents the integer number of wavelengths that the signal performs to arrive at the receiver. If f_1 and f_2 are quite similar, we can assume that the integer number of wavelengths of both signals is the same. In this way, it is possible to reduce the expression of $\Delta\varphi$ as follows:

$$\varphi_2 - \varphi_1 = 2\pi \left(\frac{d}{\lambda_2} - \frac{d}{\lambda_1} \right) \tag{2}$$

Considering that λ is defined as the propagation speed c divided by the frequency of the signal, the distance between both nodes can be approximated using the following equation:

$$d = \frac{c \cdot \Delta\varphi}{2\pi(f_2 - f_1)} \tag{3}$$

In order to improve the accuracy, our sensor nodes transmit at more than two frequencies to increase the number of available phase measurements. Thus, by averaging the available measurements, it is possible to obtain a better distance determination. For this, the sensor node transmits signals at fixed intervals of frequency defined by the variable Δf (frequency step). The first signal is transmitted at the frequency f_{start} and the last signal at f_{stop}. Δf is defined usually between 0.5Mhz and 4Mhz. In our experiments f_{start}=2403Mhz, f_{stop}=2443Mhz and Δf was set to 2Mhz.

Note, that the phase shift of the signal can be measured correctly if both transmitter and receiver are perfectly synchronized. Our sensor nodes use a method called "Active Reflector Principle" [10], which avoids this problem. The idea behind this method is to perform measurements in both directions to cancel the clock shift between both nodes.

Figure 1 shows one of our sensor nodes. The module REB233SMAD has two antennas which enables to perform 4 different POA measurements with another node. The possible combinations are node_A–antenna_1 with node_B–antenna_1, node_A–antenna_2 with node_B–antenna_1, node_A–antenna_1 with node_B–antenna_2 and node_A–antenna_2 with node_B–antenna_2. The idea of performing measurements with several antennas is to minimize the effects of multipath receptions in the distance determination process. Note that the multipath effect is different for each combination of antenna.

The firmware used in our sensor nodes obtained from Atmel, uses the following equations to combine the POA-based distance measurements from the 4 combinations of antennas and thus, obtain a final distance estimation d_f:

$$b = \frac{100cm^2}{100cm^2 + \sigma^2} \tag{4}$$

$$d_f = b \cdot d_{avg} + (1 - b) \cdot d_{min} \tag{5}$$

Here, σ and d_{avg} are the standard deviation (cm) and the average value (cm) of the 4 measurements, respectively. d_{min} is the minimum distance value obtained from the 4 combinations.

If there is a big difference between the four distance measurements (σ is large), it is assumed that the multipath effect can be strong producing positive errors in the measurements. Therefore, in this case d_{min}, which probably has a smaller error than the other values, has more importance in the estimation of the final distance d_f (see equation 5).

The internal firmware also includes different filters to avoid an erroneous estimation. Furthermore, a Distance Quality Factor (DQF) is also provided for each distance measurement. As its name suggests, this factor is correlated with the accuracy of the measurement. A low DQF value indicates that the accuracy of the performed distance estimation is poor. The DQF takes values between 0 and 100.

2.2   Distance Estimation Using RSSI

Our reference sensor node contains a register called Energy Detection (ED), which stores the strength of the incoming signal. This ED value varies from 0dBm to 83dBm. In this paper we consider the ED value as the RSSI measurement provided by the transceiver. The RF input power P_{RF} can be approximated using the expression $P_{RF} = ED - 91dBm$, which is obtained from the datasheet of the transceiver [2].

Equation 6 is used to determine the distance d between transmitter und receiver, where d_0 is a reference distance, and P_0 is the input signal power at the reference distance d_0. The parameter n is called path loss exponent, whose value is depending on the environment (normally between 2 and 4). The parameter ED_{avg} is calculated averaging the ED values referred to each antenna combination.

$$d = d_0 \cdot 10^{\frac{P_0 - P_{RF}(d)}{10 \cdot n}} = d_0 \cdot 10^{\frac{P_0 + 91 - ED_{avg}}{10 \cdot n}} \tag{6}$$

3   Evaluation of the Distance Estimation Methods

For investigating the accuracy of the described distance determination methods, we performed measurements in a park (outdoor scenario) and in the corridor of a building (indoor scenario). The height of both nodes was set to 1m and there is no obstacle between the nodes (Line-of-Sight condition).

Each distance measurement was obtained by using the measurements of the four combinations of antennas. In the POA method, equation 5 was considered to estimate the final distance. In case of RSSI, we obtained a signal strength value for each combination of antenna, Thus, we used the RSSI average value and equation 6 for the distance approximation. The calibration parameters used in the path-loss model were: $P_0 = -43dBm$, $d_0 = 1m$ and $n = 2$ [18] .

At each separation of nodes more than 300 distance measurements were carried out. The results are shown in Fig. 2 and Fig. 3. The distance error is defined as the difference between the real distance and the estimated distance. The average distance errors together with the Student-t confidence intervals (95%) are depicted in the figures.

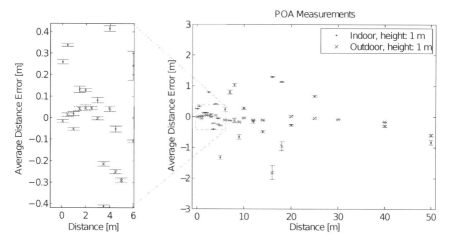

Fig. 2. Average distance error by using the POA method

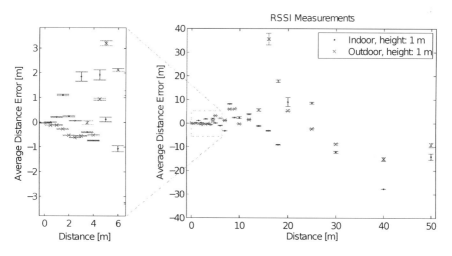

Fig. 3. Average distance error by using the RSSI method

In general, the main problem is the multipath effect, which has a negative influence on both methods. As expected, the effect is stronger in indoor environment, which leads to a worse accuracy in comparison to the outdoor scenario. In case of RSSI, the distance error kept low when the distance was shorter than 5m. From this separation, very large errors are registered. Therefore, the RSSI method should not be used in these cases.

Although the POA method requires the tranmission of several packets between both nodes, this method improves the accuracy of RSSI considerably. In all our POA-based distance measurements, the absolute average distance error was smaller than 2m. When RSSI was used, average distance errors larger than 10m were registered.

4   Evaluation of Localization Algorithms

In order to evaluate the position accuracy of *Multilateration*, *WCL* and *E-Min-Max*, a localization system was developed on a sensor network. It consists of a MN and 4 ANs which are wireslessly connected to a sensor node called "coordinator". The principal task of the coordinator is to collect the distance measurements between the MN and the ANs, and send this information to a central computer for the MN's position estimation.

At the central computer, three low-complexity 3D localization algorithms are executed. A detailed descriptions of the well-known algorithms *Multilateration* and *Weighted-Centroid Localization* can be found in [19] and [20].

The algorithm *E-Min-Max* is a improved version of the algorithm *Min-Max* presented in [19]. Using the information of the distances between the MN and the ANs, the original 2D Min-Max algorithm draws bounding squares around each AN. Then, a "definition zone" (a rectangle) is obtained by calculating logical operations (maximum and minimum value) with the coordinates of the bounding squares of the ANs. In general, the definition zone is the overlapping area of the bounding squares. The MN's position estimation is taken as the center of the definition zone.

E-Min-Max allocates weights to each corner of the definition zone and a weighted average is calculated with the coordinates of the corners and their weights. In this way, the final position estimation can be found in other places within the definition zone and not necessary at the center. A detailed description of *E-Min-Max* can be found in [21].

We used a 3D version of *E-Min-Max*. The principle of operation is the same. In a 3D scenario, bounding cubes are built around the ANs and the definition zone is a cuboid. Thus, eight weights are calculated for the corners of the definition zone. The final MN's position is considered as the weighted average value of the coordinates of the corners.

Fig. 4. Indoor scenario

We took measurements in an indoor scenario, which is a room inside a building of our university (Fig. 4). This room has many computers and tables, which produce signal reflections. In this scenario there are measurements under Line of Sight (LOS) and Non-Line of Sight (NLOS) conditions. The measurement campaign was also repeated in an outdoor scenario (LOS condition), which is a park in Dresden (see Fig. 5). Note, that here there is also signal reflections because of the ground.

Fig. 5. Outdoor scenario

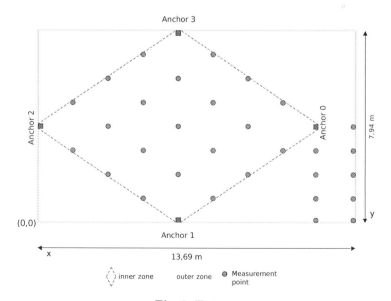

Fig. 6. Test area

Figure 6 shows the positions of the ANs and the measurement points. The test area is 7.94m wide, 13,69m long and 3.55m high. The MN was located at different points to perform measurements with the ANs. The height of the MN was set to 1.5 m. The ANs 0 and 2 were placed at a height of 2m. The heights of the ANs 1 and 3 were set to 1m.

We define an inner zone, which is the convex hull [21] created by the ANs. Here, 25 measurement points were considered. Outside the convex hull (outer zone), 10 measurement points were created. In each of these measurement points, more than 200 measurements were performed and averaged.

4.1 Protocol

A communication protocol was developed to send the distance information to the central computer.

This protocol consists of four phases, which are executed cyclically. In the "discovery" phase, i.e. phase 1, the MN broadcasts a short HELLO packet. In phase 2, the ANs that have received the packet respond within a random period (in order to minimize collisions). The MN then knows how many and which ANs are available. The ranging measurements are implemented in phase 3. In this phase, distances between the MN and each one of the anchors that have responded are sequentially estimated. Finally, in phase 4, the results are stored by the MN and sent to a fixed node of the network. If the fixed node is the coordinator, it sends this information to the central computer directly. On the contrary, if the fixed node is an AN, a multihop communication is required to deliver this information to the coordinator and the central computer. For supporting the communication between fixed nodes, we implemented the routing algorithm described in [22].

4.2 Performance Evaluation of the Position Error

Figure 7 shows the average position errors and the Student-t confidence intervals (95%) in the investigated scenarios. The position error is considered as the Euclidean distance between the real position and the estimated position. We used equation 5 and the measurements of the 4 combinations of antennas to estimate the final distance. For calculating the distance with the RSSI method, the following calibration parameter were used in this experiment: $P_0 = -57dBm$, $d_0 = 1m$ and $n = 2$.

We filtered the obtained distance values according to the DQF and the ED values. Specifically, we removed all ranging measurements, whose DQF values were smaller than 10. We also did not consider all RSSI measurements when the average ED value of the 4 combinations of antennas was smaller than 10dBm.

In general, the localization algorithms have a better performance in the inner zone compared to the outer zone. As expected, they achieve a higher accuracy level with the POA method compared to the RSSI method.

In the outdoor scenario, the distance measurements are in general more accurate that those taken in the indoor scenario.

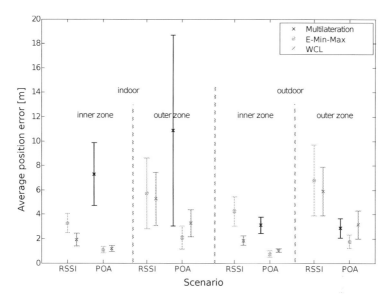

Fig. 7. Average position errors in different scenarios. The results of *Multilateration* using RSSI are out of scale and do not appear in the figure.

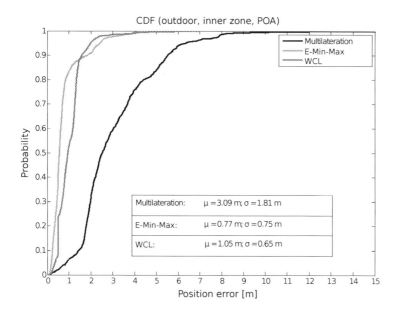

Fig. 8. Cumulative Distribution Function of the position error (outdoor, inner zone, POA method)

In many cases the position error level of *Multilateration* is so high that it offers no useful position information. Note, that the results of *Multilateration* using RSSI are out of scale and do not appear in the figure. Previous works reported about the poor robustness of *Multilateration* against large distance errors [21]. Furthermore, the relative position of the ANs plays an important role in this algorithm. If the ANs does not have an appropriate location, it means they are almost coplanar, the flip-flop problem [23] can occur degrading the position accuracy.

The position accuracy achieved by *E-Min-Max* is a little better than *WCL* when the POA method is used. The *E-Min-Max* algorithm achieves an average position error smaller than 1m in the inner zone of both scenarios. Fig. 8 shows the cumulative distribution function of the measurements of the inner zone in the outdoor scenario. Observe that it is possible to obtain a position error smaller than 1m in approx. 85% of the cases by using *E-Min-Max* .

If the RSSI method is used, *WCL* seems to be the best option. In the inner zone we can obtain an average position of about 2m, which is quite good taking into consideration, that the RSSI measurements are very dispersive.

5   Conclusion

In this paper, the distance determination methods based on RSSI and POA are evaluated by taking measurements. In the RSSI method, a calibrated path-loss model is used to determine the distance. Our results suggest that when the separation of the nodes is shorter than 5-6 meters, the distance errors are smaller than 1m in most cases. For larger separations between nodes, the distance error can achieve considerable errors (larger than 10m) and therefore the RSSI value should not be used as a distance determination method in such conditions.

The results related to the distance errors using the POA method are quite good compared with RSSI. In most cases (specially in the outdoor scenario) the distance errors were smaller than 50cm.

Additionally, a localization system was developed to evaluate the accuracy of low-complexity localization algorithms using the POA and RSSI method. For this, we created a sensor network to take distance measurements between the MN and fixed nodes. This data is sent to a central computer for the MN's position estimation. We carried out measurements in both indoor and outdoor scenarios.

In all cases, the performance of *Multilateration* was worst. This can be due to the relative positions of the ANs and the fact that this algorithm is very sensitive to large distance errors.

In our experiments, the position accuracy achieved in the inner zone (the convex hull defined by the ANs) was better than in the outer zone. If the RSSI method is used, *WCL* can achieve an average position error of about 2m inside the inner zone, which is quite acceptable taking into consideration that in many cases the RSSI measurements produce erroneous distance estimations.

The *E-Min-Max* algorithm is lightly better than *WCL* when the POA method is used for determining the distances. In the inner zone, the accuracy of both algorithms are quite similar. Although, the difference between the average position errors of both algorithms is about 1m in the outer zone. With E-Min-Max and the POA method, it is possible to achieve average position errors smaller than 1m in the investigated scenarios.

Our future work will be oriented to take more measurements in other scenarios and create a distance error model of the POA method.

Acknowledgment. The authors would like to thank to the companies Dresden Elektronik and Atmel for providing the sensor nodes used in the measurements. Specially, we thank Mike Ludwig for his feedbacks and suggestions. The authors would like to express their gratitude to Robert Baumbach and the Chair of Transport Systems Information Technology of the Technische Universität Dresden for close cooperation. This work is part of the project Cool Energy Car Communication (CECC) within the Leading-Edge Cluster Cool Silicon, which is sponsored by the German Federal Ministry of Education and Research (BMBF).

References

1. Atmel Corporation: Atmel AVR2162: REB233SMAD Hardware User Manual, Application note (2013), http://www.atmel.com/Images/doc42006.pdf
2. Atmel Corporation: Transceiver AT86RF233, Datasheet (2013), http://www.atmel.com/Images/Atmel-8351-MCU_Wireless-AT86RF233_Datasheet.pdf
3. Santos, R.A., Edwards, R.M., Seed, L.N., Edwards, A.: A location-based routing algorithm for vehicle to vehicle communication. In: Proceedings of 13th International Conference on Computer Communications and Networks, ICCCN 2004, Chicago USA, October 11-13 (2004)
4. Yun, Y.U., Choi, J.K., Hao, N., Yoo, S.J.: Location-Based Spiral Clustering for Transmission Scheduling in Wireless SensorNetworks. In: 2010 The 12th International Conference on Advanced Communication Technology (ICACT), Gangwon-Do, Korea, February 7-10 (2010)
5. Zhao, J.H., Wang, Y.C.: PosPush: A Highly Accurate Location-Based Information Delivery System, Mobile Ubiquitous Computing, Systems, Services and Technologies. In: Third International Conference on UBICOMM 2009, Sliema, Malta, October 11-16 (2009)
6. Lee, J.K., Lee, S., Kim, H.H., Lee, K.C.: Estimation of Metabolic Rate for Location-based Human Adaptive Air-conditioner in Smart Home. In: 2012 IEEE 1st Global Conference on Consumer Electronics (GCCE), Tokyo, Japan, October 2-5 (2012)
7. Lee, W., Patanavijit, V.: Performance Evaluation of TOA Estimation for Ultra-wideband System under AWGN and IEEE 802.13a Channel Model. In: 2011 Seventh International Conference Signal-Image Technology and Internet-Based Systems (SITIS), Dijon - France, (November 28, 2011)
8. Kaune, R.: Accuracy studies for TDOA and TOA localization. In: 2012 15th International Conference on Information Fusion (FUSION), Singapore, July 9-12 (2012)
9. Nanotron Technologies GmbH: Datasheet transceiver nanoLOC, http://www.nanotron.com/EN/pdf/Factsheet_nanoLOC-NA5TR1.pdf
10. Nanotron Technologies GmbH: Real Time Location Systems (RTLS), White Paper (2007)

11. Daiya, V., Ebenezer, J., Murty, S.A.V.S., Raj, B.: Experimental analysis of RSSI for distance and position estimation. In: 2011 International Conference on Recent Trends in Information Technology (ICRTIT), Chennai, India, June 3-5 (2011)
12. Wu, R.H., Lee, Y.H., Tseng, H.W., Jan, Y.G., Chuang, M.H.: Study of characteristics of RSSI signal, Industrial Technology. In: IEEE International Conference on ICIT 2008, Chengdu, China (2008)
13. Robles, J.J.: Considerations in the Design of Indoor Localization Systems for Wireless Sensor Networks. In: Lehnert, R. (ed.) EUNICE 2011. LNCS, vol. 6955, pp. 43–53. Springer, Heidelberg (2011)
14. Poujouly, S., Journet, B., Miller, D.: Laser range finder based on fully digital phase-shift measurement. In: Proceedings of the 16th IEEE Instrumentation and Measurement Technology Conference on IMTC 1999, Venice, Italy, vol. 3, pp. 1773–1776 (1999)
15. Gorostiza, E., Lazaro Galilea, L., Meca Meca, F., Salido Monzu, D., Espinosa Zapata, F., Pallares Puerto, L.: Infrared sensor system for mobile-robot positioning in intelligent spaces. Sensors 11(5), 5416–5438 (2011)
16. Wehner, M., Ricter, R., Zeisberg, S., Michler, O.: High resolution approach for phase based TOF ranging using compressive sampling. In: 2011 8th Workshop on Positioning Navigation and Communication (WPNC), Dresden, Germany, pp. S.28–S.32 (2011)
17. Atmel Corporation: Atmel AVR2042: REB Controller Base Board - Hardware User Manual, Application note (2013), http://www.atmel.com/Images/doc8334.pdf
18. Birkenmaier, J.: Development of an indoor localization system using sensor nodes. Diploma Thesis, Chair for Telecommunications, Supervisors: Jorge Juan Robles and Mike Ludwig, Dresden, Germany (2012)
19. Langendoen, K., Reijers, N.: Distributed Localization in Wireless Sensor Networks: a Quantitative Comparison. The International Journal of Computer and Telecommunications Networking - Special Issue: Wireless Sensor Networks Archive 43(4) (November 15, 2003)
20. Blumenthal, J., Reichenbach, F., Timmermann, D.: Position estimation in ad hoc wireless sensor networks with low complexity. In: Joint 2nd Workshop on Positioning, Navigation and Communication, Hannover, Germany (2005)
21. Robles, J., Supervía, J., Lehnert, R.: Extended Min-Max Algorithm for Distributed Position Estimation in Sensor Networks. In: 9th Workshop on Positioning, Navigation and Communication, Dresden, Germany (2012)
22. Robles, J., Muñoz, E.G., de la Cuesta, L., Lehnert, R.: Performance evaluation of an indoor localization protocol in a 802.15. 4 sensor network. In: 2012 International Conference on Indoor Positioning and Indoor Navigation (IPIN), Sydney, Australia (November 2012)
23. Moravek, P., Komosny, D., Simek, M.: Multilateration and Flip Ambiguity Mitigation in Ad-hoc Networks. Przegld Elektrotechniczny Selected Full Texts 88(5b), 222–229 (2012)

On the Attack-and-Fault Tolerance of Intrusion Detection Systems in Wireless Mesh Networks

Amin Hassanzadeh and Radu Stoleru

Department of Computer Science and Engineering
Texas A&M University, USA
{hassanzadeh,stoleru}@cse.tamu.edu

Abstract. Intrusion detection in Wireless Mesh Networks (WMN) has recently emerged as an important research area. The diversity in WMN hardware and applications has generated extremely diverse network types, with diverse resource levels and system and threat models. Consequently, a variety of intrusion detection systems (IDS) have been proposed by the research community, each applicable to a specific type of WMN. Although the design and implementation of specific intrusion detection mechanisms have received considerable attention, little effort has been dedicated to the attack-and-fault tolerance of IDS themselves. In this paper we propose a taxonomy that categorizes state-of-the-art IDS solutions in WMN and we investigate the attack-and-fault tolerance of IDS in this taxonomy. We first survey a series of administrative mechanisms for attack-and-fault tolerant (AFT) IDS design. Then we propose modified designs for state-of-the-art IDS solutions that are AFT. Finally, through extensive simulations, we evaluate and compare AFT designed IDS with their original designs, with respect to the IDS performance and costs.

1 Introduction

The problem of intrusion detection in wireless mesh networks (WMN), is challenging due to the limited resources (*i.e., energy, storage, and processing power*) available to participating nodes. Recently, the problem has received some attention from the research community and some solutions have been proposed [1–12]. In general, the proposed solutions can be categorized in two classes: 1) *detection engine design:* which encompasses all intrusion detection solutions that propose a detection mechanism/rule for one or few specific attacks in a particular WMN application [11,12]; and 2) *optimal monitoring mechanism:* in which, regardless of the detection engine and type of attack, the main objective is to find the optimal monitoring mechanism (e.g., IDS node placement or IDS rule assignment) that is practical based on WMN characteristics [3,4,10]. This research focuses on the second class of intrusion detection systems (IDS).

Due to the lack of concentration points in WMN where network traffic can be analyzed, research community has proposed decentralized monitoring mechanisms [7–10] for intrusion detection in WMN. A node assigned with monitoring

S. Guo et al. (Eds.): ADHOC-NOW 2014, LNCS 8487, pp. 387–401, 2014.

Table 1. Taxonomy for Traffic and Resource Aware Intrusion Detection in WMN

		Hardware Resources	
		Resourceless	Resourceful
Traffic Awareness	Traffic Agnostic	RAPID [1]	EEMON [8]
	Traffic Aware	PRIDE [3]	TRAIN [2]

task for intrusion detection is called *monitoring node* or *IDS node* (henceforth used interchangeably). The diversity in WMN hardware has resulted in *different resource levels* being allocated to monitoring tasks. Therefore, a monitoring node in WMN might be able to perform only few IDS functions (i.e., *resourceless IDS node* [3]) or a complete set of IDS functions (i.e., *resourceful IDS node* [8]). Consequently, based on the resource level in WMN, optimal monitoring mechanisms proposed by research community are either *resourceful IDS solutions* or *resourceless IDS solutions* [2].

The problem of optimal monitoring in *resourceful* WMN has been typically solved by selecting a few *monitoring nodes* that execute a complete set of IDS functions [2,8,10], and are able to make the intrusion detection decision (i.e., usually referred to as non-cooperative solutions). However, in *resourceless* WMN, resource-constrained IDS nodes are assigned few IDS functions, which results in detecting only few attacks. These nodes, in some solutions [4,6,7] exchange information for *cooperative* intrusion detection (i.e., to achieve higher detection rates) at the price of incurring high detection delay and communication overhead. *Non-cooperative* IDS were recently proposed as practical solutions for WMN. These solutions can be categorized in two classes: 1) *traffic-aware IDS:* which use the knowledge that a security administrator has about network traffic to distribute IDS functions only along routing paths [3]. Considering the distinct set of IDS functions on each node along a routing path, the entire traffic on that path is investigated by more IDS functions while none of the nodes is overloaded; and 2) *traffic-agnostic IDS:* which are based on link-coverage approach which means each node investigates the entire network traffic on the set of communication links it can monitor [1]. Hence, irrespective to the changes in WMN traffic paths, IDS nodes monitor the entire WMN traffic. In this research, we concentrate on four different non-cooperative IDS solutions in WMN (each solution is representative for a particular class of IDS) as presented in Table 1.

Although intrusion detection mechanism in WMN have received considerable focus, little attention has been paid to attacks-and-failures against/of IDS nodes. Undoubtedly, when an IDS node is compromised or faulty, it is unable to participate in intrusion detection process, thus, the intrusion detection rate will decrease and some malicious activities will remain undetected (i.e., high false negative rates). This research, inspired by similar efforts in other computer networking areas [13–17], investigates the attack-and-fault tolerance of state-of-the-art IDS solutions presented in Table 1. In order to develop AFT mechanisms for the proposed IDS solutions, we first survey related works proposed for attack or fault tolerance in all networking areas (e.g., wired, ad hoc and sensor networks). Some of the proposed solutions use redundant/backup nodes [16] to increase the network/service availability after node compromise/failure while others concentrate on camouflaging mechanisms [13, 15] to make monitoring/IDS nodes localization [14] very hard for the attacker. Furthermore, few other solutions propose fast and efficient fault detection mechanisms to detect compromised or faulty nodes [17] and recover the network from that situation [15, 16].

This research first proposes a classification for all AFT mechanisms and then concentrates on *preventive* solutions that use redundant IDS nodes to maintain high IDS availability ratio after IDS compromise/failure times. We will show that these mechanisms, at the price of higher resource consumption, increase the attack/fault tolerance level by: 1) increasing IDS availability; 2) reducing IDS compromise/failure detection time; and 3) eliminating the need for recovery actions (i.e., adopting backup nodes) [15, 16]. Taking into consideration the optimal monitoring mechanism employed by each state-of-the-art IDS solutions, we *reformulate* the optimal monitoring problem for intrusion detection in each class of IDS such that the solution becomes an AFT IDS. Through extensive simulations, the performance (e.g., intrusion detection rate) and efficiency (e.g., resource consumption) of *redesigned* and *preventive* monitoring mechanisms proposed for each class of IDS are evaluated and compared to those of the original solutions. The results show how AFT design trades off attack-and-fault tolerance levels for the amount of resources consumed by intrusion detection systems.

2   State-of-the-Art IDS Solutions

Traffic Aware and Resourceless (TW-RL): This class of IDS solutions focuses on mesh networks consisting of resource-constrained devices in which the security administrator has some knowledge about traffic paths. Since it is infeasible to perform full IDS on resource-constrained WMN nodes, a practical solution is to efficiently distribute IDS functions on the nodes along traffic paths. In such an approach, a security administrator, knowing traffic routes in the WMN, employs a traffic-aware framework that optimally places IDS functions on the nodes along routing paths. For example, it is observed [18] that when deploying WMN for disaster response, the traffic paths are always known. Furthermore, the idea of interference-load aware routing metric [19] proposed in mesh networks provides a traffic-aware framework for security administrators. PRIDE [3] is a TW-RL IDS proposed for wireless mesh networks that uses routing algorithms and monitoring tools to obtain routing information (i.e., busiest and most frequently used paths) in WMN.

Traffic Agnostic and Resourceless (TG-RL): This class of IDS solutions also concentrates on resource-constrained WMN, however, it aims at distributing IDS functions to the nodes such that they are able to monitor the entire WMN traffic irrespective to traffic path changes. RAPID [1] recently proposed a link-coverage approach for intrusion detection in TG-RL WMN. In the RAPID protocol, each node, depending on its memory resources, is assigned a subset of IDS functions to investigate the entire network traffic on the set of communication links it can monitor (i.e., its coverage area). Performing few IDS functions allows resource conservation on resourceless nodes and also increases the probability of monitoring WMN links with multiple distinct IDS functions activated on all of the nodes that can monitor the links. It is worth mentioning that for a given network size, the complexity of RG-RL is larger than that of TW-RL, as it needs to find optimal IDS function distribution for all WMN nodes, not only those located on routing paths.

Traffic Agnostic and Resourceful (TG-RF): These solutions assume that all or some of the WMN nodes are resourceful nodes [8, 10] that can perform a complete set of IDS functions to monitor the network traffic on the set of communication links in their coverage areas. TG-RF solutions assign the same set of IDS functions to a subset of MWN nodes, called monitoring nodes, where each monitoring node is responsible for a distinct part of the network. EEMON [8] is a TG-RF IDS that proposes an energy-efficient monitoring mechanism for battery-powered WMN.

Traffic Aware and Resourceful (TW-RF): This class of IDS we consider, with TRAIN [2] as a solution, focuses on resourceful WMN where the security administrator has some knowledge about traffic paths. In fact, when comparing this class with TG-RF IDS, TW-RL solutions use resourceful monitoring nodes to only monitor a subset of communication links, i.e., those located on traffic paths. Therefore, the complexity of optimal monitoring problem for intrusion detection is less than that of TG-RL and than the other two classes. This is because not only does it benefit from resourceful monitoring nodes being able to perform full IDS, but also concentrates on monitoring only few WMN links.

3   AFT Mechanisms Diagram

The IDS mechanism presented in Table 1 are not AFT (except for some special cases of the RAPID protocol, which we will explain in next sections). Therefore, if an IDS node fails (e.g., runs out of memory and crashes or its battery dies) or become compromised, part of the network will remain uncovered. This means that some WMN nodes/links become vulnerable against network attacks and that false negative rates will increase. Inspired by research in AFT design [13–17] we propose a classification which, to the best of our knowledge, is the first for AFT intrusion detection. Our proposed classification is based on the time of the action taken for AFT purposes. As shown in Figure 1, the actions are either taken before IDS attack or fault time (i.e., resulting in IDS compromise/failure) or after that.

Prevention Phase: As shown in Figure 1, prevention phase refers to the time while the IDS compromise/failure has not occurred yet. For example, a *preventive* AFT mechanism [15] may aim at increasing the risk of IDS node attack for the attacker (e.g., by using redundant monitoring node per link) or reducing the chance of node failure (e.g., by using high capacity storage or energy sources). Therefore, preventive solutions pay the AFT prices (i.e., redundant resources) at the design and implementation phase so that the IDS availability (detection

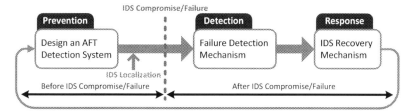

Fig. 1. A multi-phase process for designing an AFT IDS mechanism

coverage ratio) will not be affected after IDS compromise/failure. *This research focuses on preventive mechanisms and evaluates the performance of preventive AFT designs and their costs.* It is important to mention that there exists solutions focusing on IDS node camouflaging, so that IDS localization (as shown in Figure 1) becomes very hard for the attacker [13, 14].

Detection Phase: If preventive mechanisms are not used, the monitoring system must be able to detect IDS compromise/failure immediately, so the security administrator can recover the IDS mechanism quickly. The time between IDS compromise/failure and its detection by security administrator is called *detection time*. A fast and accurate detection mechanism can remarkably reduce the detection time and increase the IDS availability time. Detection mechanisms [16, 17] can be either proactive or reactive. It is worth mentioning that a preventive AFT mechanism that uses redundant monitoring nodes is already a real-time detection system since every IDS node is monitored by at least another IDS node.

Response Phase: When the IDS compromise/failure occurs and it is detected, an appropriate action is to recover the node(s) from the compromise/fault. The time between the IDS compromise/failure detection and its recovery is called *response time*. An optimal recovery mechanism minimizes the response time [15, 16]. Recovery mechanism and response time usually depend on the network topology, application, and IDS solution used in the network. We note here that although preventive solutions do not need detection and response mechanisms, it is very beneficial to consider these two mechanisms particularly for highly vulnerable WMN. This is because a preventive mechanism ultimately becomes non-preventive after a few IDS node compromises/failures.

4   AFT-Design for WMN IDS

We model a mesh network as a graph $G = (V, E)$, in which V is the set of WMN nodes $\{v_1, v_2, \cdots, v_N\}$, and $E = \{e_1, e_2, \cdots, e_Q\}$ is the set of links between them. For the traffic-aware solutions, we denote the number of nodes and links located on traffic routes by n $(n \leq N)$ and q $(q \leq Q)$, respectively. Therefore, the reduced graph $G' = \{V', E'\}$ represents the set of active nodes and links in traffic-aware WMN, where V' is the set of n active nodes $(V' \subseteq V)$, and E' is the set of q active links $(E' \subseteq E)$. The set of selected monitoring (IDS) nodes in the resourceful classes are denoted by $M = \{m_j \mid m_j \text{ is a monitoring node}\}$. We also denote the set of routing paths for the network traffic by $P = \{p_1, p_2, \cdots, p_l\}$, where $P_i^v = \{v_j \mid v_j$ is located on $p_i\}$ and $P_i^v \subseteq V'$. We denote by matrix $\mathbb{G}_{Q \times N}$ the mapping between nodes and links, i.e., $g_{hj} = 1$ iff node v_j can monitor link e_h. We also denote by matrix $\mathbb{T}_{l \times n}$ the mapping between nodes and paths, i.e., $t_{ij} = 1$ iff node j is located on path i.

We denote the residual energy and the communication load of a WMN node by b_j and c_j, respectively. Based on the maximum residual charge and communication load a node can have, both b_j and c_j are considered normalized values in range $[0, 1]$. Let $w : V \longrightarrow [0, 1]$ be a cost function that assigns a weight w_j to a node v_j based on c_j and b_j ($w_i = w(c_j, b_j) = 1/(c_j \times b_j)$), such that higher normalized c_j and b_j values result in lower weight being assigned to v_j. We also denote the set of IDS functions by $\mathcal{F} = \{f_r \mid f_r$ is a set of detection rules$\}$ with size R (i.e., $|\mathcal{F}| = R$). Let $w^f : \{\mathcal{F}\} \longrightarrow [0, 1]$ be a cost function that assigns memory load w_r^f to IDS function f_r. Consequently, vector $W^f = [w_1^f, w_2^f, \cdots, w_R^f]$ represents the amount of memory load each function in $\mathcal{F}$ imposes to the IDS node when activated on that IDS node. We use matrix $\mathbb{X}$ to show whether node v_j performs IDS function f_r (i.e., $x_{jr} = 1$) or not. Finally, vectors $\boldsymbol{\beta} = [\beta_1, \beta_2, \cdots, \beta_N]$ (*i.e., Battery Threshold*) and $\Lambda = [\lambda_1, \lambda_2, \cdots, \lambda_N]$ (*i.e., Memory Threshold*) represent the minimum energy charge required for being selected as monitor and maximum allowable memory load by IDS functions, respectively.

4.1   Resourceful IDS

EEMON [8] aims at covering all communication links while TRAIN [2] aims at covering all traffic paths, both with minimum average cost per monitoring nodes. Let S_h ($S_h \subseteq M$) be the set of selected monitoring nodes out of all possible nodes that can monitor *link* e_h, and similarly S_i' be the set of selected monitoring nodes out of all possible nodes that can monitor *path* p_i.

Therefore, the optimal monitoring problem in a battery-powered resourceful WMN (both EEMON and TRAIN) can be formulated as an integer linear program (ILP), where Constraint (2)

$$\text{minimize} \quad \sum_{v_j \in V} w_j m_j \tag{1}$$

$$\text{subject to:} \quad |S_h| \geq 1 (\text{EEMON}) \quad , \forall e_h \in E \tag{2}$$
$$|S_i'| \geq 1 \ (\text{TRAIN}) \quad , \forall p_i \in P$$
$$b_j \geq \beta_j \ (or \ b_{th}) \quad , \forall m_j \in M \tag{3}$$
$$m_j \in \{0, 1\} \tag{4}$$

indicates that every *link/path* must be covered; Constraint (3) enforces the algorithm to select the nodes with residual energy greater than a threshold. Constraint (4) means a node is either selected as a monitoring node or not.

AFT Resourceful IDS: we define δ-AFT design as an AFT IDS mechanism in which each node is monitored by $\delta + 1$ monitoring node(s) and the intrusion detection monitoring mechanism can tolerate at most δ IDS compromise/failures per link/path. Hence, in EEMON and TRAIN optimal monitoring formulations, δ-AFT design is achieved by modifying constraint (2) to $|S_h| \geq \delta$ for EEMON and $|S_i'| \geq \delta$ for TRAIN. It is worth mentioning that δ is bounded by maximum number of monitoring nodes that can potentially monitor a link/path, which is a function of network density.

4.2 Resourceless IDS

The main objective of resourceless IDS solutions is to monitor all links/paths with the maximum allowable number of IDS functions that can be performed on WMN nodes. A higher number of detection modules[1] executed on node v_j means more attack traffic can be detected on the links/paths monitored by that node. Hence, the optimal monitoring problem in resourceless WMN is formulated as the following ILP (for both PRIDE [3] and RAPID [1]):

$$\text{maximize} \quad (1/l)(\mathbf{1}^T \cdot \mathbb{T})(\mathbb{X} \cdot \mathbf{1}) \qquad \text{(PRIDE)} \qquad (5)$$
$$(1/q)(\mathbf{1}^T \cdot \mathbb{G})(\mathbb{X} \cdot \mathbf{1}) \qquad \text{(RAPID)}$$

$$\text{subject to:} \quad \mathbb{X} \cdot W^{f^T} \leq \Lambda^T \qquad\qquad\qquad\qquad (6)$$
$$(\mathbb{T} \cdot \mathbb{X})_{ir} \leq 1 \qquad \text{(PRIDE)} \qquad\qquad\qquad ,\forall i, r \quad (7)$$
$$(\mathbb{G} \cdot \mathbb{X})_{hr} \leq 1 \qquad \text{(RAPID)} \qquad\qquad\qquad ,\forall h, r$$
$$x_{jr} \in \{0,1\} \qquad \text{(PRIDE)} \qquad ,\forall v_j \in V', \forall f_r \in \mathcal{F}_j \quad (8)$$
$$x_{jr} \in \{0,1\} \qquad \text{(RAPID)} \qquad ,\forall v_j \in V, \forall f_r \in \mathcal{F}_j$$

where Constraint 6 limits the IDS memory load on every node v_j to be less than its memory threshold λ_j. Constraint 7 ensures that only one copy of each function is assigned to the nodes for each link/path. Finally, Constraint 8 means a node either performs an IDS function or not.

AFT Resourceless IDS: This class of IDS may not be able to achieve 100% link/path coverage (i.e., every link/path is monitored by all R IDS functions) due to memory constraint Λ. Suppose $\lambda_j = \lambda \, \forall v_j$, the smaller the λ is, the lower the link/path coverage will be. Therefore, if the memory threshold is very low that does not allow us to achieve 100% coverage, our IDS is always 0-AFT. When the memory threshold λ increases, it is most likely possible to achieve δ-AFT design for $\delta > 0$ in resourceless IDS. Hence, in PRIDE and RAPID, achieving higher link/path coverage rate is more important than achieving δ-AFT design.

In order to achieve δ-AFT design in this class of IDS, we have to remove Constraint 7 to ensure that more than one IDS function can be assigned to a link/path. In this case, since redundant IDS functions do not count for coverage ratio (ILP objective) [1], we need to modify the ILP objective function so that it accurately measures the link/path coverage ratio. Thus, we define function $BN :$ $\{\mathbb{Y}\} \longrightarrow \{0,1\}$ that converts y_{ij} to a binary value, i.e., if $y_{ij} = 0$, $BN(y_{ij}) = 0$, otherwise $BN(y_{ij}) = 1$. We reformulate the optimal monitoring problem for δ-AFT design of resourceless IDS classes as follows:

$$\text{maximize} \quad (1/q)(\mathbf{1}^T \cdot BN(\mathbb{T} \cdot \mathbb{X}) \cdot \mathbf{1}) \qquad \text{(PRIDE)} \qquad (9)$$
$$(1/q)(\mathbf{1}^T \cdot BN(\mathbb{G} \cdot \mathbb{X}) \cdot \mathbf{1}) \qquad \text{(RAPID)}$$

$$\text{subject to:} \quad \mathbb{X} \cdot W^{f^T} \leq \Lambda^T \qquad\qquad\qquad\qquad (10)$$
$$x_{jr} \in \{0,1\} \qquad\qquad\qquad ,\forall v_j, f_r \quad (11)$$

---

[1] RAPID and PRIDE use the concept of detection module (a group of IDS rules/functions) to reduce the complexity and increase the accuracy of the ILP [20].

The new objective function is no longer linear [1] and cannot be solved with ILP solvers. Therefore, we use Genetic Algorithm (GA), a popular and effective type of evolutionary algorithms, as used in RAPID [1] to solve the optimal monitoring problem proposed for δ-AFT design in resourceless WMN.

4.3   Solutions for AFT-Design of IDS

Although some of the solutions proposed for the optimal monitoring in state-of-the-art IDS solutions are implemented in both centralized and distributed manners, here, we only consider their centralized algorithms to compare with their centralized AFT designs. The system and attacker models considered in this research (for AFT-designs) are exactly the same as those in their original designs [1–3,8]. Similar to the original centralized solutions, the AFT-design solutions consider a WMN including mesh routers (i.e., battery powered in EEMON and TRAIN and AC-powered in RAPID and PRIDE) and a computationally powerful base station. Each router in the WMN has some local information (e.g., its communication load and its residual energy, processing/memory loads and traffic information) and periodically sends it, via a middleware and secure communication links, to the base station [1–3,8]. Based on the collected information and the δ and λ values chosen by the security administrator for resourceful and resourceless IDS, respectively, the base station then solves the optimization problem and assigns intrusion detection tasks to the nodes.

AFT-Design Resourceful IDS: Similar to original EEMON, upon collecting nodes' information, the base station uses an ILP solver (i.e., *bintprog* function of MATLAB [8]) to find the optimal set of monitoring nodes that can monitor all WMN links with $\delta + 1$ monitors. AFT-design TRAIN, as a traffic-aware solution, first removes *idle* nodes from the network, i.e., those not contributing in the traffic routing, and then optimally selects monitoring nodes (using *bintprog*) to monitor all traffic paths with $\delta + 1$ monitors. If the reduced WMN graph after removing idle nodes is disconnected, each graph component is considered as a sub-problem (to reduce the execution time) and solved separately [2].

AFT-Design Resourceless IDS: The base station in this classes performs a Genetic Algorithm to find the optimal IDS function distribution that provides maximum average link/path coverage ratio. GA solutions are encoded as bitstrings (i.e., chromosomes) of specific length and tested for fitness. In AFT-design PRIDE and RAPID formulations, matrix $\mathbb{X}$ is a solution that can be encoded as a chromosome of length $n \times R$ and the fitness (objective) value of each solution is the average link/path coverage in the WMN [1]. The genetic operations used in redesigned PRIDE/RAPID are based on operations explained in [21, 22] that their details are omitted here.

5   Performance Evaluation

5.1   Resourceful IDS

This section presents simulation results for AFT designs of two resourceful IDS solutions, EEMON and TRAIN. As shown in EEMON [8] and TRAIN [2] and by considering their problem formulations presented in Section 4, the metrics we evaluate in this section are: 1) average number of nodes selected as monitoring nodes; 2) average communication load and average residual energy charge among selected nodes as monitoring nodes, in addition to the battery threshold reduction; 3) average link coverage and intrusion detection rates; 4) time complexity and average energy consumption; and 5) a new metric called expected δ for a given δ-AFT design as we will explain it later in this section. The results are obtained from 100 random networks for each network size. We note here that 0-AFT design in simulation results means the original unmodified IDS design.

Number of Monitors: The main objective in resourceful IDS solutions is to cover the entire network links/paths with minimum number of monitoring nodes and minimum total cost. Therefore, in a δ-AFT design, as δ increases, the number of nodes that must be selected as monitoring nodes will

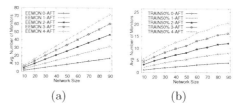

(a)   (b)

Fig. 2. Average number of monitoring nodes for different δ in: (a) EEMON; (b) TRAIN 50%

also increase (redundant monitoring nodes provide higher degree of attack and fault tolerance). Figures 2(a) and 2(b) show the average number of monitoring nodes for different δ and network sizes in EEMON and TRAIN, respectively. We note here that although TRAIN [2] evaluates this metric for different number of paths (e.g., number of paths equals to 10%, 30%, and 50% of network size), we only consider the maximum case which is number of paths equals to $0.5 \times N$ and omit the other results due to space limitations. As shown, the number of monitoring nodes linearly increases (i.e., constant percentage of nodes are selected for different N) as δ increases in both traffic-agnostic and traffic-aware solutions to provide higher levels of attack and fault tolerance. For example, more than 80% of the nodes in EEMON are selected as monitoring nodes in 4-AFT design (i.e., higher costs to achieve larger δ).

Properties of Monitoring Nodes: In EEMON and TRAIN, the cost per monitoring node is defined as a function of residual energy charge and the communication load. Therefore, the residual energy charge and the average communication load among selected node is expected to be higher than of those of non-monitoring nodes. In addition, it is possible that out of all possible nodes that can monitor a *link/path*, none of them has residual charge greater than threshold b_{th}. In this case, as mentioned in EEMON [8] and TRAIN [2], the threshold decreases until at least one of the nodes is selected. Such a threshold reduction has to be as low as possible meaning that most of the selected nodes have residual energy charge above the threshold b_{th} resulting in longer network life time.

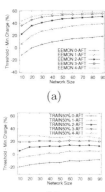

(a)

(b)

Fig. 3. [B_{th} - Minimum Charge] among selected nodes for different δ in: (a) EEMON; (b) TRAIN 50%.

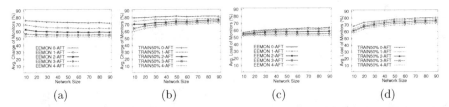

Fig. 4. Average residual charge of monitors for different δ in: (a) EEMON; (b) TRAIN 50%; Average comm. load of monitors for different δ in: (c) EEMON; (d) TRAIN 50%.

Figures 3(a) and 3(b) show the average value of $[B_{th}$ - Minimum Charge] for different δ and network sizes in EEMON (TG-RF) and TRAIN (TW-RF), respectively. Negative values mean that the minimum residual energy charge among all selected nodes is larger than the threshold and no threshold reduction has occurred. As shown, the greater the δ is, the larger the threshold reduction will be. This is because selecting more monitoring nodes (required by large δ) increases the probability of selecting low battery nodes, and consequently increases the $[B_{th}$ - Minimum Charge].

The next two metrics we consider are average residual charge and average communication load among selected nodes, as evaluated in both EEMON and TRAIN. Figures 4(a) and 4(b) depict the average residual energy charge among selected nodes (as monitoring nodes) for different δ and network sizes in EEMON and TRAIN, respectively. As depicted, the larger the δ is, the lower the average residual energy charge of monitoring nodes will be. This is because larger δ requires more monitoring nodes to achieve higher levels of attack and fault tolerance. Hence, the monitoring node selection algorithms have to select monitors among low battery nodes that decreases the average value. Figures 4(c) and 4(d) show the average communication load of selected nodes for different δ and network sizes in EEMON and TRAIN, respectively. Similar to the average residual charge, the average communication load decreases as δ increases.

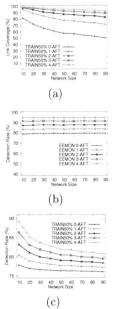

Fig. 5. (a) Average link coverage in TRAIN 50%. Average detection rate of all $40N$ Normal/Severe and Single-hop/Multi-hop attacks in: (b) EEMON; (c) TRAIN 50%.

Intrusion Detection Rates: EEMON and TRAIN aim at covering all network links and paths respectively. Average link coverage in EEMON is always 100% but TRAIN only covers a subset of links located on active routing paths. Figure 5(a) shows the average link coverage provided by TRAIN 50% when δ increases. As shown, although the original TRAIN leaves some communication links uncovered, the AFT design of TRAIN increases the average link coverage as it selects more monitoring nodes than original TRAIN. EEMON [8] considers two types of attacks, Severe (detectable by only monitoring nodes) and Normal (detectable by monitoring and non-monitoring nodes). These two attacks can be launched in single-hop and multi-hop modes. The detection rate of EEMON and TRAIN for Normal attacks, either single-hop or multi-hop, is 100% as the attack traffic is certainly monitored by a node (either monitoring or non-monitoring). In addition, Severe multi-hop attacks are also considered to be 100% detectable as both EEMON and TRAIN have at least one monitoring node that monitors multi-hop traffic. The only attack that is hard to detect is Severe single-hop attack which is only detectable by monitoring nodes.

We performed $10 \times N$ random attacks for each of 4 types (i.e., 2 types and 2 modes) for different δ in EEMON and TRAIN 50% and measured the detection rates ($40 \times N$ random attacks for each network size). Figures 5(b) and 5(c) depict the average intrusion detection rates for all combinations of Severe/Normal and single-hop/multi-hop attacks in EEMON and TRAIN, respectively. As depicted, larger δ increases the average intrusion detection rate since it results in selecting more monitoring nodes in the network that can detect Severe single-hop attacks and consequently increases the average detection rate. The lower detection rate in TRAIN (for similar δ-AFT designs as EEMON) is due to covering few paths (a subset of links) which results in selecting less monitoring nodes.

The next type of attack we consider is *EEMON and TRAIN aware* attack [2] where attacker knows which IDS solutions is used (e.g., traffic-agnostic or traffic-aware, link coverage or node coverage, etc.) but do not know what type of attack is considered to be Severe or Normal. For example, if the attacker knows that EEMON is used, he will only run single-hop attacks and if TRAIN is used, he will try to run attacks against intermediate nodes on traffic paths to avoid monitoring node on the route. Figures 6(a) and 6(b) show the average intrusion detection rates of *EEMON and TRAIN aware* attacks ($10 \times N$ random attacks for each N) in EEMON and TRAIN 50%, respectively. It is worth mentioning that EEMON, at the price of using more monitoring nodes, achieves higher detection rates than TRAIN for a given network size. Also, as δ increases, the detection rate increases too because of selecting more monitoring nodes in the network.

Time Complexity and Energy Consumption: Figures 6(c) and 6(d) show the execution time of the ILP solver when solving the optimization problem in EEMON and TRAIN, respectively. The results show the average execution time of different δ and network sizes. Generally, the execution time increases as network size (number of links/paths to be covered) increases. In addition, smaller δ increases the time complexity of the ILP solver since it reduces the solution space. As the results show, the execution time in TRAIN is always less than 0.1 seconds since it only considers traffic paths, however, the execution time in EEMON is in the order of few seconds (as it considers all communication links). We note here that higher execution times for large networks in EEMON are also because of some outliers among 100 random networks [2].

In both EEMON and TRAIN, non-monitoring nodes work in duty-cycling mode to save energy. Thus, the set of monitoring nodes changes periodically (based on the problem formulation) to extend the network life time. The current

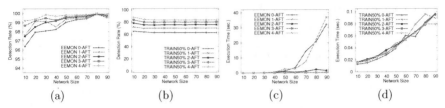

Fig. 6. Average intrusion detection rate of EEMON/TRAIN aware attacks for different δ in: (a) EEMON; (b) TRAIN 50%. Average execution time of the ILP solver for different δ in: (c) EEMON; (d) TRAIN 50%.

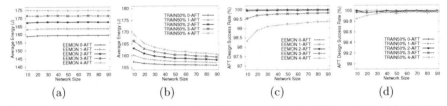

Fig. 7. Average energy consumption of 50% duty cycling for different δ in: (a) EEMON; (b) TRAIN 50%. The ratio of number of selected monitors to the expected number of monitors for different δ in: (c) EEMON; (d) TRAIN 50%.

consumption of devices used in EEMON and TRAIN (i.e., Linksys mesh routers) is 250mA, which means each device consumes 3 Watts (12V250mA). Thus, the energy consumed by each device during one minute working time (i.e., an epoch in our experiment) is 180 Joule. When duty-cycling, the energy consumption decreases depending on the duty-cycle interval. Figures 7(a) and 7(b) show the average energy consumption per node during an epoch for different δ in EEMON and TRAIN, respectively. As shown, the larger the δ is, the higher the average energy consumption will be. This is because larger δ means more nodes will work in monitoring mode and less nodes can save energy through duty-cycling.

Success Rate of δ-AFT Design: The last metric we evaluate in resourceful IDS class is the success rate of δ-AFT design in assigning $\delta + 1$ monitoring node(s) to each communication link/path. Since the number of monitoring nodes assigned to each link/path is limited by the maximum number of nodes that can cover the link/path, it is sometimes impossible to achieve δ-AFT for a given δ and network topology. In fact, the success rate of δ-AFT design in assigning $\delta + 1$ monitoring node(s) to a link depends on the network topology. We performed simulations for 100 random networks of each given network size and different δ and measured the average number of monitoring nodes per links/paths divided by δ. Figures 7(c) and 7(d) depict the success rates of δ-AFT design for different δ in EEMON and TRAIN, respectively. As one can observe, the success rate is always near 100% specially for TRAIN as it monitors less links than EEMON.

5.2   Resourceless IDS

This section evaluates the performance of resourceless IDS solutions for AFT design. As we discussed in Section 4, the main parameter in designing resourceless

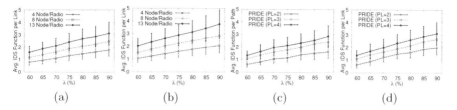

Fig. 8. The average IDS functions per link for different memory threshold (λ) and network densities in: (a) 6-Module Configuration; (b) 12-Module Configuration RAPID. The average IDS functions per path for different memory threshold (λ) and path lengths (PL) in: (c) 6-Module Configuration; (d) 12-Module Configuration PRIDE.

IDS for traffic-agnostic and traffic-aware networks is memory threshold (λ). The larger the λ is, the higher the link/path coverage will be. This is because larger λ allows nodes to execute more IDS functions which also increases the IDS function redundancy (i.e., higher levels of attack and fault tolerance). Consequently, it increases intrusion detection rates and average memory load on the nodes. Hence, in resourceless IDS, unlike resourceful IDS, we cannot change δ as a tuning parameter for AFT design, however, δ is a function of λ and network density. In other words, the security administrator gives a higher priority to link/path coverage than AFT design because for example, having two identical (redundant) IDS functions on a path is not as useful as executing two different IDS functions on the nodes along the paths. Obviously, the later provides higher path coverage (and consequently higher detection rates) than the former (i.e., lower path coverage but higher level of attack and fault tolerance).

Figures 8(a) and 8(b) show the average number of IDS functions per links in RAPID for 6-module and 12-module configurations, respectively (Note: less modules means more IDS rules in each group resulting in fewer larger groups where each of them imposes a higher memory load than a smaller module [20]). As shown, this metric is a function of memory threshold (λ) and network density. The larger the λ and network density are, the more IDS function per link (i.e., the level of attack and fault tolerance) will be. Similarly, Figures 8(c) and 8(d) depict the average number of IDS functions per paths in PRIDE for 6-module and 12-module configurations, respectively. In PRIDE, since only the nodes located on the path participate in path monitoring, the level of attack and fault tolerance is a function of path length (PL) and λ. The higher the λ and PL are, the higher the attack and fault tolerance level will be. We note here that other metrics such as intrusion detection rates and average memory loads (omitted here) in RAPID and PRIDE are exactly the same as those shown in [1,3].

6   Conclusions

In this paper, we studied the IDS attack-and-fault tolerance in wireless mesh networks (WMN). We first proposed a taxonomy of state-of-the-art IDS solutions in WMN and then investigated their attack-and-fault tolerance. Next, we showed that those solutions do not consider IDS compromise/fault scenarios. We then proposed a classification for attack-and-fault tolerant (AFT) IDS which includes prevention, detection, and recovery mechanisms in AFT design. Considering the optimal monitoring mechanism employed by each state-of-the-art IDS solution in WMN, we reformulated their optimal monitoring problems to include AFT IDS mechanisms. Through extensive simulations, the performance (e.g., intrusion detection rate) and efficiency (e.g., resource consumption) of redesigned IDS solutions were evaluated and compared to those of the original solutions.

References

1. Hassanzadeh, A., Stoleru, R., Polychronakis, M., Xie, G.: RAPID: A traffic-agnostic intrusion detection for resource-constrained wireless mesh networks. Technical report, Texas A&M University 2014-1-3 (2014)
2. Hassanzadeh, A., Altaweel, A., Stoleru, R.: Traffic-and-resource-aware intrusion detection in wireless mesh networks. Technical report, Texas A&M University 2014-1-2 (2014)
3. Hassanzadeh, A., Xu, Z., Stoleru, R., Gu, G., Polychronakis, M.: PRIDE: Practical intrusion detection in resource constrained wireless mesh networks. In: Qing, S., Zhou, J., Liu, D. (eds.) ICICS 2013. LNCS, vol. 8233, pp. 213–228. Springer, Heidelberg (2013)
4. Morais, A., Cavalli, A.: A distributed and collaborative intrusion detection architecture for wireless mesh networks. Mobile Networks and Applications (2013)
5. do Carmo, R., Hollick, M.: DogoIDS: A mobile and active intrusion detection system for IEEE 802.11s wireless mesh networks. In: HotWiSec (2013)
6. Gu, Q., Zang, W., Yu, M., Liu, P.: Collaborative traffic-aware intrusion monitoring in multi-channel mesh networks. In: TrustCom (2012)
7. Saxena, N., Denko, M., Banerji, D.: A hierarchical architecture for detecting selfish behaviour in community wireless mesh networks. Computer Communications, pp. 548 – 555 (2011)
8. Hassanzadeh, A., Stoleru, R., Shihada, B.: Energy efficient monitoring for intrusion detection in battery-powered wireless mesh networks. In: Frey, H., Li, X., Ruehrup, S. (eds.) ADHOC-NOW 2011. LNCS, vol. 6811, pp. 44–57. Springer, Heidelberg (2011)
9. Hugelshofer, F., Smith, P., Hutchison, D., Race, N.: OpenLIDS: A lightweight intrusion detection system for wireless mesh networks. In: MobiCom (2009)
10. Shin, D., Bagchi, S.: Optimal monitoring in multi-channel multi-radio wireless mesh networks. In: ACM MobiHoc (2009)
11. Glass, S., Muthukkumarasamy, V., Portmann, M.: Detecting man-in-the-middle and wormhole attacks in wireless mesh networks. In: AINA (2009)
12. Martignon, F., Paris, S., Capone, A.: A framework for detecting selfish misbehavior in wireless mesh community networks. In: Q2SWinet (2009)
13. Yu, W., Zhang, N., Fu, X., Bettati, R., Zhao, W.: Localization attacks to internet threat monitors: Modeling and countermeasures. IEEE Transactions on Computers, 1655–1668 (2010)
14. Bethencourt, J., Franklin, J., Vernon, M.: Mapping internet sensors with probe response attacks. In: USENIX Security (2005)
15. Mell, P., Marks, D., McLarnon, M.: A denial-of-service resistant intrusion detection architecture. Comput. Netw., 641–658 (2000)

16. Liu, H., Nayak, A., Stojmenovi, I.: Fault-tolerant algorithms/protocols in wireless sensor networks. In: Guide to Wireless Sensor Networks, Computer Communications and Networks, pp. 261–291 (2009)
17. Luo, X., Dong, M., Huang, Y.: On distributed fault-tolerant detection in wireless sensor networks. IEEE Transactions on Computers, 58–70 (2006)
18. Chenji, H., Hassanzadeh, A., Won, M., Li, Y., Zhang, W., Yang, X., Stoleru, R., Zhou, G.: A wireless sensor, adhoc and delay tolerant network system for disaster response. Technical report, LENSS-09-02 (2011)
19. Manikantan Shila, D., Anjali, T.: Load aware traffic engineering for mesh networks. Computer Communications, 1460–1469 (2008)
20. Hassanzadeh, A., Xu, Z., Stoleru, R., Gu, G.: Practical intrusion detection in resource constrained wireless mesh networks. Technical report, Texas A&M University 2012-7-1 (2012)
21. Hassanzadeh, A., Stoleru, R.: Towards optimal monitoring in cooperative IDS for resource constrained wireless networks. In: ICCCN (2011)
22. Hassanzadeh, A., Stoleru, R.: On the optimality of cooperative intrusion detection for resource constrained wireless networks. Computers & Security, 16–35 (2013)

Multihop Node Authentication Mechanisms for Wireless Sensor Networks[*]

Ismail Mansour[1,2], Damian Rusinek[3], Gérard Chalhoub[1,2],
Pascal Lafourcade[1,2], and Bogdan Ksiezopolski[3,4]

[1] Clermont Université, Université d'Auvergne, LIMOS, BP 10448, F-63000,
Clermont-Ferrand, France
[2] CNRS, UMR 6158, LIMOS, F-63173 Aubière, France
[3] Institute of Computer Science, Maria Curie-Sklodowska University Lublin, Poland
[4] Polish-Japanese Institute of Information Technology, Warsaw, Poland

Abstract. Designing secure authentication mechanisms in wireless sensor networks in order to associate a node to a secure network is not an easy task due to the limitations of this type of networks. In this paper, we propose different multihop node authentication protocols for wireless sensor networks. For each protocol, we provide a formal proof using Scyther to verify the security of our proposals. We also provide implementation results in terms of execution time consumption obtained by real measurements on TelosB motes. These protocols offer different levels of quality of protection depending on the design of the protocol itself. Finally, we evaluate the overhead of protection of each solution, using AQoPA tool, by varying the security parameters and studying the effect on execution time overhead of each protocol for several network sizes.

Keywords: Authentication, Wireless Sensor Network, Security, Quality of Protection, Multihop, Formal Verification.

1 Introduction

Wireless sensor networks (WSN) are more and more used in critical applications where the identity of each communicating entity should be authenticated before exchanging data in the network. The wireless nature of this technology makes it easy for intruders to try to intervene in the network activity and create any of the known attacks in WSNs [11]. Many of the current propositions focus on message authentication for ensuring data authentication and integrity, and some focus on user authentication to give access to the network for certain previously declared users. In this paper we propose a variation of different node authentication protocols that help authenticate any node in the network regardless of users.

Designing secure protocols is an error-prone task. One of the well known examples is the famous flaw found on the Needham Scroeder protocol seventeen years after its publication [19]. It clearly shows that designing secure protocols is not an easy task.

---

[*] This reasearch was conducted with the support of the "Digital Trust" Chair from the University of Auvergne Foundation.

During the last decades, several automatic tools for verifying the security of cryptographic protocols have been elaborated by several authors, like for instance Proverif[3], Avispa [25] or Scyther [4]. These symbolic tools use the Dolev-Yao intruder model [8], that considers that the intruder is controlling the network and makes the perfect encryption hypothesis[1]. The state of the art shows that formal methods are now mature and efficient enough to be used in the design of security protocol in order to avoid such logical flaws.

Another aspect which should be taken into account during WSN protocols analysis is performance which refers to the security operations. The traditional approach assumes that the best way is to apply the strongest possible security measures which make the system as secure as possible. Unfortunately, such reasoning leads to the overestimation of security measures which causes an unreasonable increase in the system load [14]. The system performance is especially important in the systems with limited resources such as wireless sensor networks or mobile devices. The solution may be to determine the required level of the protection and adjust some security measures according to these requirements. Such an approach can be achieved by means of the Quality of Protection [12,13,15] where the security measures are evaluated according to their influence on the system security.

Contributions

The originality of our work resides in the fact that it combines several aspects of security, from designing secure protocols to evaluating the implementation of our solution, going through formal automatic analysis of security and quality of protection analysis. Our contributions can be summarized in the four following points:

1. Design of multihop node authentication mechanisms.
2. Formal automatic analysis of our solutions.
3. Implementation on TelosB motes.
4. Evaluation of the quality of protection of our solutions.

Our main contribution is the design of several secure authentication protocols. In order to avoid flaws, we use Scyther [23] to prove the correctness of all our protocols automatically. We have implemented our protocols on TelosB motes in order to obtain time consumption for few nodes. From the quality of protection analysis point of view, Scyther abstracts the cost of the communication and also does not consider the computation time of cryptographic primitives. The quality of protection analysis for WSN cryptographic protocols is almost impossible to perform manually. This increases the difficulty to design secure and efficient protocols at the same time. Using our real implementation on TelosB motes, we have designed several metrics to calibrate the Automated Quality of Protection Analysis tool (AQoPA[2]). With this tool we have evaluated the quality of protection of our protocols. This analysis takes into account all security factors which affect the overall system security to determine the fastest protocol according to the level of protection that is desired by the application.

---

[1] Meaning that it is possible to obtain the plain text of an encrypted message only if the secret key is known.

[2] AQoPA is available at: http://www.qopml.org.

Related Work:

Authentication Protocols in Multihop WSNs: Very few work has been done for node authentication protocols in multihop WSNs. Most of the existing authentication protocols proposed for WSNs neglect the multihop factor. In [1], authors proposed a protocol where the base station broadcast authentication elements for in range sensor nodes to be able to authenticate new arriving nodes. In fact, they consider that any previously authenticated node can authenticate new nodes.

In [7] and [28], authors propose an authentication mechanism for users and consider that sensor nodes inside the WSN are trusted nodes. In [28], authors propose a stronger authentication protocol that ensures mutual authentication and protection against attacks from other users, which is not the case for [7].

Recently in [9], authors propose an authentication model that aims at reducing overhead for the re-authentication of sensor nodes. It is based on a ticket encrypted using a common secret key between neighbouring fixed nodes. This ticket is sent to a mobile node during the first authentication phase. This ticket is only useful when the mobile node decides to re-authenticate with this neighbour fixed node. In addition, the protocol only works well when the fixed node is in direct range with the base station, the initial authentication phase suffers from internal attacks as other sinks in the network can easily take the place of one another when they are not in communication range with the base station.

In [29], authors propose a node authentication protocol for hierarchical WSNs. The hierarchical topology is limited to a base station, cluster heads and sensor nodes. The cluster heads can reach the sensors of their clusters directly, and can also reach the base station directly. The authentication is based on hash chain functions. The proposed protocol is not resilient to insider attacks as cluster heads are trusted to forward join requests to base station. In addition, the authors did not specify how the protocol copes with a multihop topology between cluster heads and the base station.

In our proposition, we take into account the multihop factor where any node in the network is able to be authenticated by sending a request in a multihop manner towards the base station. We also consider different cases depending on the level of trust we have in intermediate nodes and their computation capacities. Finally we formally prove the security using the automatic verification tool Scyther [4].

Quality of Protection Evaluation: In the literature several quality of protection models were created for different purposes and have different features and limitations. Authors in [17] attempted to extend the security layers in a few quality of service architectures. Unfortunately, the descriptions of the methods are limited to the confidentiality of the data and are based on different configurations of the cryptographic modules. In [27], authors created quality of protection models based on the vulnerability analysis which is represented by the attack trees. The leaves of the trees are described by means of the special metrics of security. These metrics are used for describing individual characteristics of the attack. In [13], authors introduced mechanisms for adaptable security which can be used for all security services. In this model the quality of protection depends on the risk level of the analyzed processes. Authors in [20] present the quality of protection analysis for the IP Multimedia Systems (IMS). This approach presents the IMS performance evaluation using Queuing Networks and Stochastic Petri Nets. In [16], authors create the adversary-driven, state-based system security evaluation. This method quantitatively evaluates the strength of the security of the system. In [24], authors present the performance analysis of security aspects in the UML models. This approach takes as an input a UML model of the system designed by the UMLsec extension [10]. This UML model is annotated with the standard UML Profile for schedulability, performance and time, and then analysed for performance.

In [12], the Quality of Protection Modelling Language (QoP-ML) is introduced. It provides the modelling language for making abstraction of cryptographic protocols that put emphasis on the details concerning quality of protection. The intended use of QoP-ML is to represent the series of steps which are described as a cryptographic protocol. During the analysis one cannot consider only primary cryptographic operations or basic communication steps. The QoP-ML introduces the multilevel protocol analysis that extends the possibility of describing the state of the cryptographic protocol. The analysis involves the elements such as: cryptographic primitives, communication steps, information security management, key management, security policy management, legal compliance, implementation of the protocol and cryptographic algorithms as well as other factors that influence the system security. Every single operation defined by the QoP-ML is described by the security metrics which evaluate the impact of this operation on the security requirements of the system. The QoP-ML models can be automatically evaluated by the Automated Quality of Protection Analysis tool (AQoPA).

Outline: In the next section, we present five different protocols for establishing secure mutlihop communications. Then in Section 3, we use Scyther to formally prove the security of our solutions. In Section 4, we make a qualitative evaluation of our five protocols using AQoPA, before concluding the paper in the last section.

2   Multihop Authentication Protocols for WSN

We propose several protocols that allow a node to join a multihop WSN in a secure way. We distinguish two classes of protocols:

1. Direct Join to the Sink (DJS): a node joins directly through the sink.
2. Indirect Join to the Sink (IJS): a node joins the network through intermediate nodes in order to reach the sink.

We use public key Elliptic Curve Cryptography (ECC), using parameters secp160r1 and secp128r1 given by the Standards for Efficient Cryptography Group [26]. Our implementation of ECC on TelosB is based on TinyECC library [18]. More precisely we use Elliptic Curve Integrated Encryption Scheme (ECIES) the public key encryption system proposed by Victor Shoup in 2001. For all symmetric encryptions we use an optimized implementation of AES [6] with a key of 128 bits proposed by [21].

Before deployment, each node N knows the public key $pk(S)$ of the sink S and also its own pair of private and public keys, denoted $(pk(N), sk(N))$ respectively. Based on ECC, we have that $pk(N) = sk(N) \times G$, where G is a generator point of the elliptic curve. Using this material, each node N can compute a shared key with the sink S using a variation of the Diffie-Hellman key exchange without interaction between the nodes, denoted $K_{DH}(N, S)$. These computations can be done by the sink and by all nodes before deployment in order to preserve their energy.

- The sink knows its own secret key $sk(S)$ and the public key $pk(N)$ of a node N. The sink computes $K_{DH}(N, S) = sk(S) \times pk(N)$.
- Node N multiplies his secret key $sk(N)$ by the public key of the sink $pk(S)$ to get $K_{DH}(N, S)$.

Both computations give the same shared key since:

$$K_{DH}(N, S) = sk(N) \times pk(S) = sk(N) \times (sk(S) \times G) = (sk(N) \times G) \times sk(S) = pk(N) \times sk(S)$$

Notations

In what follows, we use the following notations to describe exchanged messages in our protocols:

- I: a new node that initiates the protocol,
- R: a neighbour of node I,
- S: the sink of the network (also called base station),
- J_i: the i-th intermediate node between R and S,
- n_A: a nonce generated by node A,
- $\{x\}_k$: the encryption of message x with the symmetric or asymmetric key k,
- $pk(A)$: the public key of node A,
- $sk(A)$: the secret (private) key of node A,
- $K(I, S)$: the session key between I and S,
- NK: the symmetric network key between all nodes of the network randomly generated by S,
- $K_{DH}(N, S)$: the shared symmetric key between N and S using the Diffie-Hellman key exchange without interaction described above.

2.1   Direct Join to Sink : DJS_{orig}

The protocol DJS_{orig} is the original protocol presented in [22]. It allows new nodes in range of the sink to join the network directly. We present this protocol in Figure 1. The new node I sends a direct request to S in order to establish a session key with it. The node I begins the join process by computing the symmetric key $K_{DH}(I, S)$ with the sink S. Then, node I generates a nonce n_I and adds its identity in order to form the request $\{n_I, I\}$. The request is encrypted with $K_{DH}(I, S)$ and sent to S. Upon reception, in order to decrypt the request, node S computes $K_{DH}(I, S)$ using I's identity provided by the routing protocol. Then, S verifies the identity of I^3 and generates a new session key $K(I, S)$. The join response contains n_I, the identity of S and the new symmetric session key. The response is encrypted using $pk(I)$ and is sent to I. Only I is able to decrypt the response with its secret key $sk(I)$. We note that n_I helps I to authenticate S.

---

[3] S checks if the identity of I belongs to the list of deployed nodes.

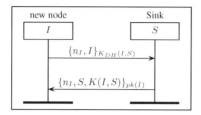

Fig. 1. DJS_{orig}: The node I joins directly the network by communicating directly with the sink S

2.2   Indirect Protocols to Join the Sink

In this section, we present four different protocols that allow a new node, out of range of S, to join the network. A new node can join the network through a neighbour node that is already authenticated in the network. The main differences between these protocols is the way the authentication of nodes between R and S is established and how messages are forwarded between them. In what follows, we describe each proposed protocol. In Table 1, we summarize the main differences between the proposed protocols.

Table 1. Operations on intermediate nodes for the Indirect Join protocols

Protocol name	Authentication	Key type	Operations on intermediate nodes			
			from R to S		from S to R	
			Encrypt	Decrypt	Encrypt	Decrypt
IJS_{orig}	no	DH with S	no	no	no	no
$IJS_{NK,dec/enc}$	yes	network key	yes	yes	yes	yes
$IJS_{K,dec/enc}$	yes	session key	yes	yes	yes	yes
$IJS_{NK,onion}$	yes	network key	yes	no	no	yes

The idea behind the different protocols is to allow the application to choose which protocol to use according to its constraints in terms of capacities, and needs in terms of security level. Using IJS_{orig} protocol is less consuming in terms of number of cryptographic operations but it assumes that all the nodes in the network are trusted nodes. $IJS_{NK,dec/enc}$ and $IJS_{K,dec/enc}$ protocols are similar in terms of number of operations but the latter is more resilient to node capture as it uses different keys along the route to the sink. As for $IJS_{NK,onion}$, it enables the network to do most of the cryptographic operations for the authentication process on the sink and thus reducing the computation time on intermediate nodes.

IJS_{orig}: This protocol is the original protocol presented in [22] and allows a new node to join the network through a neighbour node R. We present this protocol in Figure 2. The new node I sends an indirect request to S in order to establish a session key with R. The node R forwards the request to S through an intermediate nodes J_i. We note that the request and the response are just forwarded by J_i without any modifications. Node J_i is not able to decrypt any message due to the key used for encryption.

Only nodes I and S are able to decrypt the messages encrypted with $K_{DH}(I, S)$, and only R and S are able to decrypt the messages encrypted with $K_{DH}(R, S)$.

In this protocol, the authors make the assumption that intermediate nodes are trusted. Hence, it is not resilient against insider attacks executed by intermediate nodes. Indeed an intruder can play the role of any intermediate node without being detected neither by the sink nor by the new node.

In what follows, we propose three protocols that allow a new node to join the network without trusting any intermediate node. Each solution uses a different approach for solving this question and has been proven secure using Scyther.

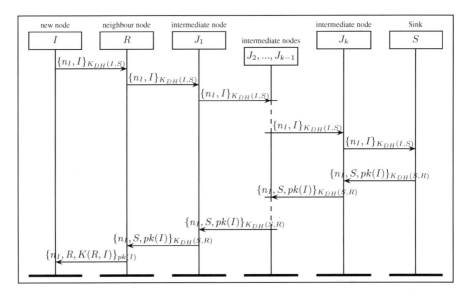

Fig. 2. IJS_{orig}: the original version. The intermediate node between R and S forwards messages without any encryption or decryption.

$IJS_{NK,dec/enc}$: The idea behind this protocol is to ensure authentication between all nodes by adding a nonce on each hop and by decrypting and encrypting exchanged messages as follows.

In Figure 3, we present $IJS_{NK,dec/enc}$ protocol. It allows new nodes to join the network through a neighbour node R using the network key for encryption/decryption on intermediate nodes. The node I sends a request containing a nonce with its own identity and the identity of R. Then, node R generates a nonce and adds it to the initial request before encrypting it with NK and forwarding it to J. Upon reception, node J decrypts the request and generates a new nonce n_J, adds it to the received request and then encrypts the result using NK. When J receives the response message, it decrypts it using NK and extracts n_J, and then forwards the response message to R while keeping n_R in the message. We note that the nonce values n_I, n_R and n_J have helped S to authenticate I, R and J respectively and make sure that the request has been forwarded by previously authenticated nodes.

This protocol is secure as proven by Scyther [23], but each intermediate node has to decrypt and encrypt a message using the same key, which is the network key. Such cryptographic operations are very resources consuming. In addition, using the same key makes a node capture attack more dangerous for it enables the attacker to decrypt the authentication process of all nodes. In the next protocol, we avoid such risk by using a session keys.

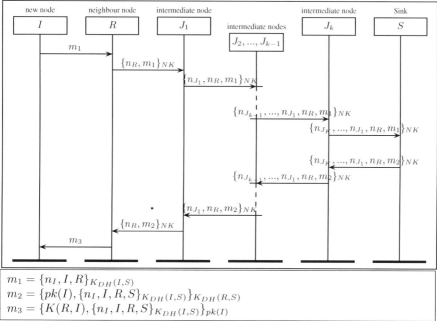

$$m_1 = \{n_I, I, R\}_{K_{DH}(I,S)}$$
$$m_2 = \{pk(I), \{n_I, I, R, S\}_{K_{DH}(I,S)}\}_{K_{DH}(R,S)}$$
$$m_3 = \{K(R, I), \{n_I, I, R, S\}_{K_{DH}(I,S)}\}_{pk(I)}$$

Fig. 3. $IJS_{NK,dec/enc}$: The intermediate nodes J_i decrypt, add a nonce value and encrypt the result message before forwarding it. It uses the network key to encrypt/decrypt this messages.

$IJS_{K,dec/enc}$: In Figure 4, we present $IJS_{K,dec/enc}$ protocol. The two main differences between $IJS_{K,dec/enc}$ and $IJS_{NK,dec/enc}$ are:

- We encrypt and decrypt the request and the response between R and S with the symmetric session key $K(J_i, J_{i+1})$ established during the previous join phases.
- We also add all identities of intermediate nodes to the initial request sent by I.

We assume that the node I is able to obtain the secure path to S from R. Indeed, the secure path is already known by R because it was able to join the network and build it using its routing protocol.

This protocol enhances the previous one by using session keys but still suffers from doing cryptographic operations on intermediate nodes. In the next protocol, we avoid overcharging intermediate nodes by doing most of the operations on the sink.

$IJS_{NK,onion}$: In Figure 5, we give a description of $IJS_{NK,onion}$ protocol which is an enhancement over $IJS_{NK,dec/enc}$ in terms of number of operations done by intermediate

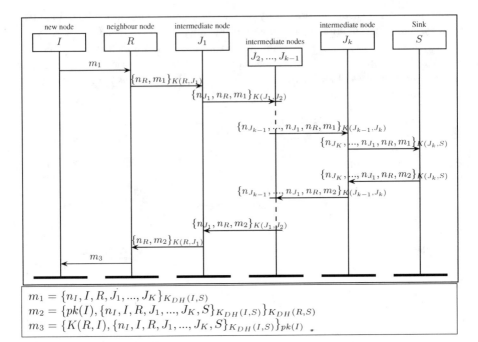

Fig. 4. $IJS_{K,dec/enc}$: The intermediate nodes J_i decrypt, add a nonce value and encrypt the result message before forwarding it. They use the session key to encrypt/decrypt this messages.

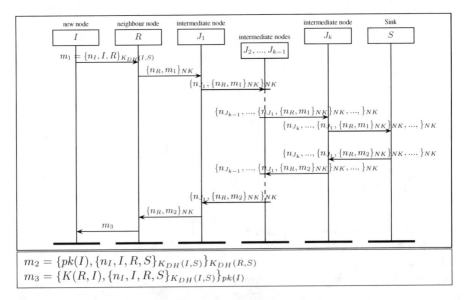

Fig. 5. $IJS_{NK,onion}$: The intermediate nodes J_i add a nonce and encrypt the request message and forward it to S

nodes. The goal is to help intermediate nodes to save time and energy. Using NK, an intermediate node J_i is able to add a nonce to the initial request and to encrypt the result before forwarding it. Upon reception, J_i is able to decrypt the response message, extract and retrieve its own nonce n_{J_i} and forward the rest of the message to R.

We note that the encryption/decryption operations that were not done by J_i are done by S. We assume that S is more efficient in computing and have more energy than the other nodes of the network.

This protocol requires less computation for intermediate nodes, but suffers from exposure due to node capture attack exactly like $IJS_{NK,dec/enc}$ protocol because the same network key is used all the way from the source node to the sink.

3 Formal Security Evaluation

Evaluating the security of cryptographic protocols is not an easy task. It is easy to design flawed protocols. During the last decades several tools have been developed to automatically verify cryptographic protocols like for instance [2,3,4]. We use Scyther [4] because it is one of the fastest tools as it has been shown in [5] and one of the most user-friendly.

3.1 Scyther Overview

Cas Cremers has developed an automatic tool called Scyther [4]. It is a free tool available on all operating systems (Linux, Mac and Windows). This tool can automatically find attacks on cryptographic protocols and prove their security for bounded and unbounded numbers of sessions. One main advantage of Scyther is that it provides an easy way to model security properties like secrecy and authentication.

3.2 Results

We verified all our protocols using Scyther for a fix bounded number of participants. More precisely, we proved the secrecy of all sensitive data exchanged (keys and nonces) and also the authenticity of the communication. Our Scyther codes are available here [23] for more information.

Moreover, for all our protocols we proved by induction the security of the protocols for any number of intermediate nodes. Each time, the base case is proven using Scyther for a small number of nodes.

- Protocol DJS: participants are one node and the sink. The verification using Scyther allows us to prove the security of our protocol.
- Protocol IJS_{orig}: Scyther found an authentication attack, where an intruder can replace any of the intermediate nodes between the new node and the sink and neither the sink nor the new node can detect its presence. This means that IJS_{orig} ensures only end-to-end authentication and fails to ensure hop-by-hop authentication. Hence, it is secure only if it is safe to send the join response through a route that was not the one used to send the join request. Indeed, in a hostile environment and in the presence of malicious nodes, it is important to be able to identify trusted nodes and be sure to route the response back through them in order to authenticate new nodes.

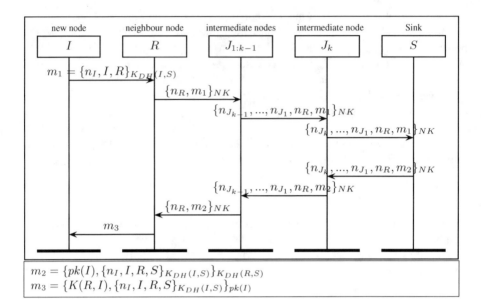

Fig. 6. $IJS_{NK,dec/enc}$: Proof by Induction

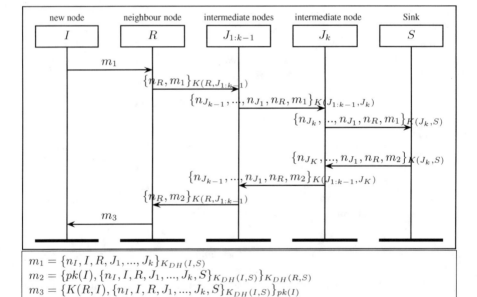

Fig. 7. $IJS_{K,dec/enc}$: Proof by Induction

– $IJS_{NK,dec/enc}$, $IJS_{K,dec/enc}$ and $IJS_{NK,onion}$: these three protocols are constructed to work for any number of intermediate nodes, for each one of them we used the same method for proving their security. They ensure end-to-end and hop-by-hop authentication. In addition, we made a proof by induction. For the initialization of our

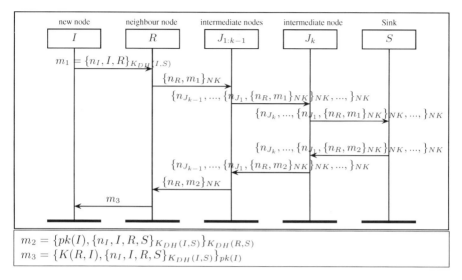

Fig. 8. $IJS_{NK,onion}$: Proof by Induction

induction, we used Scyther for proving that for 4 nodes all our protocols are secure. Then we assumed the protocols are secure for $k-1$ intermediate nodes, we showed that for k intermediate nodes they are still secure. Using the induction hypothesis, we obtain that the secrecy and authentication between I, R, J_1, ..., J_{k-1} and S is secure if S takes the place of J_k for all protocols. In order to prove the security when we add the intermediate node J_k, we consider the protocol between the following nodes I, R $J_{1:k-1}$, J_k and S (Figure 6, 7 and 8). Again using Scyther, we proved the security properties of these 5 nodes protocols.

This approach for generalizing the security of one protocol for an unbounded number of participants is a first step towards a new kind of protocols and also towards new security proofs. But it still remains a main challenge for the formal tool developers to elaborate new methods to perform such analysis automatically.

4 Quality of Protection Evaluation

The differences in our protocols come from the usage of cryptographic primitives to ensure our security goals. We modelled our protocols using QoP-ML and we used AQoPA tool to analyse them. The model can be found in the QoP-ML models library (included in the AQoPA tool). For each protocol we examine two different scenarios with different key sizes for ECIES encryption and decryption. In the first scenario, we analysed the protocols with AES algorithm in CTR mode with a 128-bit key for symmetric encryption and ECIES for public key encryption with a 128-bit key. In the second scenario, we used a 160-bit key for ECIES. In the Table 2, we provide the real execution time for all our protocols for one intermediate node J, which means that we have the following 4 nodes: I, R, J, S. These results are the averages of 20 experiments of each scenario.

We also give results of simulated execution time obtained with AQoPA tool. Notice that the time measurements slightly differ but remain within the standard deviation. This is due to the variations of execution time in the nodes during the experiments. We used AQoPA tool in order to evaluate the overall overhead of security operations for each protocol for a large number of intermediate nodes in very big networks.

Table 2. Total times of joining new node with one intermediate node

scenario 1 - ECIES - 128b key length					
Protocol name	Runtime of an actual time with S (ms)	Estimated time in AQoPA with S (ms)	Standard deviation (ms)	Estimated time in AQoPA without S (ms)	Gain %
DJS	9954.05	9920.00	123.14	3761.00	62%
IJS_{orig}	10127.32	10207.20	130.96	10071.20	1%
$IJS_{NK,dec/enc}$	10772.80	10823.16	127.40	10517.16	3%
$IJS_{K,dec/enc}$	10745.15	10823.88	125.26	10517.88	3%
$IJS_{NK,onion}$	10758.70	10823.16	126.56	10381.16	4%
scenario 2 - ECIES - 160b key length					
Protocol name	Runtime of an actual time with S (ms)	Estimated time in AQoPA with S (ms)	Standard deviation (ms)	Estimated time in AQoPA without S (ms)	Gain %
DJS	10102.35	10107.48	81.66	4113.48	60%
IJS_{orig}	10355.68	10396.60	109.13	10260.60	1%
$IJS_{NK,dec/enc}$	11072.75	11148.56	137.42	10808.56	3%
$IJS_{K,dec/enc}$	11069.20	11149.28	106.12	10809.28	3%
$IJS_{NK,onion}$	11043.05	11148.56	108.79	10638.56	4%

In Figure 9 (a), we present the execution time for all our protocols in both scenarios for 20, 40, 60, 80 and 100 intermediate nodes. Notice that the execution time for a key of 128 bits is almost equal to 160 bits. This is due to the fact that the code used is optimized for keys of 160 bits. The difference between the two scenarios become bigger when the number of intermediate node increases. Indeed, when the number of intermediate nodes increases, the number of cryptographic operations increases and the difference in execution time becomes bigger for bigger key sizes.

Note that the number of intermediate nodes gives roughly an idea about the radius of the network and not the size of the network. For example, when we evaluate a scenario with 20 intermediate nodes, it means that the furthest point of the network is 20 hops away from the sink. The total number of nodes in the network in that case will depend on the density of nodes. Keep in mind that simultaneous join request can be generated in the network and thus can take place at the same time.

It is important to notice how the time consumption of the original protocol is almost invariant when the number of intermediate nodes rises. Indeed, the main advantage of this protocol is that cryptographic operations are only done on the new node and the sink, intermediate nodes only forward the request and response without doing any additional cryptographic operation.

We also observe that $IJS_{NK,dec/enc}$ and IJS_{onion} protocols are more efficient than $IJS_{K,dec/enc}$. Indeed, for $IJS_{K,dec/enc}$ protocol, the join request has the list of all intermediate nodes starting from the first hop, whereas for $IJS_{NK,dec/enc}$ and IJS_{onion} each intermediate node adds its identifier as it forwards the requests. This makes the request message bigger for $IJS_{K,dec/enc}$ and thus needs more time for encryption and decryption along the route to the sink.

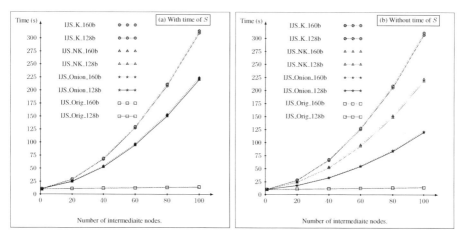

Fig. 9. Execution time of different protocols

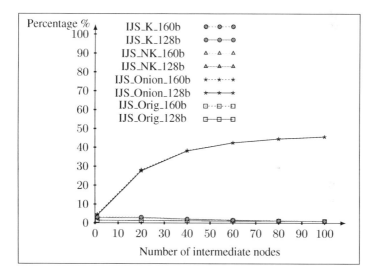

Fig. 10. Ratio of sink execution time over total execution time

Moreover, the curves for IJS_{onion} and $IJS_{NK,dec/enc}$ are very close, because the same cryptographic operations are performed by different nodes. In order to compare them, in Figure 9 (b), we did not include the time consumption at the base station for all our protocols. As expected, the protocol IJS_{onion} is more efficient than the protocols $IJS_{NK,dec/enc}$ and $IJS_{K,dec/enc}$ for the global number of cryptographic operations is less important in intermediate nodes.

In Figure 10, we present the ratio of execution time of the sink over the total execution time of our protocols given in Figure 9 (a). We clearly see that IJS_{onion} is proposed for applications where sensor nodes are energy constrained but not the base station.

5   Conclusion and Discussion

We proposed several multihop node authentication protocols for WSN. We proved the security of all of our solutions using the automatic tool Scyther. Moreover we implemented and tested all our protocols on TelosB nodes in order to evaluate the execution time of each of our solutions. Then we used AQoPA tool to perform an automatic evaluation of the overhead of protection of our solutions. Results show the cost in time consumption when the number of intermediate nodes separating the new node and the base station gets higher.

We studied different protocols that ensure different levels of security depending on the application needs. The original protocol supposes that the application does not need to use the same route for the join request and the join response. Indeed, in that case, all the nodes can participate in the routing operation for the authentication messages. This helped us significantly reduce the number of cryptographic operations. Only the new node and the sink are concerned by these operations which makes this proposal the most suitable one for very large multihop WSNs.

On the other hand, when dealing with more demanding applications, where the intermediate nodes are special nodes and have to be authenticated, more cryptographic operations are needed. We evaluated three protocols that respect that constraint. They differ, on one hand, in the resiliency against node capture attacks, and on the other, in the energy and calculation capacities assumption of the sink. With these protocols, the overhead of node authentication is very high, it reaches almost 5 minutes and 16 seconds in the most consuming scenario for 100 intermediate nodes. With the least consuming protocol, it takes around 2 minutes. Whereas the original takes around 15 seconds for authenticating a new node situated 100 hops away from the sink. The difference is significant and should be taken into account when we need to define the security needs.

We are currently working on the evaluation of key revocation and key renewal protocols for WSNs using Scyther and real testbeds on TelosB nodes. Key revocation and key renewal are very important mechanisms that need to be part of all security protocols. Our objective is to be able to achieve an acceptable security level for these protocols with the smallest number of cryptographic operations to limit the delay generated by the security overhead.

References

1. Al-mahmud, A., Akhtar, R.: Secure sensor node authentication in wireless sensor networks. International Journal of Computer Applications 46(4), 10-17 (2012), Published by Foundation of Computer Science, New York, USA
2. Armando, A., et al.: The AVISPA tool for the automated validation of internet security protocols and applications. In: Etessami, K., Rajamani, S.K. (eds.) CAV 2005. LNCS, vol. 3576, pp. 281–285. Springer, Heidelberg (2005)
3. Blanchet, B.: Automatic proof of strong secrecy for security protocols. In: IEEE Symposium on Security and Privacy, Oakland, California, pp. 86–100 (May 2004)
4. Cremers, C.J.F.: The Scyther Tool: Verification, falsification, and analysis of security protocols. In: Gupta, A., Malik, S. (eds.) CAV 2008. LNCS, vol. 5123, pp. 414–418. Springer, Heidelberg (2008)

5. Cremers, C.J.F., Lafourcade, P., Nadeau, P.: Comparing state spaces in automatic security protocol analysis. In: Cortier, V., Kirchner, C., Okada, M., Sakurada, H. (eds.) Formal to Practical Security. LNCS, vol. 5458, pp. 70–94. Springer, Heidelberg (2009)

6. Daemen, J., Rijmen, V.: The Design of Rijndael: AES - The Advanced Encryption Standard. Springer (2002)

7. Das, M.L.: Two-factor user authentication in wireless sensor networks. IEEE Transactions on Wireless Communications 8(3), 1086–1090 (2009)

8. Dolev, D., Yao, A.C.: On the security of public key protocols. In: Proceedings of the 22Nd Annual Symposium on Foundations of Computer Science, SFCS 1981, pp. 350–357 (1981)

9. Han, K., Shon, T.: Sensor authentication in dynamic wireless sensor network environments. International Journal of RFID Security and Cryptography (2012)

10. Jürjens, J.: Secure systems development with UML. Springer (2005)

11. Kavitha, T., Sridharan, D.: Security vulnerabilities in wireless sensor networks: A survey. Journal of Information Assurance and Security 5, 31–34 (2010)

12. Ksiezopolski, B.: QoP-ML: Quality of protection modelling language for cryptographic protocols. Computers & Security 31(4), 569–596 (2012)

13. Ksiezopolski, B., Kotulski, Z.: Adaptable security mechanism for dynamic environments. Computers & Security, pp. 246–255 (2007)

14. Ksiezopolski, B., Kotulski, Z., Szalachowski, P.: Adaptive approach to network security. In: Kwiecień, A., Gaj, P., Stera, P. (eds.) CN 2009. CCIS, vol. 39, pp. 233–241. Springer, Heidelberg (2009)

15. Ksiezopolski, B., Kotulski, Z., Szalachowski, P.: On qop method for ensuring availability of the goal of cryptographic protocols in the real-time systems. In: European Teletraffic Seminar 2011 (2011)

16. LeMay, E., Unkenholz, W., Parks, D., Muehrcke, C., Keefe, K., Sanders, W.H.: Adversary-driven state-based system security evaluation. In: Proceedings of the 6th International Workshop on Security Measurements and Metrics, MetriSec 2010, pp. 5:1–5:9. ACM (2010)

17. Lindskog, S.: Modeling and Tuning Security from a Quality of Service Perspective. PhD thesis, Chalmers University of Technology (2005)

18. Liu, A., Ning, N.: Tinyecc: A configurable library for elliptic curve cryptography in wireless sensor networks. In: 7th International Conference on Information Processing in Sensor Networks, pp. 245–256 (April 2008)

19. Lowe, G.: Breaking and fixing the needham-schroeder public-key protocol using fdr. Software - Concepts and Tools 17(3), 93–102 (1996)

20. Luo, A., Lin, C., Wang, K., Lei, L., Liu, C.: Quality of protection analysis and performance modeling in ip multimedia subsystem. Comput. Commun. 32(11), 1336–1345 (2009)

21. Manica, N., Saloni, M., Toldo, P.: WSN - secure comunications with AES algoritms. University of Trento - Faculty of Computer Science (2008)

22. Mansour, I., Chalhoub, G., Misson, M.: Security architecture for multi-hop wireless sensor networks. CRC Press Book (2014)

23. Mansour, I., Lafourcade, P.: Scyther code of our authentication protocols (December 2013), http://sancy.univ-bpclermont.fr/~lafourcade/scyther-code.tar

24. Petriu, D.C., Woodside, C.M., Petriu, D.B., Xu, J., Israr, T., Georg, G., France, R., Bieman, J.M., Houmb, S.H., Jürjens, J.: Performance analysis of security aspects in uml models. In: Proceedings of the 6th International Workshop on Software and Performance, WOSP 2007, pp. 91–102. ACM (2007)

25. Prérez, V.B., González, P., Cabaleiro, J.C., Heras, D.B., Pena, T.F., Pombo, J.J., Rivera, F.F.: Avispa: Visualizing the performance prediction of parallel iterative solvers. Future Generation Comp. Syst. 19(5), 721–733 (2003)

26. C. Research. Standards for efficient cryptography, sec 1: Elliptic curve cryptography (September 2000)
27. Sun, Y., Kumar, A.: Quality-of-protection (qop): A quantitative methodology to grade security services. In: ICDCS Workshops, pp. 394–399. IEEE Computer Society Press (2008)
28. Yeh, H.-L., Chen, T.-H., Liu, P.-C., Kim, T.-H., Wei, H.-W.: A secured authentication protocol for wireless sensor networks using elliptic curves cryptography. Sensors 11(5) (2011)
29. Zhang, J., Shankaran, R., Orgun, M.A., Sattar, A., Varadharajan, V.: A dynamic authentication scheme for hierarchical wireless sensor networks. In: Sénac, P., Ott, M., Seneviratne, A. (eds.) MobiQuitous 2010. LNICST, vol. 73, pp. 186–197. Springer, Heidelberg (2012)

Performance Analysis of Aggregation Algorithms for Vehicular Delay-Tolerant Networks

João N.G. Isento[1], Joel J.P.C. Rodrigues[1], Sandra Sendra[1,2],
and Guangjie Han[3,*]

[1] Instituto de Telecomunicações, University of Beira Interior, Covilhã, Portugal
[2] Universidad Politécnica de Valencia, Spain
[3] Department of Information & Communication Systems, Hohai University, China
joao.isento@it.ubi.pt, joeljr@ieee.org, sansenco@posgrado.upv.es,
hanguangjie@gmail.com

Abstract. Vehicular delay-tolerant networks (VDTNs) were proposed as an alternative network architecture for sparse vehicular communications and provide inter-networking between heterogeneous networks. The main objective of this work includes the presentation of solutions to aggregate IP packets into VDTN bundles and their performance assessment analysis. This paper presents the performance evaluation of several aggregation algorithms based on time and bundle size values with different traffic loads. Different classes of datagrams are used to evaluate the quality of service (QoS) in VDTNs. Single-class and composite-class schemes were created to optimize the aggregation of different packet classes. A laboratory testbed, called VDTN@Lab, was used to evaluate the performance of these algorithms. The hybrid algorithm and the composite-class scheme present the best performance for different types of traffic load and best priorities distribution, respectively.

Keywords: Vehicular Delay-Tolerant Networks, VDTN, Bundle Assembly, Vehicular and wireless technologies, Disruption tolerant networking, Quality of Service.

1 Introduction

Vehicular networks emerged as novel wireless network approach as a result of the advances of wireless technology and automotive industry. This kind of network can be employed in large real-world scenarios. They can be applied as the communication infrastructure in emergency scenarios [1] or to enable connectivity to rural regions [2, 3]. However, these networks introduce some single problems due to the establishment of network connectivity among vehicles and between vehicles and infrastructures. Most of these issues arise from the high mobility of

---

* This work has been partially supported by *Instituto de Telecomunicações*, Next Generation Networks and Applications Group (NetGNA), Covilhã Delegation, by National Funding from the FCT - *Fundação para a Ciência e Tecnologia* through the Pest-OE/EEI/LA0008/2013 Project.

S. Guo et al. (Eds.): ADHOC-NOW 2014, LNCS 8487, pp. 419–431, 2014.
© Springer International Publishing Switzerland 2014

vehicles, intermittent connectivity and disruption, limited transmission ranges, and large distances between network nodes. Due to these issues an end-to-end path from a source to destination often does not exists. In order to overtake the above-mentioned problems, some approaches were proposed, including Vehicular Ad-hoc Networks (VANETs), Delay Tolerant Networks (DTNs), and Vehicular Delay Tolerant Networks (VDTNs), that are considered in this work.

Vehicular ad hoc networks (VANETs) [4] follow the main characteristics of mobile ad hoc networks (MANETs) [5] where mobile nodes are vehicles such as buses or cars. The movement of mobile nodes is limited to roads and the network performance is highly affected by traffic flow. Two types of communications can be performed, vehicle to vehicle and vehicle to infrastructure. This road infrastructure is based on node placement along the roads improving connectivity and service arrangement.

In order to provide connectivity between heterogeneous networks in high mobility scenarios, delay tolerant networks (DTNs) [6] can deal with considerable aspects such as sparse and intermittent connectivity, low transmission reliability and it is assumed that an end-to-end link in not available. DTN layered architecture considers a new layer, called Bundle layer, under the Application layer allowing a link between heterogeneous networks and distinct transmission rates. This Bundle layer is responsible to assembling several application data packets into one or more protocol data units, called bundles. Through this layer, the store-carry-and-forward paradigm allows the bundle transmission over DTN nodes.

Vehicular delay-tolerant networks (VDTNs) [7] were proposed as vehicular communication solution that uses vehicles as a network infrastructure to provide connectivity to a wide range of scenarios, offering a low-cost asynchronous communications, variable delays, and bandwidth constraints. VDTNs consider three different types of nodes: terminal, relay, and mobile nodes. Terminal nodes, usually placed at the edge of the network (both fix and mobile nodes), are considered the access points to the VDTN network. Mobile nodes move along roads, following a random or a predicted movement, allowing data for being carried between terminal nodes. In certain scenarios mobile nodes may act as terminal nodes by generating and receiving data. Finally, relay nodes are considered to increase the network connectivity and the number of contact opportunities. The increase of contact opportunities will allow the increase of bundle delivery probability, which leads to a decrease of the bundle average delay. VDTNs, unlike DTNs, place the bundle layer under the network layer, in order to aggregate incoming IP packets into large size packets, called data bundles. VDTNs also support a store-carry-and-forward paradigm to overcome several issues that arise from disconnection and intermittence. An illustration of these layers interaction can be seen at Figure 1. Although all these features, VDTNs still have to overlap some problems related to network connectivity. These problems arise mainly from the conjunction of the high mobility of vehicles and its velocity. These two aspects will have great impact in the network performance given constant changes on the network topology and short contact durations. This motivates the use of

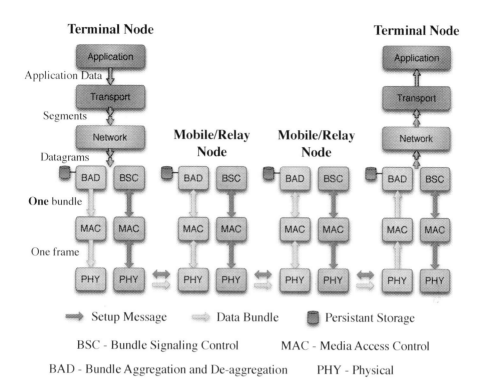

Fig. 1. Messages and Data Bundles transmissions between VDTN nodes, from a source to a destination

several approaches in order to maximize the use of the few available resources that improve the network performance. Examples of these approaches are the use of fragmentation mechanisms [8] or bundle aggregation/de-aggregation algorithms.

Essentially, the idea behind aggregation algorithms on VDTNs and bundle placement under the network layer comes from the optical burst switching networks (OBS) approach and it was used with success on this technology [9]. In OBS, Internet protocol (IP) datagrams are assembled into large data packets, called bursts. These burst are generated taking into account an amount of data with similar characteristics (e.g. same destination, application, or priority). In order to study the impact of aggregation on VDTNs, this paper presents, discusses, and analyses the performance of several bundle aggregation/de-aggregation algorithms for VDTNs.

This paper addresses the problem of data aggregation and de-aggregation on VDTNs. Then, it proposes a set of aggregation and de-aggregation algorithms for VDTNs. Several approaches are considered taking into account the time and size parameters. The bundle generation process is based on a timer unit and/or a threshold that defines the maximum number datagrams per bundle. The proposed timer-based solution considers a time interval for bundle generation while threshold-based approach has a predefined maximum number of datagrams per bundle. A third proposal is based on a hybrid solution with the combination of the time and threshold parameters, trying to find the best performance for bundling. The main contributions of the paper are the following:

– Creation of bundle assembly algorithms with QoS support for VDTNs;
– Comparison analysis and validation of the proposed algorithms;
– Performance evaluation and demonstration through a laboratory testbed.

The remainder of the paper is organized as follows. Section 2 elaborates on the related literature about data aggregation algorithms, mainly, focusing on vehicular communications. Section 3 presents the proposed aggregation algorithms considering also QoS support mechanisms, while Section 4 presents and describes the laboratory VDTN testbed. The performance assessment analysis of the aggregation algorithms for VDTNs are presented in Section 5. Finally, Section 6 concludes the paper and points some directions for future work.

2   Related Work

In this section, several aggregation solutions to real problems in vehicular communications are addressed. These problems show how important the aggregation algorithms are to improve the performance of protocols, services, and applications in vehicular networks.

Mobile ad-hoc networks face security risks and energy consumptions issues due to limits resources and its infrastractureless network environment. To perform a monitoring and detection in a MANET the authors in [10] develop both lossless and lossy aggregation mechanisms to reduce the energy cost and bandwidth consumption while preserving the detection accuracy. Lossless aggregation intent to reinforce the detection information, in which the de-aggregated information contains the same data as the original one. In the lossy mechanism, the decompressed data contains most of the original information accepting some loss of fidelity. Real-world experiments and simulations are conducted to assess the effectiveness of this solution.

VoIP traffic over mobile ad-hoc networks faces several issues related to the mobility of the nodes and routing protocols introducing latency and packet loss rate. Authors in [11] present several solutions to improve the quality of VoIP calls in MANET environment. One of these solutions relates to aggregate several VoIP frames within a single packet. Due to this feature, and depending on the voice codec used, the performance of the VoIP traffic was improved. Several simulations were performed using the NS-2 simulator with the *NS2Voip++* extension and each simulation had the duration of 4000 seconds.

K-hop bandwidth aggregation [12, 13] is proposed for enhance cooperative video streaming over hybrid vehicular ad-hoc networks and 3G/3.5G cellular network. In this approach each member (vehicle) in the same group is known to each other and share the same path and the same destination. The video requests are sent over 3G/3.5G network. However, the bandwidth in this network is very limited and not support higher quality video. In order to improve the quality of video playback the requester can ask to other members, called helpers, to cooperatively download the requested video through their cellular network. Then, all video "fragments" are sent to the requester over ad hoc network. k-hop aggregation solves the problem of discovering the best helpers and how the data is sent to the requester. This approach can get better quality video that cooperative video streaming without the K-hop aggregation scheme.

In a vehicular ad hoc network, one of the main issues is the problem of the message routing, where the location of the destination node is unkown due to the high node mobility. One of the best solution for this problem is to use location service protocols. However, these protocols require large volume of signaling overhead which can result in network traffic congestion and high resource consumption. Region-based Location Service Management Protocol (RLSMP) [14] uses message aggregation enhanced by geographical clustering to minimize the volume of the signaling overhead. In this protocol, each vehicle frequently check its geographical cluster while moving, without any additional communication or delay. RLSMP uses message aggregation to bypass the drawbacks of independent updating and querying messages. The messages are aggregated so that data is grouped, resulting in more efficient bandwidth usage and minimal communication overhead. This protocol guarantees scalability with the increase of the number of nodes in the network.

In [15], authors propose an adaptive packet aggregation for header compression. Aggregation packets in a vehicular environment can cause redundant information and, therefore, high volume of traffic. This algorithm, first, perform static header compression, and then aggregate a dynamic number of packets according to the delay constraints. Using the adaptive procedure, the system will adaptively change the number of aggregated packets according to the current end-to-end delay. Simulation results, using the NS-2 simulator, showed that the proposed aggregation algorithm outperforms the existing schemes in terms of end-to-end delay and packet delivery rate.

In [16–18], the authors intent to study the effect of aggregation of DTN bundles on space communications. This work tries to find if aggregation of multiple DTN bundles has advantages over the default approach of "one bundle per block". In space communications, bundles are passed as "service data units" to Licklider Transmission Protocol (LTP) for transmission. LTP is an end-to-end protocol for deep space communications. These service data unites are encapsulated in LTP blocks. Then, each LTP block is divide into LTP data segments, according the maximum transmission unit (MTU) size. To reduce the amount of ACK traffic, multiple DTN bundles are aggregated into a single LTP block. The aggregation of multiple bundles has significant advantages over the default approach.

The complete failure of telecommunications infrastructures is a common characteristic when a disaster occurs. A DTN-based solution is presented in [19] in order to aggregate as much information as possible from an area of interest (AoI) within a disaster zone. Users using mobile phones generate messages with disaster-related information and aggregate them with their respective coverage areas into a new message to minimize the overall message collection delay. To prevent duplicate message from users, this system a filter is constructed to drop possible duplicated messages. Through simulation, using a real geographical map, this solution achieved a smaller delay in message delivery.

All of these aforementioned solutions have contributed to the conception and design of the proposed VDTN aggregation algorithms. All of these aggregation schemes show the influence that the packet aggregation has in vehicular communications.

3   Aggregation and De-aggregation Algorithms

Bundle aggregation algorithms,, proposed in [20, 21], collect contributions from the burst assembly schemes of optical burst switching networks (OBS). In OBS networks, burst assembly is the mechanism of assembling incoming packets from the higher layer into bursts at the ingress edge node of the OBS network. As data units arrive from the higher layer, they are stored in queues according to their destination and class. The burst assembly algorithm place these packets into bursts based on some assembly approach. Based on this available method for OBS, VDTN bundle aggregation approach follows a similar process. Figure 2 illustrates the schematic mechanisms of the bundle assembly unit. In this case, incoming IP packets (datagrams) arrive at the Routing Module (RM) that distribute in function of their destination node and places them in the corresponding queue (q_0, q_1, ..., q_N) at the Bundle Assembly Module, based on packets destination. If available, the class of each packet is also considered and packets are distributed on sub-queues (c_0, ..., c_M). Based on the assembly mechanism, data bundles are generated at this module and sent to network core when a contact opportunity are available.

The main aspect in bundle assembly is the generation criteria for determining when a VDTN bundle should be created. There are two main types of bundle aggregation techniques: timer and threshold based algorithms. A combination of these two algorithms, called hybrid algorithm, is considered in this work. All details are explained below.

Timer-Based Algorithms. Usually, a bundle is created and injected into VDTN network at periodic time intervals following time-based algorithms. These algorithms use a timer to indicate an assembly cycle of each queue. There is a fixed threshold (T_N) that act as the primary criterion to create a bundle. Thereby, they provide almost uniform gaps between successive bundles from the same queue into a VDTN network. However, the bundle length changes with the offered load.

Threshold-Based Algorithms. In threshold-based algorithms, a bundle is created and injected into a VDTN once the assembly data length reaches or exceeds the given threshold (L_N). Threshold-based algorithms generate bundles at

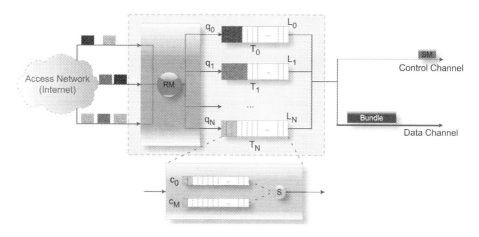

Fig. 2. llustration of the bundle assembly process

non-periodic time intervals. A length threshold is used as the primary criterion to create a bundle. The threshold is placed as a limiting parameter on the minimal length or the number of contained packets. In these algorithms, once the threshold is reached, all the packets in the queue i are assembled into a bundle.

Hybrid Algorithms. The above two types has some shortcomings under certain conditions. For example, the threshold-based algorithms may will experience a long assembly waiting time when the input traffic is low. Timer-based algorithms solve this problem, however, under high traffic conditions, it will have a larger variance of bundle length. Both types of algorithms are too rigid and not adapt their parameters according to input traffic conditions. Then, some hybrid solution have been proposed. These algorithms can dynamically 2 the time, threshold, or both values, according to the real-time input traffic situation.

Bundle Assembly Framework for QoS Support. The primary issues appear to decide which class of packets and how many packets of each class to put into a same bundle. To provide QoS support, the bundle assembly policies should take into account the number of packet classes as well as the number of bundle priorities supported at the VDTN core. In this work it is assumed the number of classes and the number of priorities is the same.

Let N be the number of input packet classes at the edge and let M be the number of bundle priorities supported in the VDTN core. Given packet classes and bundle priorities, the objective is to meet the QoS requirements by defining a set of bundle types, which specify how packets are aggregated, and by assigning an appropriate bundle priority to each bundle type. In this model, the length of a bundle is defined by the number of packets on it. This process works on the sub-queues of each queue of packets at the bundle assembly unit. Basically, two main approaches exists as shown in Figure 3.

Packet
Class

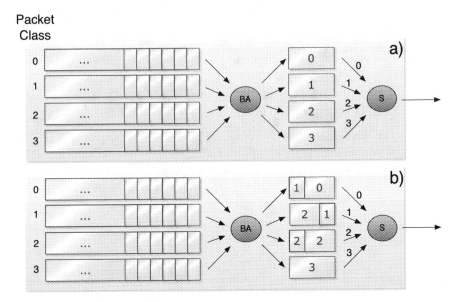

Fig. 3. Proposed aggregation algorithms with QoS support: (a) Single-class bundle with $N = M = 4$; (b) Composite-class bundle with $N = M = 4$

In Figure 3 (a) is shown the Single-Class Bundle with $N = M$ algorithm. In this case, a bundle type can be created such that each bundle contains a single class of packets. The priority of a bundle will directly correspond to a specific class of packets contained in the bundle. When the timer or threshold is reached, depending on current algorithm, the bundles are created.

The Composite-Class Bundle with $N = M$, illustrated in Figure 3 (b), allows different classes of packets being placed in the same bundle. The packets are placed in the bundle following a decreasing order of class, such that the higher class packets are located at the head of the bundle. If a connection failure occurs, the bundle is fragmented and the higher class packets are transmitted first. The bundle priority is equal to higher class packet contained in that bundle.

De-aggregation. As above-mentioned, the aggregation process occurs at the source terminal nodes that generate bundles for VDTN network. Then, when these bundles arrive at the destination terminal node, each one will be de-aggregated into their original datagrams. These disassembled datagrams are delivered to the upper layers till the Application Layer or forwarded to the external network (e.g. Internet). The de- aggregation process is shared by all the aggregation algorithms.

4   VDTN Laboratorial Testbed

The VDTN@Lab testbed was created in a laboratory environment, and its design allows the update of different hardware and software components with minimal impact on the others. The testbed considers the three types on VDTN nodes: terminal, relay, and mobile nodes. Terminal nodes act as the access points to the VDTN network and may be fixed or mobile. Stationary relay nodes are located at crossroads and are equipped with store-and-forward capabilities. Being at crossroads, the relay nodes increase the number of contact opportunities in scenarios with low node density [22]. Mobile nodes, which interact with relay nodes, are vehicles opportunistically exploited to collect and disseminate data bundles through the network. They move along roads carrying data that must be delivered to the terminal nodes.

Fig. 4. Illustration of network nodes interactions: a) mobile node interacting with a relay nodes; b) mobile nodes interacting with each other; c) mobile node interacting with a terminal node; d) mobile nodes interacting with other nodes and following paths.

To emulate the terminal and relay nodes, several laptops and desktops are used. To emulate mobile nodes (e.g vehicles) LEGO MINDSTORMS NXT robotic cars and a netbook computer are used. LEGO NXT robots are programmed with several mobility models (e.g., bus movement or random movement across roads), allowing performance assessment studies under different movement patterns. Figure 4 illustrates the interactions between all network nodes, and the mobility patterns followed by the mobile nodes. In order to allow communication between networks nodes, several software modules are deployed into network nodes. These software modules implement the above-presented concepts of the VDTN architecture, as well as several routing protocols and scheduling and dropping policies. They also provide several functionalities to emulate specific network constraints (e.g. range, storage).

5   Performance Analysis

The aggregation algorithms have been evaluated over different input traffic conditions. Input traffic (datagrams) is created in the same application of terminal nodes and this application generates different types of traffic load (considering low, medium, and high). In such a case of a low traffic load, datagrams are created in a time interval of 1 up to 3 seconds; medium traffic load in the time interval of the 0.5 up to 1.5 seconds, and the high traffic load in the time interval of the 0.25 up to 0.75 seconds. These datagrams have random destination terminal nodes and its size is uniformly distributed and average length is 1250 bytes. Data bundles have a time-to-live (TTL) fixed on 10 minutes. The timer and threshold values changes among different testbed experiments in order to evaluate the performance of bundle assembly. For the timer-based algorithm the timer value is fixed on 120 seconds. Threshold-based algorithm threshold value is equal to 150 datagrams. The hybrid algorithm has both the timer and threshold values. Three terminal nodes, two relay nodes and four mobile nodes are considered in this work. Figure 5 show the average aggregation delay as function of traffic loads (easy, medium, and high).

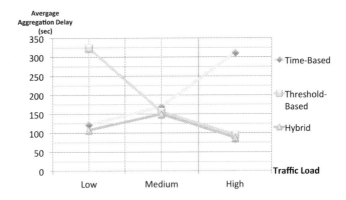

Fig. 5. Performance evaluation of proposed algorithms: Average aggregation delay as function of traffic load for different traffic loads (low , medium, and high).

For this study, all the aggregation algorithms above described are used. As may be seen, the Hybrid Algorithm presents the best results and gains of approximately 14, 20, and 225 (for low, medium, and high traffic loads) when compared to the Timer-based Algorithm. Comparing Hybrid Algorithm with the Threshold-based Algorithm, it presents gains of approximately 216, 6, and 7 seconds. Under low traffic load, timer-based algorithms are the best technique for bundle generation and, under high traffic level, threshold-based algorithms have a better behavior for improvement of VDTN bundle aggregation. Thus, it is natural that Hybrid Algorithm, based on the two approaches, presents the best performance.

For the QoS schemes, we intent to prove that the QoS requirements of incoming packets not change with bundle assembly. For this, a uniform distribution of classes number was calculated. Equation (1) presents a possible distribution. Value $p(i)$ represent the percentage of packets with class i. $p(j)$ represent the percentages of previous classes with $i \geq j \leq n - 2$.

$$p(i) = (\frac{100 - (\sum_{j}^{n-1} p(j))}{i + 2}) \times 2 \tag{1}$$

Incoming packets were defined with four classes: Class 0 (more priority), Class 1, Class 2, and Class 3 (less priority). By (1), the distribution is about 40%, 30%, 20%, and 10% respectively. It is assumed that incoming packets arrival at the terminal node according to a Poisson process and average packet length is about 1250 Bytes. The four schemes above described has been considered on the experiments: Single-Class Bundle with $N = M = 4$ (Single 4:4), Composite-Class Bundle with $N = M = 4$ (Composite 4:4).

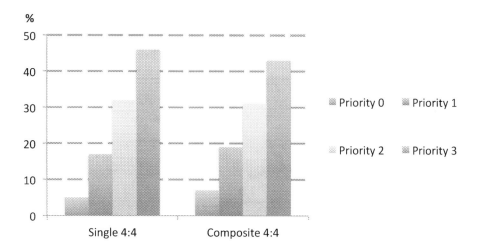

Fig. 6. Performance evaluation of proposed algorithms: Distribution of bundle priorities after aggregation algorithms

Figure 6 shown the distribution, in percentage, of priorities applied in VDTN network core. The gray lines represent the optimal values of these percentages given by (1). In general, four schemes produce an acceptable distribution of priorities, along with the Composite schemes having better results. This behavior occurs because the aggregation of different classes in a single bundle.

6   Conclusion and Future Work

This paper studied the mechanisms of aggregation and de-aggregation for vehicular delay-tolerant networks (VDTNs) and presented several algorithms evaluating their impact on the overall performance of VDTNs emulated in a laboratory testbed, called VDTN@Lab. The main goal of the study was the evaluation of different aggregation mechanisms, including approaches with QoS support, to find the best approach that should improve the bundle aggregation delay and the best priorities distribution. It was concluded that the Adaptive Hybrid Algorithm is the best bundle assembly algorithm under any traffic load condition. With QoS support, the Composite-Class Bundle presents the best results. As a future research is important to study the behavior of the edge network and know if this QoS framework produces significant changes of traffic.

References

1. Asplund, M., Nadjm-Tehrani, S., Sigholm, J.: Emerging information infrastructures: Cooperation in disasters. In: Setola, R., Geretshuber, S. (eds.) CRITIS 2008. LNCS, vol. 5508, pp. 258–270. Springer, Heidelberg (2009)
2. Networking for Communications Challenged Communities: Architecture, Test Beds and Innovative Alliances, http://www.n4c.eu/
3. Wizzy Digital Courier - leveraging locality, http://www.wizzy.org.za/
4. Maqsood, A., Khan, R.: Vehicular ad-hoc networks. International Journal of Computer Science 9(1), 401–408 (2012)
5. IETF MANET Working Group, http://datatracker.ietf.org/wg/manet/charter/
6. Cerf, V., Burleigh, S., Hooke, A., Torgerson, L., Durst, R.: Delay-tolerant networking architecture. RFC 4838 (2007)
7. Isento, J., Rodrigues, J., Dias, J., Paula, M., Vinel, A.: Vehicular delay-tolerant networks - a novel solution for vehicular communications. IEEE Intelligent Transportation Systems Magazine 5(4), 10–19 (2013)
8. Dias, J., Rodrigues, J., Isento, J.: Performance assessment of fragmentation mechanisms for vehicular delay-tolerant networks. EURASIP Journal on Wireless Communications and Networking (1), 195 (2011)
9. Rodrigues, J.: Optical Burst Switching Networks: Architectures and Protocols. Fundação Nova Europa, Universidade da Beira Interior Portugal (2008)
10. Zhang, D., Ge, L., Hardy, R., Yu, W., Zhang, H., Reschly, R.: On effective data aggregation techniques in host-based intrusion detection in manet. In: IEEEConsumer Communications and Networking Conference (CCNC 2013), pp. 85–90 (January 2013)
11. Mysirlidis, C., Galiotos, P., Dagiuklas, T., Kotsopoulos, S.: Performance quality evaluation with voip traffic aggregation in mobile ad-hoc networks. In: IEEE 14th International Symposium and Workshops on a World of Wireless, Mobile and Multimedia Networks (WoWMoM 2013), pp. 1–6 (June 2013)
12. Huang, C.-M., Yang, C.-C., Lin, H.-Y.: A bandwidth aggregation scheme for member-based cooperative networking over the hybrid vanet. In: IEEE 17th International Conference on Parallel and Distributed Systems (ICPADS 2011), pp. 436–443 (December 2011)

13. Huang, C.-M., Lin, T.-H., Tseng, K.-C.: Bandwidth aggregation over vanet using the geographic member-centric routing protocol (gmr). In: 12th International Conference on ITS Telecommunications (ITST 2012), pp. 737–742 (November 2012)
14. Saleet, H., Basir, O.: Location-based message aggregation in vehicular ad hoc networks. In: IEEE Globecom Workshops, pp. 1–7 (November 2007)
15. Wang, T.-P., Chen, Y.-C.: Adaptive packet aggregation for header compression in vehicular wireless networks. In: IEEE 13th International Conference on High Performance Computing and Communications (HPCC 2011), pp. 935–939 (September 2011)
16. Wang, R., Wei, Z., Zhang, Q., Hou, J.: Ltp aggregation of dtn bundles in space communications. IEEE Transactions on Aerospace and Electronic Systems 49(3), 1677–1691 (2013)
17. Wei, Z., Wang, R., Zhang, Q., Hou, J.: Aggregation of dtn bundles for channel asymmetric space communications. In: IEEE International Conference on Communications (ICC 2012), pp. 5205–5209 (June 2012)
18. Hu, J., Wang, R., Zhang, Q., Wei, Z., Hou, J.: Aggregation of dtn bundles for space internetworking systems. IEEE Systems Journal 7(4), 658–668 (2013)
19. Fajardo, J., Yasumoto, K., Shibata, N., Ito, M.: Dtn-based data aggregation for timely information collection in disaster areas. In: IEEE 8th International Conference on Wireless and Mobile Computing, Networking and Communications (WiMob 2012), pp. 333–340 (October 2012)
20. Isento, J., Dias, J., Rodrigues, J., Chen, M., Lin, K.: The effect of bundle aggregation on the performance of vehicular delay-tolerant networks. In: IEEE 8th International Conference on Mobile Adhoc and Sensor Systems (MASS 2011), pp. 813–818 (October 2011)
21. Isento, J., Dias, J., Rodrigues, J., Chen, M., Lin, K.: Performance assessment of aggregation and de-aggregation algorithms for vehicular delay-tolerant networks. In: IEEE 8th International Conference on Mobile Adhoc and Sensor Systems (MASS 2011), pp.158–160 (October 2011)
22. Paula, M., Isento, J.N., Dias, J., Rodrigues, J.: A real-world VDTN testbed for advanced vehicular services and applications. In: IEEE 16th International Workshop on Computer Aided Modeling and Design of Communication Links and Networks (CAMAD 2011), pp. 16–20 (2011)

VEWE: A Vehicle ECU Wireless Emulation Tool Supporting OBD-II Communication and Geopositioning

Óscar Alvear, Carlos T. Calafate, Juan-Carlos Cano, and Pietro Manzoni

Department of Computer Engineering (DISCA),
Universitat Politècnica de València, Spain
oalvear@gmail.com, {calafate,jucano,pmanzoni}@disca.upv.es

Abstract. Almost all the vehicles built during the last decade integrate an On Board Diagnostic (OBD-II) interface, through which it is possible to monitor and manage multiple operational parameters. In the past few years, Bluetooth OBD-II devices have been introduced in the market to facilitate connection to mobile devices. With the increased use of these devices, many applications for real-time control and monitoring of different parameters are being developed in the automotive sector. This infrastructure has opened a broad research area related to "smart driving". The main problem in the development and testing of these applications is the need to debug and validate them using different vehicles, under different configurations and scenarios. Our proposal to address this problem is VEWE: Vehicle Wireless Emulator offering realistic vehicle dynamics, which allows testing the correctness and the performance of mobile applications using off-the-shelf computers, thereby speeding up the development time and lowering costs. Another useful functionality of VEWE is the emulation of Geo-positions in Android systems through a GPS-Emulator. By combining both functionalities, VEWE provides a complete and flexible development environment for mobile vehicular applications.

1 Introduction

Mobile devices have experienced a technological breakthrough in recent years, evolving towards high performance terminals with multi-core microprocessors, being smartphones a clear representative exponent of this trend. In addition, the On Board Diagnostics (OBD-II) [1] standard, available since 1994, has recently become an enabling technology for in-vehicle applications due to the appearance of Bluetooth OBD-II connectors [2]. These connectors enable a transparent connectivity between the mobile device and the engine's Electronic Control Unit (ECU).

The range of possibilities that arise when combining cars and smartphones is endless, allowing, for example, diagnosing the car via mobile devices which assume the tasks that are typically performed by the On Board Unit (OBU) of the vehicle, sending the collected data to a platform where diagnosis and vehicle

S. Guo et al. (Eds.): ADHOC-NOW 2014, LNCS 8487, pp. 432–445, 2014.

maintenance can be done, detecting possible failures automatically, or developing solutions to automatically analyze driver behavior and suggest corrective actions when necessary.

Currently, it is already possible to find several smartphone-based applications that rely on OBD-II communications [3]. However, the development of these applications is costly since the developer has to deal with real ECUs from different manufacturers in order to test and debug the applications, usually requiring taking the vehicle for short test trips. To avoid this requirement, and to speed-up the development process, in this paper we propose VEWE, a vehicle wireless emulator that includes map-based mobility modeling, and a simulated engine ECU accessible through a wireless interface. The functionality provided by VEWE allows the developers to test OBD-II based smartphone applications as if moving in a real vehicle.

This paper is organized as follows: in section 2 we present some related works, evidencing how our work differs from previous ones. In section 3 we provide an overview of the OBD-II standard. The VEWE solution is introduced in section 4, and technical details are provided in section 5. Finally, section 6 presents the main conclusions of this paper.

2   Related Works

Currently we can find a broad range of smartphone applications able to communicate with a vehicle's ECU to provide enhanced services to drivers in the scope of Intelligent Transportation Systems (ITS) area.

Torque [4] is a well known Android application that allows monitoring all sorts of parameters available through the OBD-II interface (e.g. vehicle speed, engine RPM, Fuel pressure), offering the user a comfortable visualization by allowing to personalize the actual size and position of data displays.

Teng et al. [5] implement an Android-based mobile device platform able to read data on the vehicle ECUs; results are graphically displayed as a virtual instrument on the mobile panel.

Meseguer et al. proposed DrivingStyles [6], a solution combining an Android-based application and a web platform that is able to determine the type of road where the driver is circulating, as well as his driving habits. Its main goal is to help promoting a safer and more ecological driving style by making drivers more conscious about their behavior on the road.

Zaldivar et al. [7] proposed an Android-based application that monitors the vehicle through the On Board Diagnostics (OBD-II) interface, being able to detect accidents and sending details about the accident to pre-defined destinations through either e-mail or SMS; these tasks are immediately followed by an automatic phone call to the emergency services.

Tahat et al. [8] developed an Android application able to monitor the vehicle's fuel consumption and other vital electromechanical parameters. Data can also be sent to the vehicle's manufacturer maintenance department, allowing to detect and predict vehicle faults while moving.

The aforementioned research works highlight the interest in developing smartphone applications that interact with vehicles. Nevertheless, since available OBD-II emulation tools are too simple and lack GPS integration [9,10], developers should perform test driving to evaluate their proposals, which is expensive and time consuming. The platform presented in this paper aims to simplify and accelerate development by modeling vehicle dynamics in detail and allowing to jointly emulate OBD-II communications and geopositioning, thus meeting the basic requirements of mobile application developers working in this field.

3   The OBD-II Standard

The On-board Diagnostic (OBD) standards [11] were developed in the USA to detect car engine problems that can provoke an increase of $CO2$ gas emission levels beyond acceptable limits. To achieve this purpose, the system is constantly monitoring the different elements related to gas emissions, including engine management functions, being a powerful tool to diagnose problems on vehicles' electrical systems. When a failure is detected, the system must store it in its memory so that technicians may analyze it later on.

The first OBD standard, known as OBD-I, defined just a few parameters to monitor, and did not establish a specific emission level for vehicles. Thus, failures resulted in just a visual warning to the driver and the storage of the error. The second generation of OBD, known as OBD-II, standardizes different elements such as the connector used for diagnostic, the electrical signaling protocols, and the message format. Additionally, it defines a list of parameters that can be monitored, assigning a specific code to each parameter. A detailed list of DTCs (Diagnostic Trouble Codes) is also defined in the standard (see [11]).

Several operating modes are defined by the OBD-II standard to provide an easier interaction with the system, and achieve the desired functionality. Most automobile manufacturers have introduced additional operation modes that are specific to their own vehicles, thus offering a full control of the available functionality.

The European version of the OBD-II standard, known as EOBD, is mandatory for all gasoline and diesel vehicles since 2001 and 2003, respectively. Despite it introduces some small improvements, EOBD strongly resembles OBD-II, sharing the same connectors and interfaces.

Figure 1 shows an example of both male and female OBD-II connectors. In particular, the male connector shown in the figure is part of a Bluetooth-enabled OBD-II device that offers a bridge between the vehicle's internal bus and a smartphone using a Bluetooth connection.

3.1   Communication Protocols

Although the physical interface is well defined, the communications protocol varies depending on the manufacturer. Different protocols are available: (i) SAE J1850 PWM (Pulse-Width Modulation) [12], (ii) SAE J1850 VPW (Variable

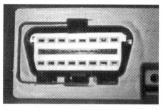 

(a) Female connector   (b) Male connector.

Fig. 1. Example of a) an in-vehicle OBD-II female connector, and b) a Bluetooth-enabled OBD-II device with male connector

Pulse Width) [12], (iii) ISO 9141-2 [13], (iv) ISO 14230 KWP2000 (Keyword Protocol 2000) [1], and (v) ISO 15765 CAN [11]; these protocols present significant differences between them in terms of the electrical pin assignments. Notice that most vehicles implement only one of these protocols. For instance, Chrysler uses the ISO 9141-2 protocol, General Motors uses SAE J1850 VPW, and Ford uses SAE J1850 PWM.

3.2 Diagnostic Trouble Codes (DTCs)

Diagnostic Trouble Codes were standardized in document ISO 15031-6 [14], and allows engine technicians to easily determine why a vehicle is malfunctioning using generic scanners. The proposed format assigns alphanumeric codes to the different causes of failure, although extensions to the standard are allowed to support manufacturer-specific failures.

3.3 OBD Message Formats

The OBD system was designed to offer a flexible communications system. Message delivery among different devices requires defining the type of message to be delivered, along with the transmitter and the receiver devices. The adoption of different message priorities is also supported in order to make sure that critical information is processed first.

However, depending on the protocol using by each vehicle, the format of this message may vary slightly (see figure 2). Notice that both frame formats allow up to 7 data bytes, and they also include a checksum field in order to detect any transmission errors.

3.4 OBD-II PIDs

The OBD-II PIDs (OnBoard Diagnostics Parameter IDs) are identification codes of the different parameters that can be measured in a vehicle. The OBD-II standard, defined by the Society of Automotive Engineers as SAE J1979, defines

(a) Frame format adopted by the SAE J1850, ISO 9141-2 and ISO 14230-4 standards.

(b) Frame format adopted by the ISO 15765-4 (CAN) standard.

Fig. 2. OBD frame formats

Table 1. OBD-II operation modes

Mode	Description
01	Show current data
02	Show freeze frame data
03	Show stored Diagnostic Trouble Codes
04	Clear Diagnostic Trouble Codes and stored values
05	Test results, oxygen sensor monitoring (non CAN only)
06	Test results, other component/system monitoring (Test results, oxygen sensor monitoring for CAN only)
07	Show pending Diagnostic Trouble Codes (detected during current or last driving cycle)
08	Control operation of on-board component/system
09	Request vehicle information
0A	Permanent Diagnostic Trouble Codes (DTCs) (Cleared DTCs)

some parameters, although vehicle manufacturers usually introduce their own codes.

As shown in table 1, the SAE J1979 standard defined some operation modes for accessing OBD-II information.

The SAE J1979 standard also defines several OBD-II PIDs for the different modes of operation. We focus on mode 01 since it provides access to the current vehicle data. Numerical values are sent and received in hexadecimal format. Table 2 shows some codes along with their description and the parameters required to interpret them.

The "PID" column indicates the code identifier, the "Size" column indicates the length in number of bytes, the "Description" column describes the code, columns "Min" and "Max" indicate the minimum and maximum values, respectively, column "Units" indicates the units in which the code is read, and finally the "Formula" column shows which formula must be applied to interpret the received value.

Table 2. OBD-II PIDs

PID	Size	Description	Min	Max	Units	Formula
00	4	Supported PIDs				
05	1	Engine coolant temperature	-40	215	°C	A-40
0A	1	Fuel pressure	0	765	kPa (gauge)	A*3
0C	2	Engine RPM	0	16,383.75	rpm	((A*256)+B)/4
0D	1	Vehicle speed	0	255	km/h	A

When communicating via OBD-II two codes are required: one indicating the mode, and another one to specify the monitored parameter. For example, when requesting the vehicle speed, the identifier is "010D" where "01" indicates the mode, and "0D" the speed parameter. The response to this request would be "410D XX"; notice that the first digit identifier is changed from "0" to "4" to indicate an answer, and "XX" is the returned value.

In the "Formula" column, letter A represents the first byte returned, letter B represents the second byte, letter C the third byte, and so on.

4   VEWE: Vehicle ECU Wireless Emulator

VEWE is a solution developed to simulate vehicular behavior, and offer access to the vehicle's ECU parameters through a Bluetooth OBD-II interface. By using VEWE, the developers of OBD-II based smartphone applications are able to test them as if they were moving in a real vehicle, thereby speeding up the development time and lowering costs.

VEWE is able to generate all the parameters of a moving vehicle, and it allows the user to control the vehicle's mobility within a map through a joystick, as well as manually setting new parameters and have control over the Bluetooth communications. Map information is obtained from OpenStreetMap [15].

The VEWE platform is composed of three components: the main application, called VEWE server, and two Android based applications: GPSEmulator and OBDIICapture. Together, they provide a complete test system for Android-based mobile vehicular applications.

VEWE has the following features:

Multiplatform. It was developed in Java, using the Bluecove library for Bluetooth connectivity [16] and JMapViewer library for the map viewing functionality, making it a multiplatform system fully functional on both Windows and Linux operating systems.

Friendly graphical interface. It has a friendly and intuitive graphical user interface, developed using the Java Swing library. The user simply "turns on" the vehicle, and then it controls its path over a map (taken from Open-StreetMap) using a joystick.

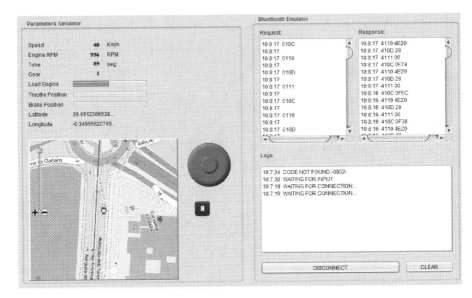

Fig. 3. VEWE Server application including the vehicle mobility simulator (left) and the OBD-II/Bluetooth connection manager (right)

Flexible. You can add or remove different simulated parameters, as well as specify the formula for calculating their value; other input parameters can be referenced in such a formula.

Geocoding simulation. Using the GPSEmulator application, we can emulate geopositions on any Android system based on data generated by VEWE in real-time. GPS coordinates are transferred between the VEWE Server and the GPSEmulator via Bluetooth.

Control of Bluetooth/OBD-II communications. The applications displays, in real-time, all the information exchanged between the simulator and the mobile device, as well as all the events generated therein. This is useful to debug OBD-II based applications being developed.

4.1 VEWE Server

The VEWE Server is the main Java application, being responsible for simulating the behavior of a moving vehicle, its engine Electronic Control Unit (ECU), and its location. For this endeavor it relies on different components: (1) Vehicle Mobility Simulator, (2) Map Manager, (3) Engine ECU, and (4) an OBD-II Emulator.

Concerning the Vehicle Mobility Simulator (1), this component is responsible for simulating a real vehicle, and dynamically determining the value of all relevant parameters, including those registered in the engine ECU. Among these parameters we can find:

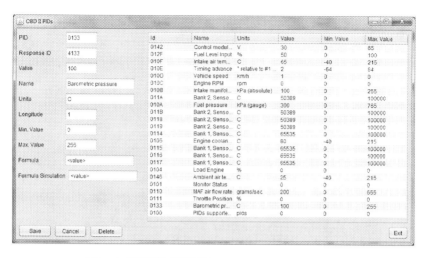

Fig. 4. VEWE OBD-II parameter management

- **Speed:** Updated according to the simulated vehicle speed.
- **Engine RPM:** Indicates the engine speed in terms of Revolutions Per Minute at any time.
- **Time:** Registers the time since the simulation started.
- **Gear:** Registers the current gear being used by the vehicle.
- **Engine Load:** Percentage value indicating the current load of the engine.
- **Throttle Position:** Percentage value indicating the relative position of the throttle pedal.
- **Brake Position:** Percentage value indicating the relative position of the brake pedal.
- **Latitude/Longitude:** Register the current vehicle coordinates.

The Map Manager (2) retrieves the actual street map from the OpenStreetMap platform [15], and provides a visual representation of the vehicle's position; such position is updated throughout time, and can be served to clients upon request.

Concerning the Engine ECU (3), it is a component used to store the value of all relevant PIDs being simulated, and the values stored can be requested through requests coming from the OBD-II interface. The VEWE Server also provides an OBD-II PID management interface (see figure 4), which contains a list of all parameters being simulated. From this interface it is possible to manually manipulate all the parameters handled by the Engine ECU.

The attributes associated to each parameter are:

OBD-II PID. Identifies the PID parameter and has a length of 4 hexadecimal digits, where the first two digits indicate the mode, and the following two PID identifier.

Response ID. Indicates the response ID for that specific PID, usually by changing the first digit from "0" to "4" .

Value. Indicates the value of the PID at that time. When a new parameter is introduced, this attribute indicates the default parameter value.

Name. Indicates the name or description of the PID.

Units. Specifies the unit associated to the PID value.

Length. States the size, in number of bytes, for each PID.

Min Value. Indicates the minimum value allowed for the PID.

Max Value. Indicates the maximum value allowed for the PID.

Formula. Indicates the formula used to transform the PID value into OBD-II format.

Simulation Formula. Indicates the formula used for calculating the PID value based on other PID values (if applicable).

Finally, regarding the Bluetooth OBD-II interface emulator (4), it is the component is responsible for controlling the Bluetooth connection. In particular, it creates and maintains the OBD-II Bluetooth connection, and registers all data sent through the established Bluetooth channel. The registered data is split into three parts:

1. Request: Shows the data requests received from clients.
2. Response: Displays the system responses returned to clients.
3. Logs: Registers important events related to the connection, such as connection status, missing codes, etc.

Besides the VEWE server application described above, our solution also includes two Android components - GPSEmulator and OBDIICapture -, which have quite different goals. These are described below.

4.2   GPSEmulator

Concerning GPSEmulator, it is used to emulate a (fake) location on Android systems (see Figure 5.a). Basically it creates a service which retrieves simulated geo-positioning data from the VEWE-Server via Bluetooth, and then injects the simulated locations on the Android system; other sources of localization data, such as GPS, WiFi or GSM-based triangulation, are disabled to avoid conflicts. This way the simulated coordinates are assumed as real by other running applications.

The application interface is quite simple. When the service starts, the Bluetooth device connected to VEWE is selected for coordinate retrieval; when the service is stopped, the Bluetooth connection to the VEWE server is closed, and emulated georeferencing ends.

4.3   OBDIICapture

With regard to OBDIICapture (see Figure 5.b), it is a different Android component designed to support the manual introduction of AT commands and OBD-II PID requests, and through which data from a real Bluetooth OBD-II interface is captured. It is useful to analyze the initial message exchange when using real

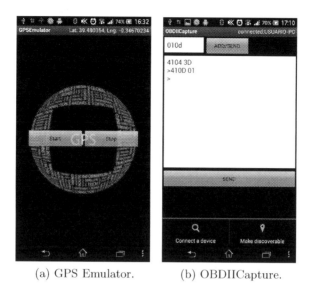

(a) GPS Emulator.   (b) OBDIICapture.

Fig. 5. VEWE Android Components

OBD-II interfaces by providing a log service, and to validate the message output of the VEWE server. In addition, it was also useful to measure the response times and the maximum message rate in a real environment, allowing to tune VEWE so as to achieve a similar behavior.

5   VEWE Implementation Details

VEWE was built using a layered architecture through which the different features of the system become independent, allowing to tweak each of them independently, thereby simplifying the development and maintenance of the system.

The VEWE server is a multithreaded system where separate threads are responsible for handling: (i) data generation, (ii) the graphical environment, and (iii) Bluetooth communications. Communication between threads is achieved through the shared variable paradigm, so that changes introduced by one thread are automatically detected by other threads.

VEWE follows a client-server paradigm, where an Android-based client application queries a Java-based server to retrieve the values of the different parameters stored in the emulated engine ECU (see figure 6). To this purpose it relies on a Bluetooth connection between the client and the server, upon which an OBD-II serial connection is established.

Below we provide details about the vehicle simulator, the emulation of GPS coordinates, and the OBD-II parameter database.

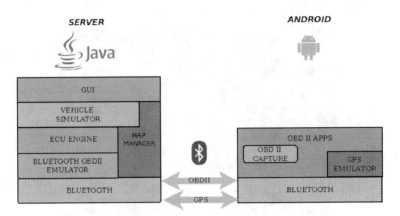

Fig. 6. VEWE architecture

5.1   The Vehicle Simulator

The vehicle simulator is a key element in the VEWE Server, being responsible for modeling vehicle mobility according to the user input, the vehicle characteristics, and the terrain profile. To that purpose it relies on the acceleration calculator module which, by taking into account the different forces affecting the vehicle's mobility - traction, friction, aerodynamic and gravity - determines the acceleration value (see figure 7) using realistic vehicle dynamics. Based on the acceleration value, it then updates the vehicle position on the map, as well as the desired parameters on engine's ECU (e.g. speed, RPM, gear).

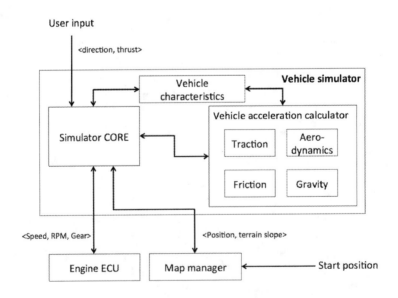

Fig. 7. Overview of the vehicle simulator

Real-time updating of the vehicle position on the map provides the user with a feeling of control over the simulated vehicle using the joystick available on screen. Notice that, by modifying the vehicle characteristics element, different types of vehicles can be modeled.

5.2 Emulating GPS Coordinates

The Map Manager component obtains updated vehicle coordinates from the vehicle simulator described above. In particular, vehicle positions are updated by taking as reference: (i) the current vehicle position, (ii) the orientation, and (iii) the distance traversed; the latter is calculated by combining the vehicle's speed with the inter-sample times.

The Map Manager component is able to provide geopositioning information to external applications through a Bluetooth channel. In our framework, we developed an Android application whose only purpose is to generate fake positions based on the information retrieved from the Map Manager via Bluetooth (see Figure 8). This application, called *GPS Emulator*, uses the *mock location* functionality provided in the Android API to introduce the fake locations into the system; to avoid interferences, it also cancels all other sources of localization information, including GPS, WiFi, or cellular-based positioning. This way, all running applications that register for localization services will receive the mock locations generated by the VEWE Server Map Manager component.

By making the speed calculated using GPS coordinates match the speed value returned by the OBD-II interface, we are able to provide consistent data to application developers.

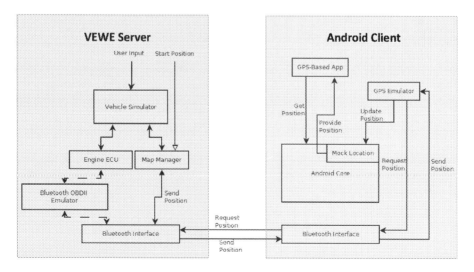

Fig. 8. Emulated position update procedure

5.3   Emulated OBD-II Interface

Providing an emulated OBD-II interface accessible via Bluetooth is one of the main goals of the VEWE platform. The OBD-II interface emulator is able to maintain a serial connection with a client, and adequately process AT commands and PID requests as would occur when using real OBD-II devices.

In order to retrieve the actual PID values, the OBD-II layer communicates with the Engine ECU element, responsible for storing and maintaining all PIDs supported. The Engine ECU handles a data structure (see Figure 9) capable of storing all the vehicle parameters (OBD-II Codes) as well as AT parameters.

The OBD-II interface emulator then converts the retrieved valuers to the appropriate format for delivery through the serial port.

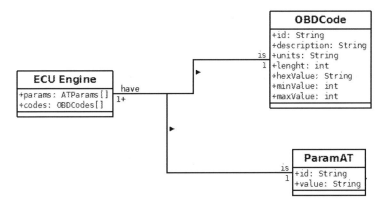

Fig. 9. Structure of the OBD-II parameter database

6   Conclusions

The evolution path towards sophisticated solutions in the ITS sector include smartphone/vehicle integration approaches. Based on current technologies, such approaches require smartphones to communicate with the vehicle's ECU through the OBD-II interface by relying on wireless connectors, typically Bluetooth-based.

In the literature different solutions are starting to emerge that make use of smartphone/vehicle integration using wireless OBD-II interfaces. However, developing such solutions is costly and time consuming, typically requiring several test drives to properly debug and tune the functionality of the applications being developed. In this paper we provide an efficient solution to this problem by introducing a platform able to emulate a Bluetooth-based OBD-II connection, along with the GPS coordinates of the vehicle. To this aim we provide VEWE, which is able to simulate the mobility of a vehicle controlled by a user using realistic acceleration/deceleration dynamics, registering data in a simulated engine ECU accordingly.

Overall, the proposed solution is expected to boost application development in the ITS area, reducing the development effort and costs, and making it feasible for developers without access to vehicles to develop this type of applications as well.

Acknowledgments. This work was partially supported by the *Ministerio de Ciencia e Innovación*, Spain, under Grant TIN2011-27543- C03-01.

References

1. International Organization for Standardization, ISO 14230-1:1999: Road vehicles, Diagnostic systems, Keyword Protocol 2000 (1999)
2. Elm Electronics - Circuits for the Hobbyist, Obd to rs232 interpreter (2013)
3. Martinez, S., Meseguer, J., Zaldivar, J., Calafate, C., Cano, J.-C., Manzoni, P., Fogue, M., Martinez, F.: Smartphones as the keystone for leveraging the diffusion of ITS applications. In: 9th ITS European Congress (2013)
4. Hawkins, I.: Torque: OBD2 Performance and Diagnostics for your Vehicle (2014), http://torque-bhp.com/
5. Teng, H.-F., Wang, M.-J., Lin, C.-M.: An implementation of android-based mobile virtual instrument for telematics applications. In: 2nd International Conference on Innovations in Bioinspired Computing and Applications (IBICA), pp. 306–308 (2011)
6. Meseguer, J.E., Calafate, C.T., Cano, J.C., Manzoni, P.: Drivingstyles: A smartphone application to assess driver behavior. In: 2013 IEEE Symposium on Computers and Communications (ISCC), pp. 535–540 (July 2013)
7. Zaldivar, J., Calafate, C., Cano, J.-C., Manzoni, P.: Providing accident detection in vehicular networks through OBD-II devices and Android-based smartphones. In: 2011 IEEE 36th Conference on Local Computer Networks (LCN), pp. 813–819 (2011)
8. Tahat, A., Said, A., Jaouni, F., Qadamani, W.: Android-based universal vehicle diagnostic and tracking system. In: IEEE 16th International Symposium on Consumer Electronics (ISCE), pp. 137–143 (2012)
9. Freematics, Freematics OBD-II Emulator (2013)
10. Briggs, G.: OBDSim (2013)
11. International Organization for Standardization, ISO 15765: Road vehicles, Diagnostics on Controller Area Networks (CAN) (2004)
12. SAE International - Vehicle Architecture For Data Communications Standards, Class B Data Communications Network Interface (2006)
13. International Organization for Standardization, ISO 9141-2:1994/Amd 1:1996 (1996)
14. International Organization for Standardization, ISO 15031-6: Road vehicles – Communication between vehicle and external equipment for emissions-related diagnostics – Part 6: Diagnostic trouble code definitions (2010)
15. The OpenStreetMap Project (2014), http://www.openstreetmap.org/
16. Bluecove Team, Bluecove (2008), http://bluecove.org/

Density Map Service in VANETs City Environments

Pratap Kumar Sahu[1], Abdelhakim Hafid[1], and Soumaya Cherkaoui[2]

[1] Dept. of Computer Science and Operation Research,
University of Montreal, Montreal (QC), Canada
{sahupk,ahafid}@iro.umontreal.ca
[2] Dept. of Electrical and Computer Engineering,
University of Sherbrooke, Sherbrooke (QC), Canada
Soumaya.Cherkaoui@USherbrooke.ca

Abstract. Vehicle density information is crucial for efficient functioning of many vehicular applications including emergency notification, driver assistance and infotainment applications. This information is used for evacuation planning in accident scenarios, finding alternate routes in case of road congestion, and providing stable routing paths for uninterrupted internet connection. In a city scenario, it is a tedious task to collect and share large volume of data containing density information. In this paper, a mechanism is proposed to create a density map for city environments. A hierarchy (i.e. tree) is established, where a node represents a road segment and the density is used to determine the height of the node; the root node represents the road segment having highest node density. The purpose of this tree is to multi-level collection and aggregation of density information at till the root node is reached. Then, the aggregated density information (i.e. density map) is forwarded down the hierarchy. Simulation results show that the proposed mechanism allows highly accurate computing of density map while generating low network overhead.

Keywords: Congestion, Curve-fitting, VANET.

1   Introduction

With the advancement of wireless technology and development in the automobile industry, the current road transportation involves intelligent vehicles. Over the last several years, vehicular ad Hoc networks (VANETs)[1][2] have drawn considerable attention from research community because of the promising services and applications provided through the cooperation of equipped vehicles. The major goal of vehicular Ad hoc networks is to increase the safety of passengers [23] and enhance the driving experience by providing services, such as collision warning, lane change assistance, intelligent navigation and road traffic control.

Real-time monitoring, control and management of road traffic are significant necessities in megacities. Awareness of vehicle density is essential for effective traffic management operations [4] [27]; Knowledge of vehicle density is useful in determining the congestion status of a road and allows drivers to navigate through

S. Guo et al. (Eds.): ADHOC-NOW 2014, LNCS 8487, pp. 446–460, 2014.
© Springer International Publishing Switzerland 2014

congestion free routes. Moreover, the performance of VANET communication protocols (e.g., broadcast, routing and content sharing) varies greatly according to vehicle density. Thus, vehicle density information at any place and at any instant allows VANET protocols to adapt their behavior in different density situations to sustain the required performance.

In traditional vehicle density estimation techniques [3][7-8], the information collected from road side sensors (magnetic loop detectors) or surveillance cameras are sent to a central server where the information is processed and disseminated either proactively or in response to a query sent by a driver. Because of the high cost involved in installation and maintenance, road sensors are deployed near junctions or in selected highways; thus, it is difficult to estimate vehicle density of the entire city. These devices are also limited in capacity; for example, it is difficult for a surveillance camera to perform complex image processing and analysis in real-time. Since, vehicles are equipped with GPS and wireless transceivers, vehicle-to-vehicle (V2V) communication can be leveraged to estimate vehicle density in a decentralized manner. As the network bandwidth is limited, it is necessary to develop mechanisms that estimate vehicle density of a whole city without inducing overhead in the network.

In this paper, we propose a novel service, called Density Map Service (DMS), for vehicular networks, with the objective of providing precise density information to vehicles. The main challenge in realizing the service is to provide real time information while satisfying end-to-end-delay requirements and minimizing overhead. It is known that spatial density information is huge to deal with. Therefore, we use a comparison based aggregation mechanism [9][17] to substantially compress the density information. To lower end-to-end delay, a single virtual gateway[1] should be identified to store local density information for each road segment. Basically, a hierarchy (i.e. tree) is established where a node represents a road segment and the height of a node represents the density of the corresponding road segment. The root node represents the road segment having highest node density. At each level, the density of a node is combined with the density of nodes at the next lower level. The combined information is expressed in the form of an equation using curve-fitting method. Once the root node receives the desired density map, it is transmitted down the hierarchy, eventually distributing the density information in the entire network.

Density information is very useful for any routing protocol to take routing decisions [10] [13][19][28]. Denser paths are considered to be better connected routes and routing decisions are made on connectivity. To be able to evaluate the novelty of our proposal we choose a routing protocol (i.e. BAHG) [13], which is evaluated in presence and absence of density map service.

The rest of the paper is organized as follows. A brief discussion of related works is presented in Section-2. Section-3 describes the proposed mechanism. Section-4 contains performance evaluation. Finally, a summary is provided in Section-5.

---

[1] A single vehicle or a group of vehicles stationed around a position to act as a gateway to store data; when any of these vehicles leave the position, new vehicles take over.

2   Related Works

To collect traffic density information of a road segment, various methods are adopted in Intelligent Transportation Systems. Traditionally, two techniques are used to measure the traffic density in a neighbourhood: (i) through periodic beacon exchanges: beacons are periodically broadcasted by all vehicles with position information [10][20][21][22]. Thereafter, selected vehicles collect one-hop density information and forward it to their neighbours; and (ii) using sensing devices: various sensing equipment are used to measure the density of a road segment[3][7][8]; they include magnetic loop detectors, surveillance cameras, and radar speed guns. In [3][7][8], traffic data is collected using sensors placed at either ends of a road segment and various methods are presented to estimate the vehicle density. In [3], using a Markov model, Singh et al. attempted to solve the density estimation problem using the arrival rate of vehicles. Yuan et al. [7] proposed a Lagrangian state estimator-based approach to estimate vehicle density. An extended and modified approach based on Kalman filter is adopted for density estimation in [8]. Buch et al. [6] proposed an interesting use of camera sensing techniques for urban traffic monitoring; video cameras are installed at light poles to collect vehicle count information. In [11], three vision sensors were used to compute the number of vehicles. In [12], cumulative road acoustics were used in estimating road traffic density and the impact of noise on the estimation was analyzed. Though it requires all vehicles to be equipped with wireless transceivers, density estimation achieved through periodic beaconing is pretty accurate in one hop. It is a difficult task to extend the accuracy in density information to the multi-hop. Magnetic loop detectors are often considered to be expensive and installation is a cumbersome task. Also, surveillance cameras and radar speed guns face similar issues. Apart from that, analysis of images captured through cameras involves huge processing and communication cost. Also, sensors are usually confined to some geographic locations which limit the capability of sensors to capture events (i.e. accident, traffic congestion, road condition and weather condition) for the entire network.

   The distribution of density information is very similar to location services in vehicular networks [5] [14] [15]. The location services can be classified into three major categories: (i) centralized, where a central server usually takes care of entire system; (ii) decentralized, where the location information is held on distributed entities; and (iii) query-based, where the source probes the destination position through flooding. In a centralized mechanism (e.g., [10][25]), as the central server has the core responsibility to provide location service, memory, computational capabilities and bandwidth can be easily exhausted; also, installation of such sensing devices, which are connected to a central server is quite expensive. On the other hand, in a decentralized system (e.g., [14] [24]), the moving entities hold the location information. Since, vehicular networks consist of fast moving vehicles; there is always a possibility that a vehicle gets lost in the crowd carrying a lot of location information. In the Map-Based location service (MBLP) [26], the area is divided into hierarchy of squared regions, where a waypoint stores and provides location information of each vehicle in the square; a hierarchy of waypoints stores location information of the entire area. The random selection of waypoints and mobility of vehicles selected for waypoints make it inappropriate for vehicular networks. The query based location service proposed in RLS [15] uses controlled flooding to get the real time position information of a vehicle. This location service is quite useful to take routing decision if the source or the destination changes the position noticeably. Usually, a vehicle moves to a new position quite often from its original position once the connection is established between the source and the destination pair for data exchanges in routing. Therefore, further position updates generate unreasonable network overhead.

The major difference among location service and density estimation is that density information is shared for a region (i.e. road segment) whereas in location service, a position information in shared. The amount of data involved in density map service would be very high. For a centralized location service, the amount of data to be communicated to the server would be astronomical and make it infeasible for vehicular networks. Similarly, for a distributed hierarchy based location service, the data transfer time would be very long and the topology may change by the time the source collects the information. For query based systems, though collecting density information of a road segment involves small communication cost, it is infeasible to collect density information of a large area. Therefore, it would not be possible to use a location service to provide density information. Apart from that, location services are not free from shortcomings.

3   Proposed Scheme

In this paper we propose a novel service for vehicular networks, namely a 'Density Map Service", or D-MAP with the objective of providing precise density information to vehicles. The main challenge in realizing the service is to provide real time information while satisfying end-end-delay requirements and lower the message overhead. It is known that spatial density information is very huge to deal with. Therefore, we used a comparison based semantic aggregation mechanism to substantially compress the density information. To lower the end-to-end delay, a virtual gateway should be identified to store local density information. Basically, a hierarchy (i.e. tree) is established where a node represents a road segment and the height of a node represents the density of the corresponding road segment. The root node represents the road segment having highest node density. At each level, the density of a node is combined with the density of nodes at the next lower level. The combined information is expressed in the form of an equation using curve-fitting method. Once the root node receives the desired density map, it is transmitted down the hierarchy, eventually distributing the density information in the entire network.

3.1   Definitions, Hierarchy Constructions and Procedures

Virtual Gateway: To collect information of each road segment, a storage node must be selected. A single vehicle or a group of vehicles stationed around a position to act as a gateway to store data; when any of these vehicles leave the position, new vehicles take over. We name the storage node as the virtual gateway.

Border Nodes: Two vehicles, on either end of the road segment must provide density information to the virtual gateway. These vehicles which pass the density information to the virtual gateway are called border nodes.

Border Region: A border node operates in a region adjacent to the intersection from which it has the visibility to other border nodes of adjacent road segments. This region is called border region.

Membership Level-0: Initially, all road segments are assigned a membership level of 0.

Membership Level-1: For edge e_i , the membership level M_{e_i} is set to 1 if $\exists$ an edge $e_j \in N_a(e_i)$ s.t. $d_{e_j} < d_{e_i}$ and *Membership Level* $M_{e_i} \neq 1$ and $M_{e_j} \neq 1$, where d_{e_i} and d_{e_j} are the densities of edges e_i and e_j respectively.

Membership Level ≥ 1: For edge e_i , membership level M_{e_i} is set to $L+1$ if $\forall$ edge $e_j \in N_b(e_i)$, $d_{e_i} > d_{e_j}$ and $M_{e_j} = L$, where d_{e_i} and d_{e_j} are the densities of edges e_i and e_j respectively.

Procedure (Membership Level 1)
    When a road segment is having higher average density than its adjacent neighboring road segments and none of the road segments has any membership level, then the road segment with higher density has its membership level change to 1 and

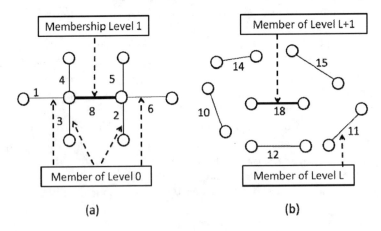

(a)   (b)

Fig. 1. (a). Membership Level 1   **Fig. 1.** (b). Higher Level Membership >1

all its neighboring road segments become its children. The process is repeated for each road segment which neither has membership level 1 nor is a child of a road segment with membership level 1. In Fig 1(a), all adjacent neighboring segments have lower density than the middle road segment (bold in Fig. 1(a)).

Procedure (Membership Level >1)

When a road segment has higher average density than its non-adjacent neighboring road segments and all such road segments have the same membership level (i.e. L), then the road segment with highest density has its membership level change to L+1 and all its neighboring road segments become its children. The process is repeated for each road segment which neither has membership level L+1 nor is a child of a road segment with membership level L+1. In Fig 1(b), all non-adjacent neighboring segments are having lower density than the middle road segment.

3.2 Segment-Wise Virtual Gateway Creation and Density Information Collection

Before collecting density information, a storage site should be selected to keep density information of each road segment. As vehicles are mobile, it is infeasible to select a single node for entire duration of time for storage purpose. Thus, the virtual gateway and border nodes are assigned the responsibility of the road segment. In Fig 2, vehicles A and G represent border nodes and D the virtual gateway; vehicle A selects forwarding node C and sends the density information and C forwards the density information to the next hop (i.e. D) which is the virtual gateway here. Apparently, vehicle C sums up the density up to D and the density it has received from A (i.e. 4). Similarly, G sends density information to E which sends it to G.

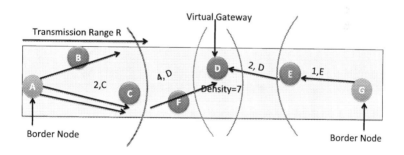

Fig. 2. Density information Collection and Storage

3.3 Selection of Border Nodes and Virtual Gateway

Density information collection and caching are two important requirements in density map creation. To collect information of each road segment, two border nodes and a storage node are assigned. We name the storage node as the virtual gateway. The job of the border nodes is to pass the neighborhood density information to the virtual gateway as shown in Fig 2. When no node is chosen to be the virtual gateway, the node closest to the mid-point of a road segment declares itself as the virtual gateway through its beacon message. When a virtual gateway is away from mid-point of the road segment by a distance R/2, it denounces its virtual gateway status through the beacon message. Here, R is the transmission range of a vehicle. A new node declares itself as the virtual gateway when an old node denounces its virtual gateway status.

A border node should be a node which is one of the closest nodes to the intersection. If a road segment is connected to an intersection, then the end of the road segment including the intersection is called border region whose length is R. A node declares itself as the border node if it is closest to the mid-point of the border region and it denounce its border node status if its predicted next position is not in the border region at the time of its next beacon sending. When a road segment length is less than R, then no border node is selected and the virtual gateway takes care of the density information collection.

3.4   Aggregation Based Data Collection and Density Share Algorithm

In the data collection procedure, two pieces of information, namely average density and standard deviation of all road segments, are collected at the root (i.e. the road segment having highest level) of the hierarchy. In this procedure, each road segment at a given level of the hierarchy collects information from its child road segments; then, it combines the collected information with its own average density and standard deviation and sends the combined information to its parent. Information collection starts at road segments at level 0; each of these road segments sends its average density and standard deviation information to its parent (road segment at level 1). Upon receipt of this information, a road segment at level 1combines its own information with the information received from its children and sends the combined information to its parent (road segment at level 2). In this way, the information is sent upward in the hierarchy.

The main challenge in this procedure lies in sending the combined information. As shown in the hierarchy, a road segment at level L has to send its own information and information of all road segments present in the sub-tree rooted at it. It is clear that, as the level increases, the forwarding of information creates huge communication overhead and leads to significant amount of collisions. As a consequence, it results in very low packet delivery ratio at the nodes at intermediate levels. To address this issue, we propose a data aggregation scheme [9][17] in which, a road segment performs aggregation on the information received from its children and its own information. The aggregation operation is described as follows.

At a given node at level 1, the density information (average or standard deviation) of its children can be sorted in decreasing order. Accordingly, an ordering of the segments can be determined based on the sorted density information. If this information is plotted against the segments ordering, then a curve showing the decreasing trend of the density is obtained. The curve can be represented by a polynomial of some order; we use polynomial least square curve fitting method to obtain a polynomial of order >2. Since, there are two pieces of information (average density and standard deviation) two polynomials are obtained. Let $P(x)$ and $Q(x)$ denote j^{th} order polynomials that represent the decreasing trends of average densities and standard deviations respectively; they are defined as follows:

$$P(x) = a_0 + a_1 x + a_2 x^2 + \ldots\ldots + a_j x^j \qquad (1)$$

$$Q(x) = b_0 + b_1 x + b_2 x^2 + \ldots\ldots + b_j x^j \qquad (2)$$

where x denotes the order of the road segment. Let n denotes the number of road segments whose information needs to be aggregated. Let m_i denotes the average density of the road segment having order i. The mean square error for $P(X)$ is given as follows:

$$err^2 = \sum_{i=1}^{n} (m_i - (a_0 + \sum_{k=1}^{j} a_k x_i^k)) \qquad (3)$$

We used the technique described in [18] to minimize err^2 and obtain the coefficients in (1) and (2).

Thus, at level 2, a node receives one or more polynomials (e.g., for average density) from its children;   before sending density information to its parent at level 3, the node merges all these polynomials to one polynomial (one for average and one for standard deviation); more specifically, it starts by extracting density information from the received polynomials and then sorting this information in decreasing order; it also computes an ordering of the segments according to the sorted density information. Finally, it computes two polynomials, as described above, and sends them to its parent.

We introduce a packet, called UP-DP (Upward Density Packet) to carry the average density and standard deviation information. The fields those are included in an UP-DP, namely the number of levels, highest membership level, total number of segments at each level, the ids of those segments, average density and standard deviation. At level 0, a road segment includes its average density and standard deviation in UP-DP to be sent to its parent at level 1. But, for all other levels, an UP-DP to be sent to the upper level contains the coefficients that describe the polynomial of average density and the polynomial of standard deviation in addition to other fields. Since, the decreasing trend of average density is same as that of standard deviation, the segment ordering remains same and is included in UP-DP. The benefit of the proposed aggregation scheme is that the average density and standard deviation of multiple road segments are sent to the next upper level in the hierarchy using just two polynomials instead of a huge data set.

When the root receives polynomials from its children road segments, it merges the polynomials to form a single polynomial that contains the information of the entire network. More specifically, the root forms two final polynomials, one containing the average density and the other containing the standard deviation of all road segments in the network.

In Fig 3 (a), all road segments are initially assigned a membership level of 0. On running the procedure to assign membership level 1, in the first phase most of the road segments become either promoted to level 1 or become the children of road segments having level 1. It took three iterations to complete this process. Fig 3 (b), in two iterations, level 2 membership could be computed. In Fig 3(c), membership level 3 and the root segment is obtained.

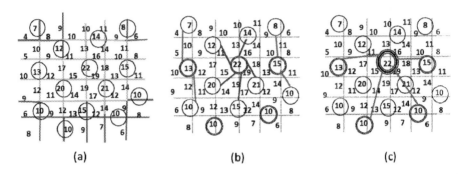

(a)   (b)   (c)

Fig. 3. Road Segments with their (a). Membership Level 1, (b) Membership Level 2 and (c) Membership Level 3.

3.5   Density Information Extraction

The two final polynomials formed at root of the hierarchy at the end of the collection/aggregation procedure embed the network-wide average density and standard deviation information. In order to allow each road segment to construct the density map, the coefficients of these final polynomials as well as the sorted list of segments identifiers are distributed to all road segments. A new packet, called downward density packet (DN-DP), is created to send this information. The propagation of DN-DP is accomplished by sending it down the hierarchy. When DN-DP is received by road-segments at the lowest level, the information in DN-DP is processed to construct the density map. In particular, segment ordering is used to extract the average density and standard deviation of each road segment from the received polynomials. For example, let $P(x)$ and $Q(x)$ denote the final polynomials that embed the average density and standard deviation for the entire network; for a road segment of order i, its density and standard deviation is obtained by computing the value of $P(i)$ and $Q(i)$ respectively.

Fig. 4, shows an example of the operation of the density aggregation algorithm; for clarity, we show only the average density aggregation. As shown in Fig. 4, road segment P (and Q) collects average density from each of its children. Upon receipt of this information, P sorts them in descending order as shown in Fig. 4. Then, it prepares the corresponding fitted-curve and forwards it to node E. On getting data from multiple children, namely P and Q, E extracts the average density of each node, in the sub-tree rooted at E, from the 2 polynomials it received; then, it applies the density aggregation algorithm to produce a polynomial and forwards it to node H as shown in Fig. 4.

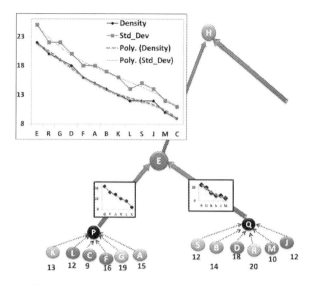

Fig. 4. Density information collection and aggregation

4   Performance Evaluation

In this section, we discuss the performance of the proposed density map algorithm. We used ns-2 [16] network simulator to conduct all simulations. The proposed algorithm is evaluated on varying network density and the distance between the source and destination. The purpose is first to evaluate the accuracy of the density service, and second to evaluate how such a service helps the performance of a routing protocol that uses this service. The simulation parameters are as follows:

Detour Factor (%): In the routing protocol BAHG [13], proactively a path is predicted from the source to the destination. If an entirely new path or the part of path is chosen to be new, that change in percentage is called the detour factor.

Packet Delivery Ratio (%): It is ratio of total number of packets received at the destination to the total number of packets generated by the source.

Control Overhead (%): It is the ratio of the control packets in bytes over the total packets in bytes sent in the entire networks.

Table 1. Simulation Parameters

Parameters	Values
Simulation Area	4000m X 4000m
Number of Nodes	100-500
Radio Range	300m
Data Rate	2 Mbps
Channel Model	Two Ray Ground
Number of Beacons/Second	1
Vehicle Speed	10m/s-30 m/s
Number of Routing Connections	10
Number of lanes	2

4.1   Simulation Environment

In the simulation, we consider an area of 4000m X 4000m. For this prototype model we choose a grid structure with 64 intersections, 8 intersections on each row. To evaluate the efficiency of the density map service, we investigate the performance of BAHG routing protocol [13] which uses this service. The performance of this routing protocol is compared with having density map service and in the absence of the density map service. The number of source-destination pairs for the routing is chosen to be 10.

4.2   Results and Discussion

In Fig 5, detour factor is presented for different node densities. When a predicted path is not followed, the change in path increases detour factor. In this figure, the BAHG routing protocol is taken into consideration for the evaluation. It is found that, with the help of D-MAP scheme, BAHG performs much better. However, detour factor decreases with increase in node density. With increase in node density the connectivity increases which reduces detouring of the routing packets. BAHG does a proactive routing planning without being aware of any density information and it is the reason that BAHG with D-MAP performs better.

In Fig 6, detour factor is presented for various source-destination distances. A longer route has higher chance of detouring. Therefore, with increased source-destination distance the detour factor increases. BAHG performs with the help of D-MAP scheme because this scheme helps in finding better connected paths.

Fig 7 and Fig 8 present the packet delivery ratio for various node densities and for different source-destination distances respectively. In Fig 7, the difference in packet delivery ratio between the BAHG with D-MAP and BAHG without BAHG is higher than Fig 8. This indicates that node density has impacts heavily on packet delivery ratio. With higher node density a more connected path can be chosen on getting suggestion from density map service. This improves overall packet delivery ratio which is evident from Fig 7. On the contrary the distance between the source and destination does not impact much. Though small, the change in packet delivery ratio in Fig 8 is attributed to length of a route. In Fig 8 it is seen that the higher the length of a path, the lower the packet delivery ratio without the suggestion of density map service.

In Fig 9, the ratio of the control overhead to the total network overhead is evaluated on various node densities. The beacon message is higher in size for D-MAP services. Therefore, the BAHG with D-MAP service is having higher control overhead through very small. This could be a good bargain if we achieve higher packet delivery ratio with slight increase in control overhead. With increase in node densities the control overhead increases as the overall data packet transmission is kept the same.

In Fig 10, the ratio of the control overhead to the total network overhead is evaluated on different source destination distances. If the length of a routing path increases, the number of hops from the source to the destination increases. However, the number of beacon messages remain the same as the node density is kept the same. Therefore, the control overhead decreases with source-destination distance. As we mentioned earlier, the size of the control messages (i.e. predominantly beacon messages) is higher than normal beacon messages. Therefore, BAHG with D-MAP services have higher control overhead.

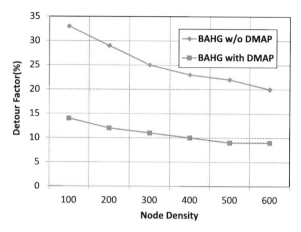

Fig. 5. Detour Factor Vs Node Density with Avg. SRC-DST Distance of 3km

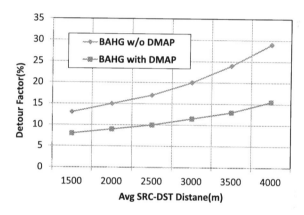

Fig. 6. Detour Factor Vs SRC-DST distance with node density of 400 nodes

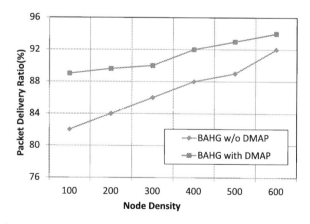

Fig. 7. Packet Delivery Ratio Vs. Node Density with Avg. SRC-DST Distance of 3km

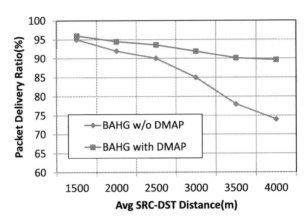

Fig. 8. Packet Delivery Ratio Vs. SRC-DST distance with node density of 400 nodes

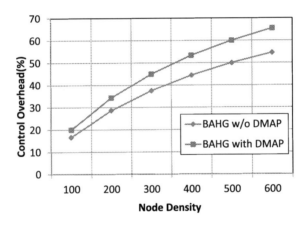

Fig. 9. Control Overhead Vs. Node Density with Avg. SRC-DST Distance of 3km

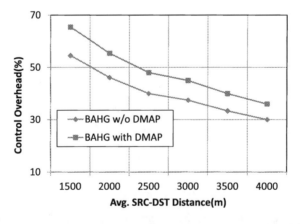

Fig. 10. Control Overhead Vs. SRC-DST distance with node density of 400 nodes

5   Conclusion

Vehicular density map provides better planning for safety message dissemination, driver assistance information and infotainment applications. Apart from that it helps in evacuation planning when an accident occurs, can avail drivers an alternate route information of a congested path, and can provide uninterrupted internet connection through a stable routing path. Using our D-MAP mechanism every vehicle is informed about the density of all road segments in the network. This is achieved through a hierarchy of density tree. A semantic data aggregation is done at the root of the tree using curve-fitting methods and this data is exchanged periodically with each segment which is part of the tree. Through our simulations, it is evident that extra control overhead is not added to adopt the density map service though it can provide better route planning which can be proved by better packet delivery ratio. In future, we plan to propose a recommendation system for vehicular traffic to minimize the overall congestion.

References

[1] Morgan, Y.L.: Managing DSRC and WAVE Standards Operations in a V2V Scenario. International Journal of Vehicular Technology (2010)

[2] Hartenstein, H., Laberteaux, K.P.: A tutorial survey on vehicular ad hoc networks. IEEE Communications Magazine, 164–171 (2008)

[3] Singh, K., Baibing, L.: Estimation of traffic densities for multilane roadways using a markov model approach. IEEE Transactions on Industrial Electronics 59(11), 4369–4376 (2012)

[4] Artimy, M.M., Robertson, W., Phillips, W.J.: Assignment of dynamic transmission range based on estimation of vehicle density. In: Proceedings of ACM International Workshop on Vehicular Ad hoc Networks (2005)

[5] Brahmi, N., Boussedjra, M., Mouzna, J., Cornelio, K.V.: An improved Map-based Location Service for Vehicular Ad Hoc Networks. In: Proceedings of IEEE 6th International Conference on Wireless and Mobile Computing, Networking and Communications (2010)

[6] Buch, N., Velastin, S.A., Orwell, J.: A review of computer vision techniques for the analysis of urban traffic. IEEE Trans. on Intelligent Transportation Systems 12(3), 920–939 (2011)

[7] Yuan, Y., van Lint, J.W.C., Wilson, R.E., van Wageningen-Kessels, F., Hoogendoorn, S.P.: Real-time Lagrangian traffic state estimator for freeways. IEEE Trans. Intell. Transp. Syst. 13(1), 59–70 (2012)

[8] van Hinsbergen, C.P.I.J., Schreiter, T., Zuurbier, F.S., van Lint, J.W.C., van Zuylen, H.J.: Localized extended kalman filter for scalable realtime traffic state estimation. IEEE Trans. Intell. Transp. Syst. 13(1), 385–394 (2012)

[9] Xiang, L., Luo, J., Rosenberg, C.: Compressed data aggregation Energy-efficient and high-fidelity data collection. IEEE/ACM Transactions on Networking PP(99), 1 (2013)

[10] Jerbi, M., Senouci, S.-M., Rasheed, T., Ghamri-Doudane, Y.: Towards efficient geographic routing in urban vehicular networks. IEEE Trans. Veh. Technol. 58(9), 5048–5059 (2009)

[11] Meng, C., Weihua, Z., Barth, M.: Mobile traffic surveillance system for dynamic roadway and vehicle traffic data integration. In: Proceedings of 14th International IEEE Conference on Intelligent Transportation Systems (ITSC), pp. 771–776 (2011)

[12] Tyagi, V., Kalyanaraman, S., Krishnapuram, R.: Vehicular traffic density state estimation based on cumulative road acoustics. IEEE Transactions on Intelligent Transportation Systems PP(99), 1–11 (2012)

[13] Sahu, P.K., Wu, E.H., Sahoo, J., Gerla, M.: BAHG: Back-Bone-Assisted Hop Greedy Routing for VANET's City Environments. IEEE Transactions on Transportation Systems 14(1), 199–213 (2013)

[14] Kieß, W., Füßler, H., Widmer, J.: Hierarchical location service for mobile ad-hoc networks. ACM SIGMOBILE Mob. Comput. Commun. Rev. 8(4), 47–58 (2004)

[15] Käsemann, M., Füßler, H., Hartenstein, H., Mauve, M.: A reactive location service for mobile ad hoc networks. Dept. Comput. Sci., Univ. Mannheim, Mannheim, Germany. Tech. Rep. TR-14-2002 (November 2002)

[16] The Network Simulator-ns-2, http://www.isi.edu/

[17] Rajagopalan, R., Varshney, P.: Data-aggregation Techniques in Sensor Networks: A Survey. IEEE Commun. Surveys Tutorials 8(4), 48–63 (2006)

[18] Björck, Å.: Numerical Methods for Least Squares Problems. SIAM (1996) ISBN978-0-89871-360-2

[19] Ding, Y., Xiao, L.: SADV: Static-node-assisted adaptive data dissemination in vehicular networks. IEEE Trans. Veh. Technol. 59(5), 2445–2455 (2010)

[20] Sommer, C., Tonguz, O.K., Dressler, F.: Traffic information systems: Efficient message dissemination via adaptive beaconing. IEEE Commun. Mag., Autom. Netw. Appl. Ser. 49(5), 173–179 (2011)

[21] Rezende, C., Boukerche, A., Ramos, H.S., Loureiro, A.A.F.: A Reactive and Scalable Unicast Solution for Video Streaming over VANETs. IEEE Transactions on Computers (2014)

[22] Chuang, M.C., Chen, M.C.: DEEP: Density-Aware Emergency Message Extension Protocol for VANETs. IEEE Trans. on Wireless Communications (October 2013)

[23] Xu, Q., Mark, T., Ko, J., Sengupta, R.: Vehicle-to-vehicle safety messaging in DSRC. In: Proc. VANET, pp. 19–28 (October 2004)

[24] Li, J., Jannotti, J., De Couto, D.S.J., Karger, D.R., Morris, R.: A scalable location service for geographic ad hoc routing. In: Proc. ACM MOBICOM, pp. 120–130 (2000)

[25] The CitySense Sensor Network Project, http://www.citysense.net

[26] Boussedjra, M., Mouzna, J., Bangera, P., Pai, M.M.M.: Map-Based Location Service for VANET. In: Proceedings of the International Conference on Ultra Modern Telecommunications, pp. 1–6 (2009)

[27] Mao, R., Mao, G.: Road traffic density estimation in vehicular networks. IEEE Wireless Communications and Networking Conference, WCNC (2013)

[28] Frank, R., Giordano, E., Cataldi, P., Gerla, M.: TrafRoute: A Different Approach to Routing in Vehicular Networks. In: VECON 2010, Niagara Falls, Canada (2010)

Author Index